*Excel*

# SNOWBALL MATHS 7

Get the results you want

Moonie Kim

PASCAL PRESS

# *Excel* Snowball

# MATHS YEAR 7

## *Get the Results You Want!*

Just as a snowball becomes larger as it rolls down a hill and collects more snow, so too will students increase their knowledge as they work through ***Excel*** **Snowball Maths Year 7**.

This book has been developed by a high-school Maths teacher who extensively trialled the work in his classrooms over a seven-year period. His students' confidence increased significantly while using the units from this book, leading to success in their schoolwork and tests.

Not just a write-in workbook, ***Excel*** **Snowball Maths Year 7** is also a self-learning book:

- **Questions have embedded hints and instructions** to show students the necessary steps to solve them—see page 42.
- **Working lines are carefully laid out with plenty of room for students to write**, which will guide them step by step through each solution—see page 49.
- **Concepts are introduced slowly and methodically** to ensure students move onto the next level of questions only when they have mastered the previous level.
- **One-page unit reviews** are included to help students efficiently revise their work for each chapter.

This book is written for Year 7 students who want:

- ✓ an introduction to the Year 7 Maths course
- ✓ a quick and efficient revision of the course
- ✓ support in better understanding Maths concepts
- ✓ to fill the gaps in their Year 7 Maths knowledge
- ✓ to boost confidence in their mathematical ability.

### ABOUT THE AUTHOR

**Moonie Kim** BSc BEd (University of Sydney) is a high-school Mathematics teacher with over ten years teaching experience. He has taught all levels of Mathematics, including HSC Mathematics Extension 2, in schools with a high number of EALD (English as an additional language or dialect) students.

# Table of Contents

## Chapter 1: Integers



## Chapter 2: Indices



## Chapter 3: Fractions



## Chapter 4: Fractions, decimals and percentages



## Chapter 5: Algebra

UNIT 1

**INTEGERS**

# Introducing integers

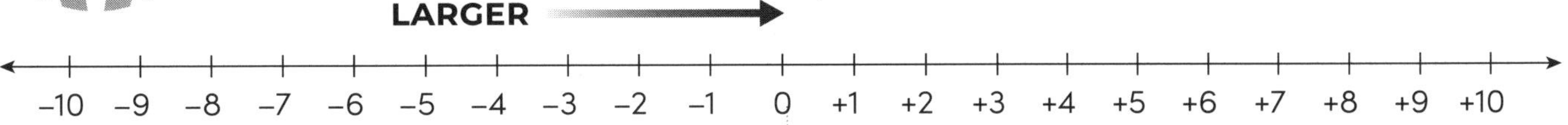


**1** Write the correct symbol (either < or >) to show the larger number:

| | | |
|---|---|---|
| **a** 1 ☐ 4 | **b** 3 ☐ 1 | **c** 4 ☐ 9 |
| **d** 5 ☐ 6 | **e** 5 ☐ 2 | **f** 3 ☐ 7 |
| **g** 0 ☐ 5 | **h** 1 ☐ 0 | **i** 4 ☐ 0 |
| **j** 6 ☐ 0 | **k** 3 ☐ 0 | **l** 0 ☐ 2 |
| **m** 0 ☐ −5 | **n** −1 ☐ 0 | **o** −4 ☐ 0 |
| **p** −6 ☐ 0 | **q** −3 ☐ 0 | **r** 0 ☐ −2 |
| **s** −4 ☐ 3 | **t** 5 ☐ −7 | **u** −6 ☐ 9 |
| **v** −2 ☐ 4 | **w** 5 ☐ −4 | **x** −8 ☐ 3 |

**2** Write the correct symbol (either < or >) to show the larger number:

| | | |
|---|---|---|
| **a** −4 ☐ 4 | **b** −5 ☐ 5 | **c** 1 ☐ −1 |
| **d** 3 ☐ 3 | **e** −2 ☐ 2 | **f** 8 ☐ −8 |
| **g** −1 ☐ −7 | **h** −5 ☐ −9 | **i** −6 ☐ −4 |
| **j** −3 ☐ −2 | **k** −4 ☐ −5 | **l** −7 ☐ −10 |
| **m** 2 ☐ 5 | **n** 7 ☐ 0 | **o** −4 ☐ 0 |
| **p** −6 ☐ 5 | **q** −2 ☐ 2 | **r** −9 ☐ −2 |
| **s** −11 ☐ −12 | **t** 5 ☐ −6 | **u** 0 ☐ −3 |

**3** Write the following sets of integers in descending order (largest to smallest):

| | | |
|---|---|---|
| **a** 1, 9, 5 ______ | **b** 5, 0, 3 ______ | **c** 2, 7, 4 ______ |
| **d** 3, 5, 9 ______ | **e** 4, 0, 6 ______ | **f** 7, 1, 3 ______ |
| **g** 3, −1, 6 ______ | **h** −2, 5, 1 ______ | **i** −3, 0, 3 ______ |
| **j** −2, 5, 2 ______ | **k** 4, −3, 0 ______ | **l** −6, 3, 7 ______ |
| **m** −3, −1, 6 ______ | **n** −2, −5, 4 ______ | **o** −1, 3, −6 ______ |
| **p** −7, 2, −2 ______ | **q** −1, −7, 1 ______ | **r** −8, −3, 0 ______ |
| **s** −6, −4, −2 ______ | **t** −3, −5, −1 ______ | **u** −5, −2, −6 ______ |
| **v** 0, −4, −2 ______ | **w** −5, −7, −2 ______ | **x** −6, −4, −5 ______ |
| **y** −10, 0, −9 ______ | **z** 8, −7, −6 ______ | **za** −9, −10, −8 ______ |

INTEGERS

# Introducing integers

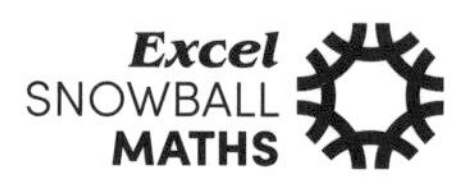


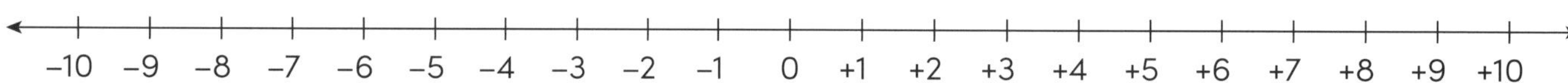


**1** A counter is currently on +4 position.

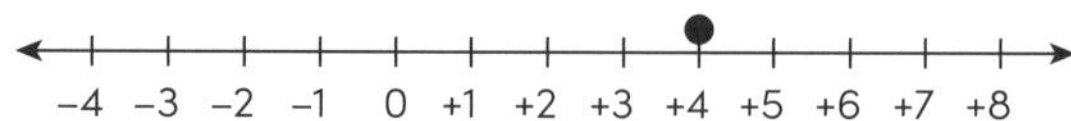


Where will it be if it is moved:

- **a** 3 places to the right? ______
- **b** 4 places to the left? ______
- **c** 2 places to the right? ______
- **d** 4 places to the right? ______
- **e** 7 places to the left? ______
- **f** 3 places to the right then 4 places to the left? ______
- **g** 2 places to the left then 2 places to the right? ______
- **h** 4 places to the left then 2 places to the right? ______

**2** A peg is currently on −3 position.

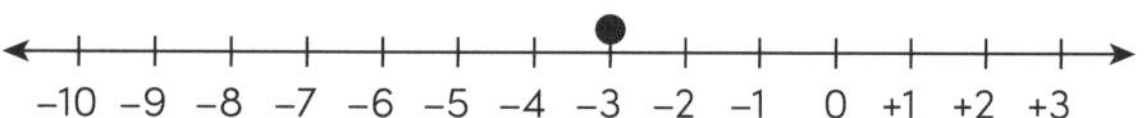


Where will it be if it is moved:

- **a** 4 places to the right? ______
- **b** 2 places to the left? ______
- **c** 5 places to the right? ______
- **d** 2 places to the right? ______
- **e** 7 places to the left? ______
- **f** 2 places to the right then 5 places to the left? ______
- **g** 3 places to the left then 1 place to the right? ______
- **h** 5 places to the left then 3 places to the right? ______

**3** A counter is currently on +5 position.

Find where it will be if it is moved:

- **a** 2 places forward? ______
- **b** 4 places backwards? ______
- **c** 1 place forward? ______
- **d** 3 places forward? ______
- **e** 6 places backwards? ______
- **f** 3 places back then 5 places back? ______
- **g** 2 places forward then 6 places back? ______
- **h** 1 place forward then 4 places back? ______

**4** The temperature is currently −2 °C.

What will the temperature be if it:

- **a** rises 3 degrees? ______
- **b** falls 5 degrees? ______
- **c** rises 4 degrees? ______
- **d** falls 3 degrees? ______
- **e** falls 9 degrees? ______
- **f** rises 2 degrees then falls 5 degrees? ______
- **g** falls 1 degree then rises 7 degrees? ______
- **h** falls 4 degrees then rises 6 degrees? ______

INTEGERS

# Introducing integers

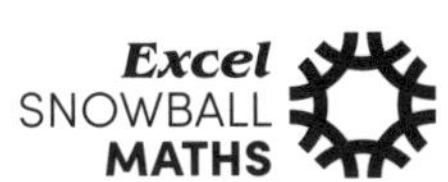

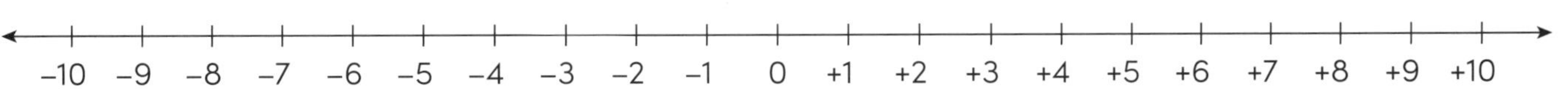


**1** A hiker is currently on +1 position.

Find where the hiker will be if he hikes:

- **a** 4 steps up →
- **b** 3 steps down ←
- **c** 2 steps up
- **d** 6 steps down
- **e** 5 steps up then 8 steps down
- **f** 4 steps down then 7 steps up
- **g** 4 steps down then 6 steps down
- **h** 1 step up then 9 steps down

**2** Manahil's current bank account is –$3.

What will be her account balance if she:

- **a** is charged $7? ←
- **b** deposits $5? →
- **c** is charged $6?
- **d** deposits $10?
- **e** is charged $8?
- **f** deposits $4 and is then charged $9?
- **g** is charged $5 and then charged another $3?
- **h** is charged $7 and then deposits $6?

**3** Adam is currently 3 floors above ground level (+3). On what floor will he be if he goes:

- **a** 2 floors up?
- **b** 4 floors down?
- **c** 3 floors up?
- **d** 5 floors down?
- **e** 2 floors up then 6 floors down?
- **f** 3 floors down then 5 floors up?
- **g** 1 floor up then 6 floors down?

**4** Donna is playing a game where the points score can be positive or negative. She has 3 points. How many points will she have if she:

- **a** loses 4 points?
- **b** gains 2 points?
- **c** loses 7 points?
- **d** loses 10 points?
- **e** gains 2 points then loses 7 points?
- **f** gains 5 points then loses 10 points?
- **g** loses 4 points then loses another 4 points?

**INTEGERS**

# Adding integers

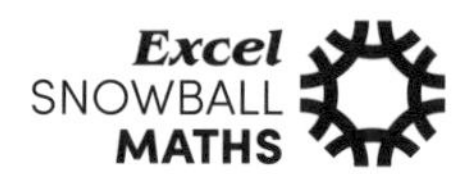


**1** Evaluate the following:

**a** $-3 + 4 =$ ____________

Move 4 places to the $\overrightarrow{\text{RIGHT}}$.

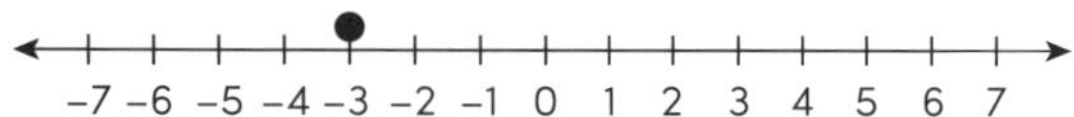


**b** $-1 + 2 =$ ____________

Move 2 places to the $\overrightarrow{\text{RIGHT}}$.

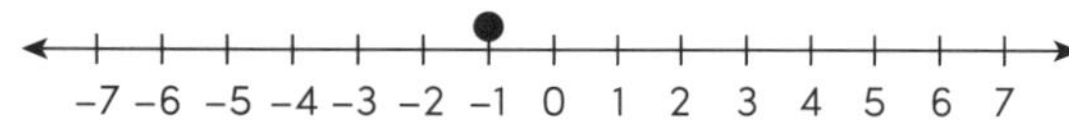


**c** $-5 + 3 =$ ____________

Move 3 places to the RIGHT.

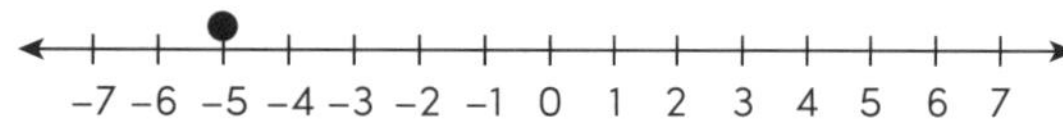


**d** $-4 + 6 =$ ____________

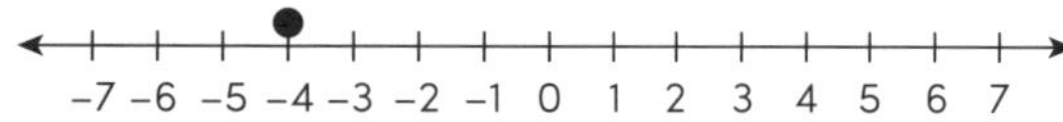


**e** $-6 + 4 =$ ____________

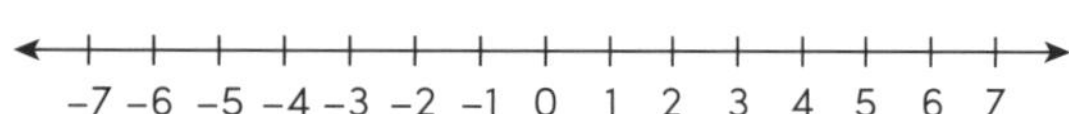


**f** $-2 + 6 =$ ____________

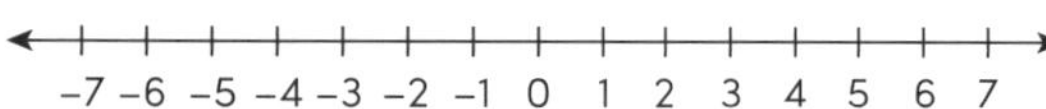


**g** $-5 + 7 =$ ____________

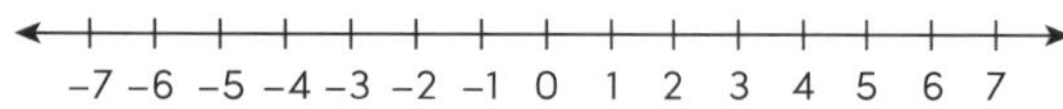


**h** $-7 + 5 =$ ____________

–7 –6 –5 –4 –3 –2 –1 0 1 2 3 4 5 6 7

**i** $-4 + 7 =$ ____________

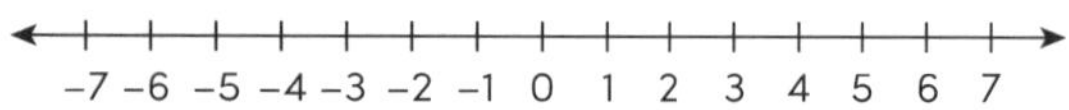


**j** $-3 + 3 =$ ____________

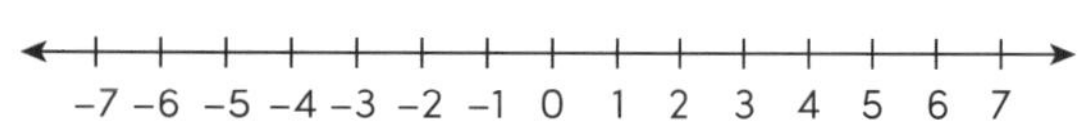


**k** $-6 + 5 =$ ____________

–7 –6 –5 –4 –3 –2 –1 0 1 2 3 4 5 6 7

**2** Evaluate:

**a** $-7 + 3 =$ ____________

**b** $-4 + 5 =$ ____________

**c** $-2 + 8 =$ ____________

**d** $-3 + 9 =$ ____________

**e** $-1 + 5 =$ ____________

**f** $-7 + 2 =$ ____________

**g** $-6 + 8 =$ ____________

**h** $-2 + 4 =$ ____________

**i** $-8 + 3 =$ ____________

**j** $-5 + 8 =$ ____________

**k** $-6 + 2 =$ ____________

**l** $-3 + 7 =$ ____________

**m** $-2 + 2 =$ ____________

**n** $-5 + 5 =$ ____________

**o** $-7 + 10 =$ ____________

**p** $-6 + 10 =$ ____________

**q** $-4 + 11 =$ ____________

**r** $-3 + 10 =$ ____________

**s** $-14 + 12 =$ ____________

**t** $-17 + 13 =$ ____________

**u** $-2 + 10 =$ ____________

**v** $-9 + 10 =$ ____________

# INTEGERS
## Adding integers

1 Evaluate the following:

a $3 + (-4) =$ ________
Move 4 places to the LEFT.

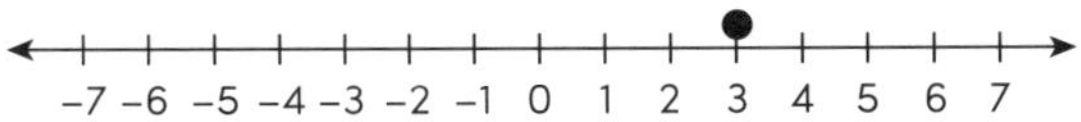

b $5 + (-6) =$ ________
Move 6 places to the LEFT.

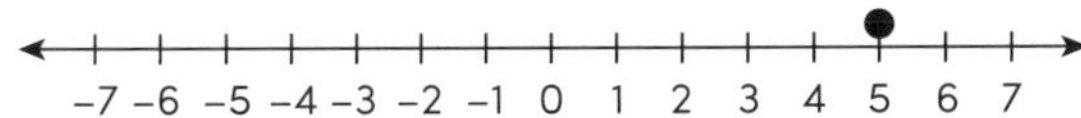

c $4 + (-2) =$ ________
Move 2 places to the LEFT.

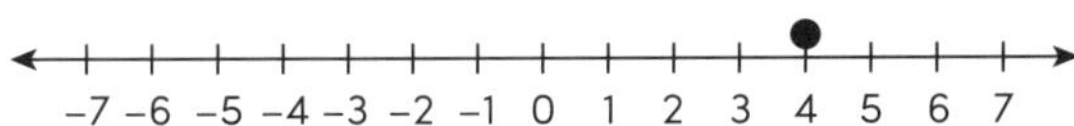

d $1 + (-4) =$ ________

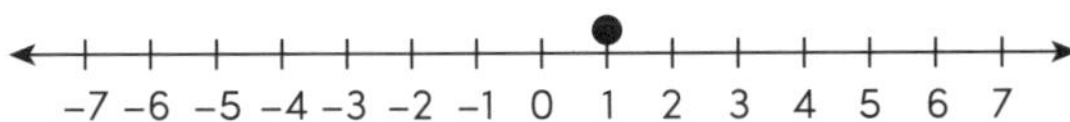

e $2 + (-6) =$ ________

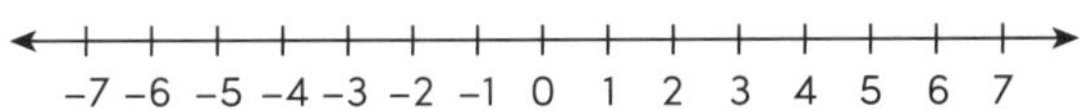

f $4 + (-3) =$ ________

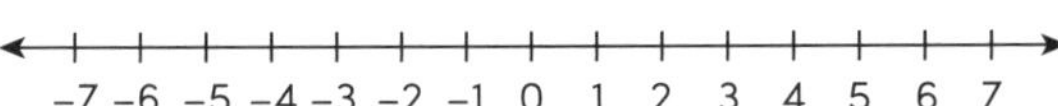

g $5 + (-8) =$ ________

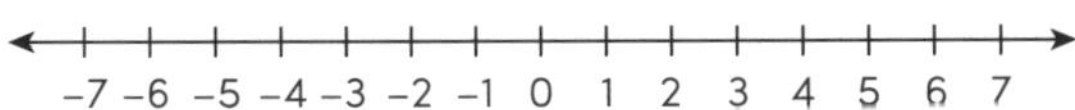

h $2 + (-4) =$ ________

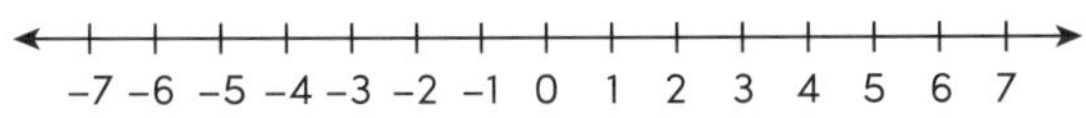

i $6 + (-5) =$ ________

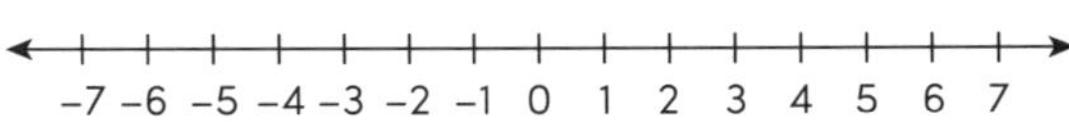

j $3 + (-8) =$ ________

–7 –6 –5 –4 –3 –2 –1 0 1 2 3 4 5 6 7

k $1 + (-6) =$ ________

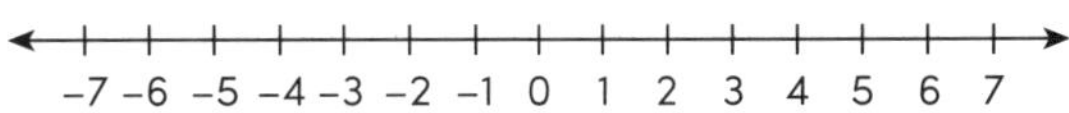

2 Evaluate:

a $4 + (-8) =$ ________

b $3 + (-6) =$ ________

c $2 + (-6) =$ ________

d $5 + (-5) =$ ________

e $3 + (-3) =$ ________

f $4 + (-10) =$ ________

g $2 + (-10) =$ ________

h $7 + (-6) =$ ________

i $2 + (-8) =$ ________

j $2 + (-9) =$ ________

k $5 + (-7) =$ ________

l $9 + (-8) =$ ________

m $8 + (-5) =$ ________

n $4 + (-6) =$ ________

o $5 + (-10) =$ ________

p $2 + (-8) =$ ________

q $6 + (-6) =$ ________

r $1 + (-2) =$ ________

s $7 + (-7) =$ ________

t $13 + (-8) =$ ________

u $15 + (-10) =$ ________

v $8 + (-12) =$ ________

UNIT 6

INTEGERS

# Adding integers

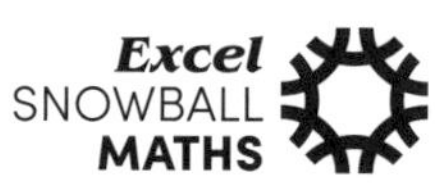

**1** Evaluate the following:

**a** $-3 + (-3) =$ ________
Move 3 places to the LEFT.

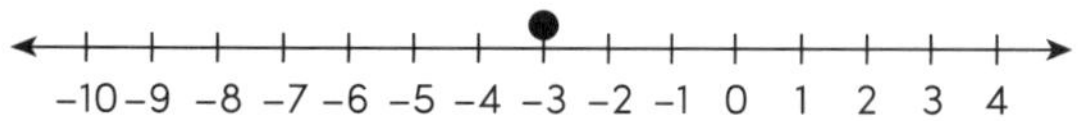

**b** $-2 + (-5) =$ ________
Move 5 places to the LEFT.

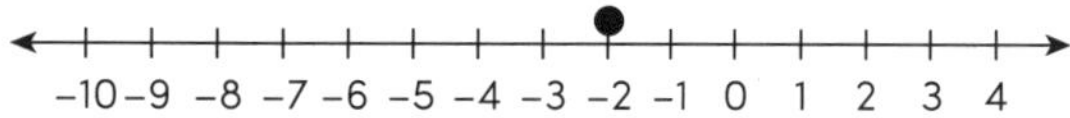

**c** $-4 + (-3) =$ ________
Move 3 places to the LEFT.

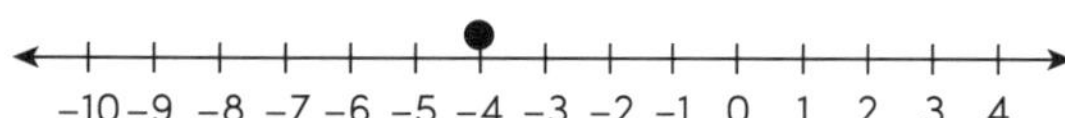

**d** $-3 + (-1) =$ ________

**e** $-5 + (-3) =$ ________

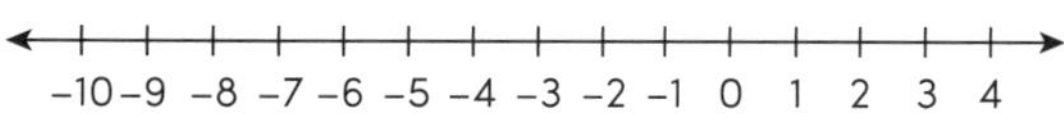

**f** $-6 + (-3) =$ ________

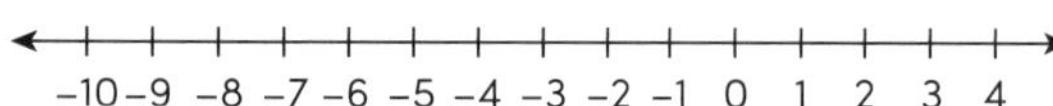

**g** $-1 + (-7) =$ ________

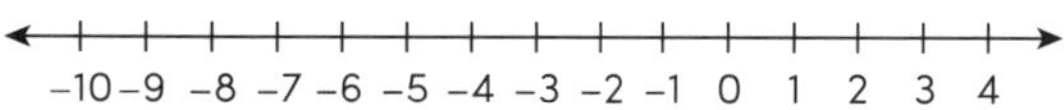

**h** $-5 + (-4) =$ ________

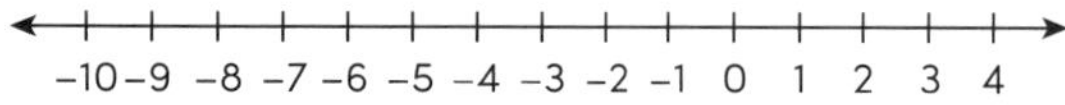

**i** $-2 + (-6) =$ ________

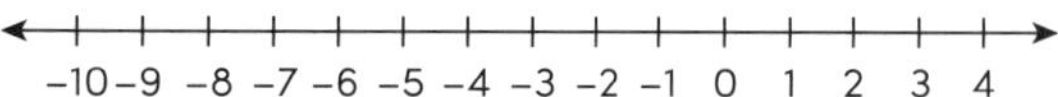

**j** $-1 + (-5) =$ ________

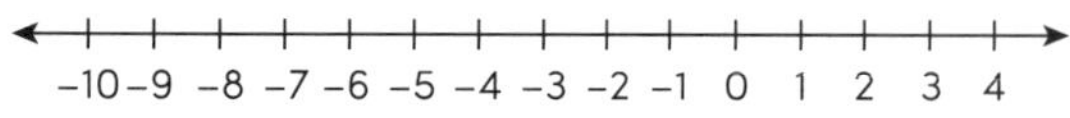

**k** $-4 + (-4) =$ ________

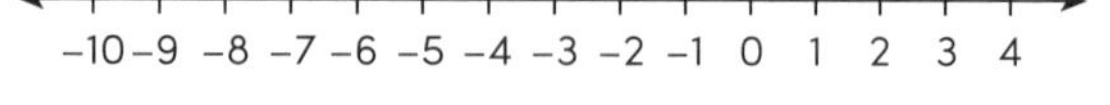

**2** Evaluate:

**a** $-4 + (-2) =$ ________

**b** $-1 + (-6) =$ ________

**c** $-2 + (-4) =$ ________

**d** $-5 + (-1) =$ ________

**e** $-2 + (-2) =$ ________

**f** $-6 + (-2) =$ ________

**g** $-7 + (-1) =$ ________

**h** $-3 + (-4) =$ ________

**i** $-5 + (-3) =$ ________

**j** $-2 + (-3) =$ ________

**k** $-6 + (-4) =$ ________

**l** $-8 + (-2) =$ ________

**m** $-7 + (-2) =$ ________

**n** $-4 + (-7) =$ ________

**o** $-5 + (-10) =$ ________

**p** $-3 + (-8) =$ ________

**q** $-4 + (-10) =$ ________

**r** $-5 + (-8) =$ ________

**s** $-7 + (-6) =$ ________

**t** $-13 + (-8) =$ ________

**u** $-6 + (-9) =$ ________

**v** $-3 + (-7) =$ ________

**INTEGERS**

# Subtracting integers

**1** Evaluate the following:

**a** 3 – 8 = ________

Move 8 places to the LEFT.

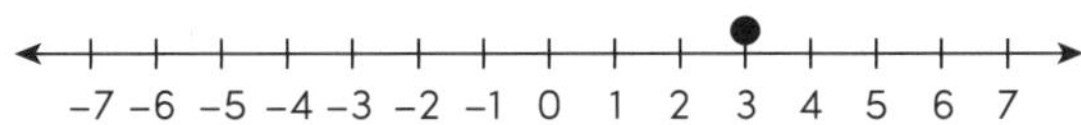


**b** 1 – 2 = ________

Move 2 places to the LEFT.

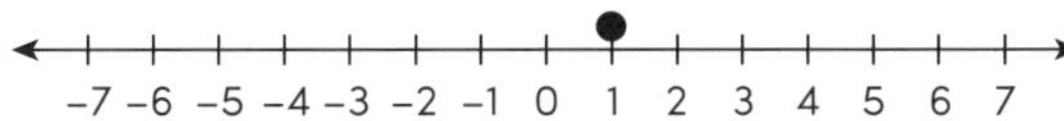


**c** 5 – 7 = ________

Move 7 places to the LEFT.

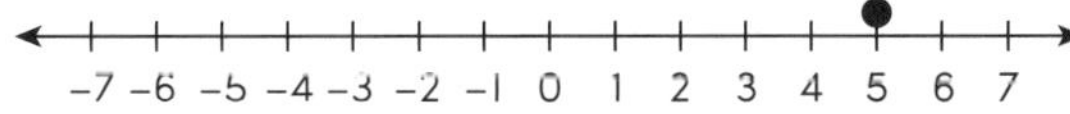


**d** 4 – 6 = ________

**e** 6 – 9 = ________

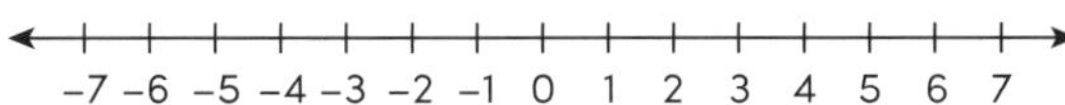


**f** 1 – 7 = ________

–7 –6 –5 –4 –3 –2 –1 0 1 2 3 4 5 6 7

**g** 4 – 8 = ________

–7 –6 –5 –4 –3 –2 –1 0 1 2 3 4 5 6 7

**h** 4 – 9 = ________

–7 –6 –5 –4 –3 –2 –1 0 1 2 3 4 5 6 7

**i** 1 – 8 = ________

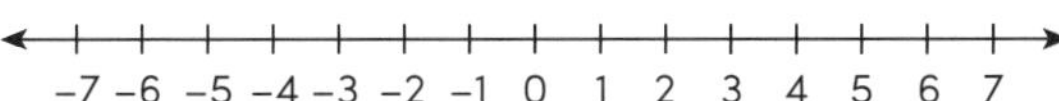


**j** 3 – 8 = ________

–7 –6 –5 –4 –3 –2 –1 0 1 2 3 4 5 6 7

**k** 6 – 10 = ________

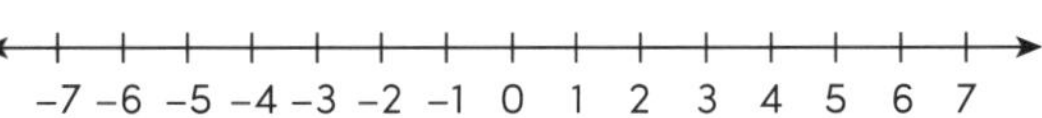


**2** Evaluate:

**a** 2 – 8 = ________

**b** 2 – 4 = ________

**c** 3 – 5 = ________

**d** 4 – 10 = ________

**e** 4 – 7 = ________

**f** 5 – 10 = ________

**g** 1 – 6 = ________

**h** 7 – 10 = ________

**i** 3 – 4 = ________

**j** 2 – 6 = ________

**k** 2 – 9 = ________

**l** 1 – 5 = ________

**m** 1 – 3 = ________

**n** 5 – 8 = ________

**o** 2 – 3 = ________

**p** 2 – 10 = ________

**q** 5 – 6 = ________

**r** 7 – 9 = ________

**s** 1 – 10 = ________

**t** 9 – 12 = ________

**u** 10 – 13 = ________

**v** 12 – 15 = ________

UNIT 8

**INTEGERS**

# Subtracting integers

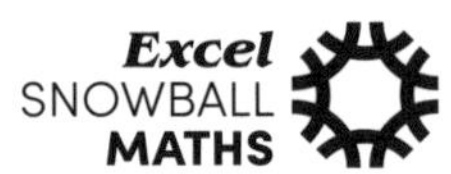


**1** Evaluate the following:

**a** −3 − 4 = ________

Move 4 places to the LEFT.

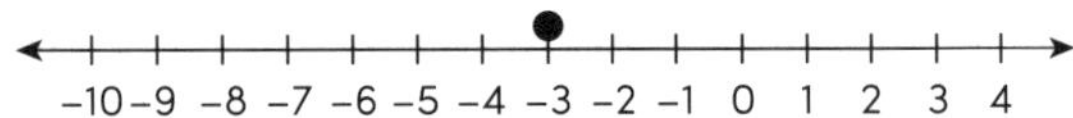


**b** −1 − 5 = ________

Move 5 places to the LEFT.

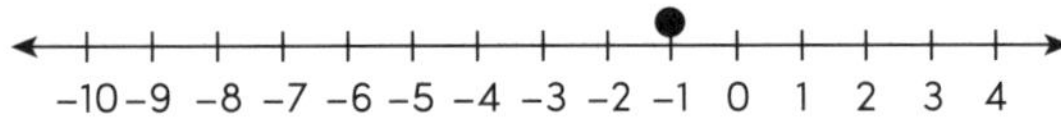


**c** −2 − 3 = ________

Move 3 places to the LEFT.

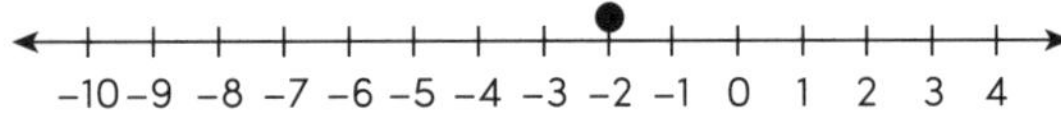


**d** −4 − 3 = ________

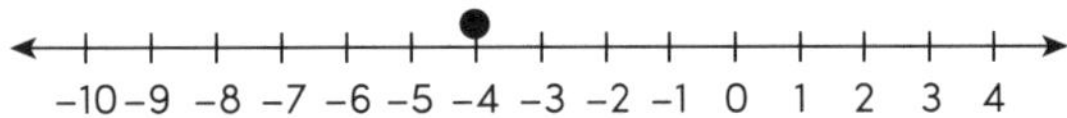


**e** −2 − 5 = ________

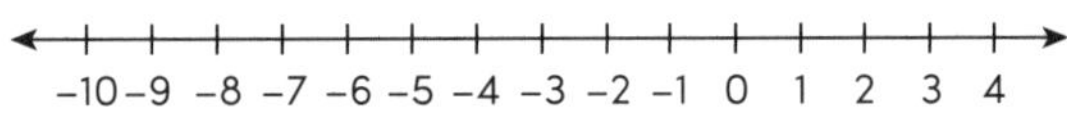


**f** −3 − 6 = ________

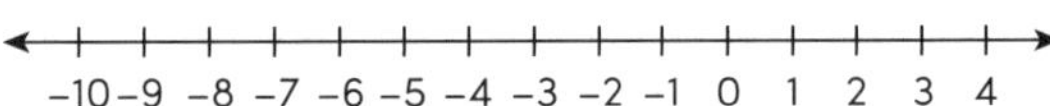


**g** −2 − 6 = ________

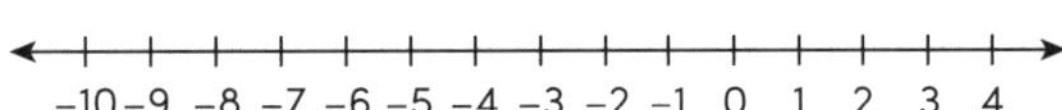


**h** −5 − 2 = ________

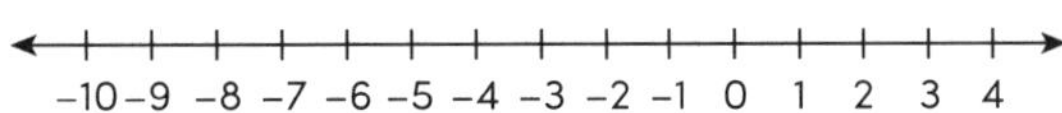


**i** −4 − 2 = ________

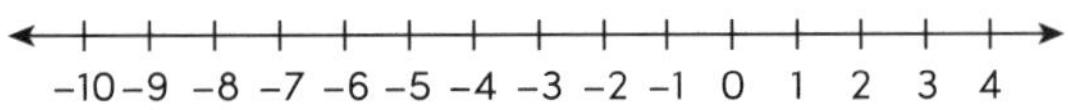


**j** −2 − 8 = ________

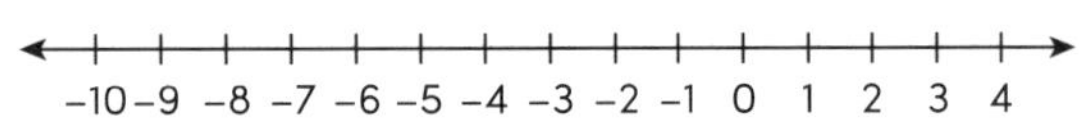


**k** −3 − 3 = ________

−10 −9 −8 −7 −6 −5 −4 −3 −2 −1 0 1 2 3 4

**2** Evaluate:

**a** −2 − 4 = ________

**b** −2 − 2 = ________

**c** −1 − 3 = ________

**d** −3 − 5 = ________

**e** −3 − 1 = ________

**f** −4 − 4 = ________

**g** −4 − 5 = ________

**h** −3 − 2 = ________

**i** −1 − 4 = ________

**j** −5 − 4 = ________

**k** −5 − 3 = ________

**l** −4 − 1 = ________

**m** −1 − 2 = ________

**n** −7 − 1 = ________

**o** −1 − 1 = ________

**p** −7 − 2 = ________

**q** −8 − 4 = ________

**r** −5 − 5 = ________

**s** −9 − 2 = ________

**t** −16 − 4 = ________

**u** −13 − 3 = ________

**v** −17 − 2 = ________

UNIT 9

## INTEGERS
# Subtracting integers

**1** Evaluate the following:

**a** $3 - (-4)$
$= 3 \boxed{+} 4$
= ______

**b** $1 - (-6)$
$= 1 \boxed{+} 6$
= ______

**c** $2 - (-4)$
= ____ ☐ ____
= ______

**d** $5 - (-4)$
= ____ ☐ ____
= ______

**e** $10 - (-3)$
= ______
= ______

**f** $0 - (-5)$
= ______
= ______

**g** $6 - (-4)$
= ______
= ______

**h** $7 - (-6)$
= ______
= ______

**i** $9 - (-7)$
= ______
= ______

**j** $15 - (-4)$
= ______
= ______

**2** Evaluate the following:

**a** $-3 - (-5)$
$= -3$ ☐ $5$
= ______

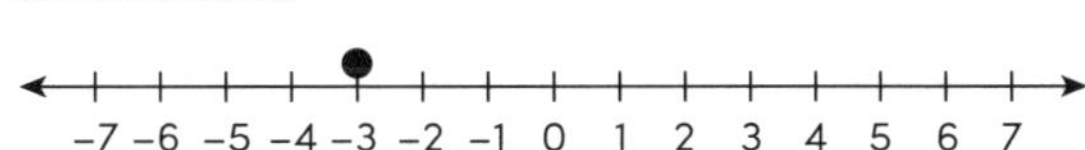


**b** $-4 - (-7)$
$= -4$ ☐ $7$
= ______

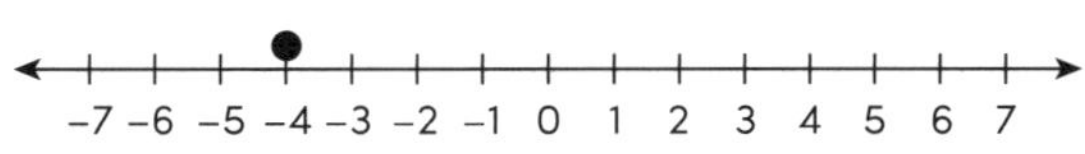


**c** $-7 - (-8)$
$= -7$ ☐ $8$
= ______

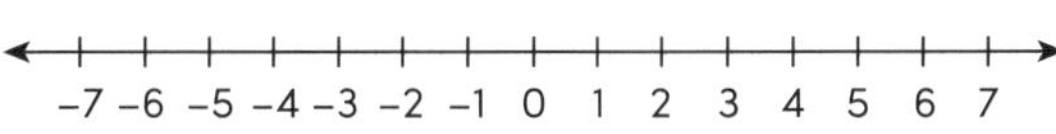


**d** $-2 - (-4)$
= ____ ☐ ____
= ______

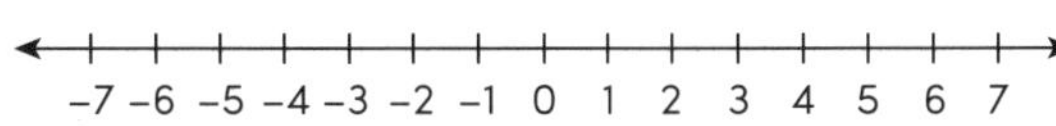


**e** $-6 - (-4)$
= ____ ☐ ____
= ______

−7 −6 −5 −4 −3 −2 −1 0 1 2 3 4 5 6 7

**f** $-7 - (-3)$
= ____ ☐ ____
= ______

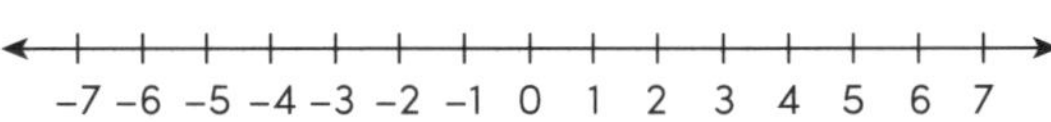


**g** $-5 - (-2)$
= ______
= ______

**h** $-4 - (-1)$
= ______
= ______

−7 −6 −5 −4 −3 −2 −1 0 1 2 3 4 5 6 7

**i** $-6 - (-8)$
= ______
= ______

**j** $-7 - (-2)$
= ______
= ______

−7 −6 −5 −4 −3 −2 −1 0 1 2 3 4 5 6 7

## INTEGERS

# Integer problems

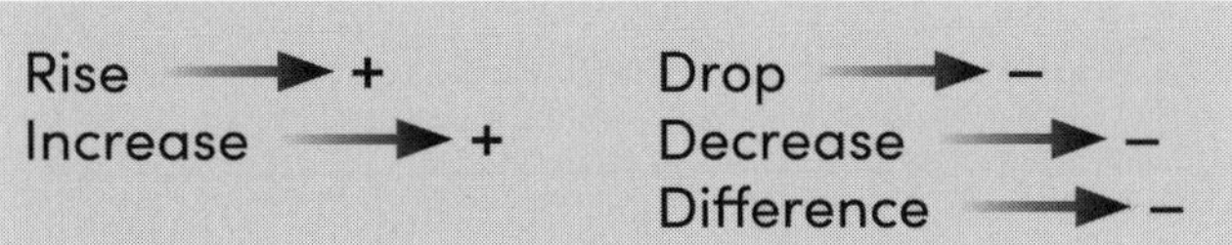


**1** Solve the following problems with or without a calculator:

**a** What is the temperature if it rises 15 °C from −13 °C?

−13 ☐ (rise) ________ = ________ °C

**b** What is the temperature if it drops 12 °C from −8 °C?

−8 ☐ (drops) ________ = ________

**c** What is the temperature if it drops 4 °C from −15 °C?

________ (from) ☐ ________ = ________

**d** What is the temperature if it rises 18 °C from −16 °C?

________ (from) ☐ ________ = ________

**e** The temperature increased by 16 °C from −9 °C. What was the final temperature?

________________

**f** The temperature decreased by 15 °C from −6 °C. What was the final temperature?

________________

**2** Solve:

**a** From the ground floor, a lift went down 2 floors then up 7 floors. What floor is the lift now at?

−2 ☐ ______ = ________

**b** From the ground floor, a lift went down 5 floors then up 9 floors. What floor is the lift now at?

−5 ☐ ______ = ________

**c** From the ground floor, a lift went down 8 floors then up 6 floors. What floor is the lift now at?

________________

**d** From the ground floor, a lift went down 4 floors then up 10 floors. What floor is the lift now at?

________________

**e** From the ground floor, a lift went up 12 floors then down 15 floors. What floor is the lift now at?

________________

**f** From the ground floor, a lift went up 21 floors then down 26 floors. What floor is the lift now at?

________________

**3** Solve:

**a** Sydney is 10 hours ahead of London, while Mexico is 6 hours behind London. What is the time difference between Sydney and Mexico?

10 ☐ −6 = ________

**b** Paris is 1 hour ahead of London, while New York is 5 hours behind London. What is the time difference between Paris and New York?

1 ☐ −5 = ________

**c** Bangkok is 7 hours ahead of London, while Alaska is 8 hours behind London. What is the time difference between Bangkok and Alaska?

________________

UNIT 11

INTEGERS

# Integer problems

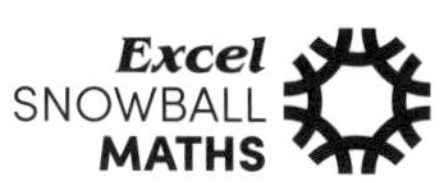

**1** Solve:

**a** Athens is 3 hours ahead of London, while Detroit is 4 hours behind London. What is the time difference between Athens and Detroit?

______________________

______________________

**b** Hong Kong is 8 hours ahead of London, while Costa Rica is 6 hours behind London. What is the time difference between Hong Kong and Costa Rica?

______________________

______________________

**c** Indiana is 5 hours **behind** London, while Alaska is 8 hours behind London. What is the time difference between Indiana and Alaska?

______________________

______________________

______________________

**d** Samoa is 11 hours behind London, while Detroit is 4 hours behind London. What is the time difference between Samoa and Detroit?

______________________

______________________

______________________

**e** Hawaii is 10 hours behind London, while Mexico is 6 hours behind London. What is the time difference between Hawaii and Mexico?

______________________

______________________

______________________

Deposit → +     Withdraw → –

**2** Solve:

**a** Jason's bank balance is currently $130. If he withdraws $180, what will his balance be?

______ (currently) ☐ ______ = ______

**b** David's bank balance is currently –$250. If he deposits $390, what will his balance be?

______ (currently) ☐ ______ = ______

**c** Sam's bank balance is currently –$550. If he deposits $800, what will his balance be?

______________________

______________________

**d** Fred's bank balance is currently $680. If he withdraws $700, what will his balance be?

______________________

______________________

UNIT 12

**INTEGERS**

# Multiplying and dividing integers

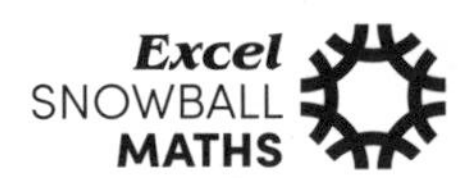


**1** Evaluate the following:

**a** $-5 \times 3$ = ______ **b** $-2 \times 4$ = ______ **c** $-9 \times 4$ = ______

**d** $-6 \times 8$ = ______ **e** $3 \times (-5)$ = ______ **f** $7 \times (-2)$ = ______

**g** $9 \times (-4)$ = ______ **h** $6 \times (-5)$ = ______ **i** $-3 \times 10$ = ______

**j** $4 \times (-7)$ = ______ **k** $9 \times (-8)$ = ______ **l** $-5 \times 8$ = ______

**2** Evaluate the following:

**a** $-2 \times (-7)$ = ______ **b** $-1 \times (-20)$ = ______

**c** $-10 \times (-4)$ = ______ **d** $-3 \times (-10)$ = ______

**e** $-5 \times (-11)$ = ______ **f** $-6 \times (-10)$ = ______

**g** $-8 \times (-9)$ = ______ **h** $-9 \times (-7)$ = ______

**i** $-3 \times (-8)$ = ______ **j** $-7 \times (-7)$ = ______

**k** $-4 \times (-4)$ = ______ **l** $-5 \times (-5)$ = ______

**3** Evaluate the following:

**a** $(-5)^2 = -5 \times (-5) =$ ______

**b** $(-3)^2 = -3 \times (-3) =$ ______

**c** $(-6)^2 =$ ______ $\times$ ______ $=$ ______

**d** $(-7)^2 =$ ______ $\times$ ______ $=$ ______

**e** $(-2)^2 =$ ______

**f** $(-8)^2 =$ ______

**g** $(-10)^2 =$ ______

**h** $(-9)^2 =$ ______

**i** $(-4)^2 =$ ______

**j** $(-1)^2 =$ ______

**k** $(-11)^2 =$ ______

**l** $(-12)^2 =$ ______

**4** Evaluate the following:

**a** $-15 \div 3$ = ______ **b** $-8 \div 4$ = ______ **c** $-36 \div 4$ = ______

**d** $-48 \div 8$ = ______ **e** $15 \div (-5)$ = ______ **f** $14 \div (-2)$ = ______

**g** $36 \div (-4)$ = ______ **h** $30 \div (-5)$ = ______ **i** $-30 \div 10$ = ______

**j** $28 \div (-7)$ = ______ **k** $72 \div (-8)$ = ______ **l** $-40 \div 8$ = ______

**5** Evaluate the following:

**a** $-63 \div (-7) =$ ______ **b** $-20 \div (-20) =$ ______

**c** $-42 \div (-6) =$ ______ **d** $-36 \div (-9) =$ ______

**e** $-55 \div (-11) =$ ______ **f** $-45 \div (-9) =$ ______

**g** $-72 \div (-9) =$ ______ **h** $-90 \div (-10) =$ ______

**i** $-32 \div (-4) =$ ______ **j** $-49 \div (-7) =$ ______

**k** $-16 \div (-4) =$ ______ **l** $-25 \div (-5) =$ ______

UNIT 13

INTEGERS

# Multiplying and dividing integers

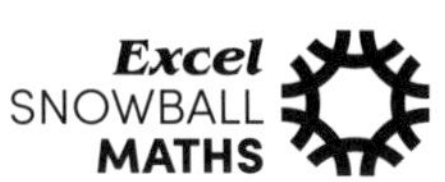

**1** Evaluate the following:

**a** $\frac{-12}{4}$ = ________

**b** $\frac{-16}{2}$ = ________

**c** $\frac{-20}{2}$ = ________

**d** $\frac{-24}{6}$ = ________

**e** $\frac{25}{-5}$ = ________

**f** $\frac{32}{-8}$ = ________

**g** $\frac{21}{-7}$ = ________

**h** $\frac{42}{-6}$ = ________

**i** $\frac{27}{-9}$ = ________

**j** $\frac{-16}{-4}$ = ________

**k** $\frac{-20}{-4}$ = ________

**l** $\frac{-30}{-10}$ = ________

**m** $\frac{-12}{-2}$ = ________

**n** $\frac{-45}{-5}$ = ________

**o** $\frac{-64}{-8}$ = ________

**2** Find the unknown value:

**a** $-5 \times ? = -15$

$? = -15 \div -5$

$=$ ________

**b** $-2 \times ? = -8$

$? = -8 \div -2$

$=$ ________

**c** $-9 \times ? = -36$

$? = -36 \div$ ________

$=$ ________

**d** $-6 \times ? = -48$

$? = -48 \div$ ________

$=$ ________

**e** $-3 \times ? = -27$

**f** $-7 \times ? = -14$

**g** $-4 \times ? = -36$

**h** $-10 \times ? = -30$

**i** $-8 \times ? = -40$

**j** $-7 \times ? = -28$

**k** $? \times 4 = -20$

**l** $? \times 6 = -18$

**m** $? \times 5 = -30$

**n** $? \times 7 = -42$

**o** $? \times (-8) = -56$

**p** $? \times (-3) = -27$

**q** $6 \times ? = -30$

**r** $5 \times ? = -45$

**s** $-8 \times ? = -40$

**t** $? \times 9 = -72$

UNIT 14

**INTEGERS**

# Order of operations

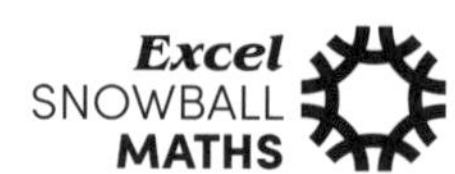

–10 –9 –8 –7 –6 –5 –4 –3 –2 –1 0 +1 +2 +3 +4 +5 +6 +7 +8 +9 +10

**1** Evaluate the following:

**a** $3 - 4 \times 2$
$= 3 -$ ______
$=$ ______

**b** $4 - 3 \times 2$
$= 4 -$ ______
$=$ ______

**c** $6 - 4 \times 2$
$= 6$ ______
$=$ ______

**d** $7 - 5 \times 2$
$= 7$ ______
$=$ ______

**e** $10 - 6 \times 2$
$=$ ______
$=$ ______

**f** $8 - 4 \times 3$
$=$ ______
$=$ ______

**g** $9 - 5 \times 2$
$=$ ______
$=$ ______

**h** $12 - 5 \times 3$
$=$ ______
$=$ ______

**i** $8 - 6 \times 2$
$=$ ______
$=$ ______

**j** $7 - 4 \times 2$
$=$ ______
$=$ ______

**k** $10 - 3 \times 5$
$=$ ______
$=$ ______

**l** $6 - 7 \times 1$
$=$ ______
$=$ ______

**m** $8 - 4 \times 4$
$=$ ______
$=$ ______

**n** $12 - 5 \times 4$
$=$ ______
$=$ ______

**2** Evaluate the following:

**a** $-3 - 4 \times (-2)$
$= -3 +$ ______
$=$ ______

**b** $-5 - 3 \times (-2)$
$= -5 +$ ______
$=$ ______

**c** $-4 - 5 \times (-2)$
$= -4$ ______
$=$ ______

**d** $-5 - 4 \times (-2)$
$= -5$ ______
$=$ ______

**e** $-6 - 5 \times (-2)$
$=$ ______
$=$ ______

**f** $-8 - 6 \times (-2)$
$=$ ______
$=$ ______

**g** $-10 - 4 \times (-3)$
$=$ ______
$=$ ______

**h** $-8 - 4 \times (-3)$
$=$ ______
$=$ ______

**i** $-10 - 3 \times (-5)$
$=$ ______
$=$ ______

**j** $-6 - 2 \times (-5)$
$=$ ______
$=$ ______

**k** $-12 - 7 \times (-2)$
$=$ ______
$=$ ______

**l** $-15 - 5 \times (-4)$
$=$ ______
$=$ ______

**m** $-20 - 6 \times (-4)$
$=$ ______
$=$ ______

**n** $-18 - 3 \times (-6)$
$=$ ______
$=$ ______

**3** Evaluate the following:

**a** $3 + 4 \times (-2)$
$= 3 + (-$ ______ $)$
$=$ ______

**b** $4 + 3 \times (-2)$
$= 4 + (-$ ______ $)$
$=$ ______

**c** $7 + 5 \times (-2)$
$= 7$ ______
$=$ ______

**d** $2 + 6 \times (-2)$
$= 2$ ______
$=$ ______

**e** $3 + 5 \times (-2)$
$=$ ______
$=$ ______

**f** $7 + 6 \times (-2)$
$=$ ______
$=$ ______

**g** $10 + 4 \times (-3)$
$=$ ______
$=$ ______

**h** $12 + 5 \times (-3)$
$=$ ______
$=$ ______

**i** $9 + 6 \times (-2)$
$=$ ______
$=$ ______

**j** $8 + 4 \times (-3)$
$=$ ______
$=$ ______

**k** $15 + 6 \times (-3)$
$=$ ______
$=$ ______

**l** $11 + 3 \times (-5)$
$=$ ______
$=$ ______

**m** $12 + 4 \times (-5)$
$=$ ______
$=$ ______

**n** $24 + 5 \times (-6)$
$=$ ______
$=$ ______

UNIT 15

INTEGERS

# Order of operations

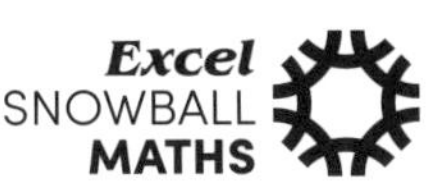

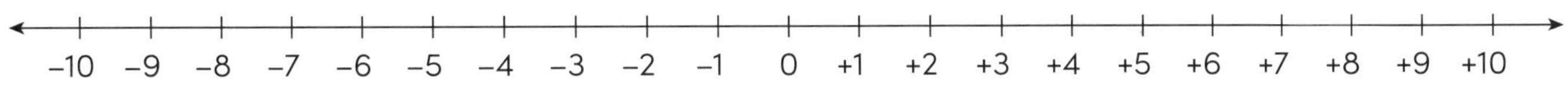


**1** Evaluate the following:

**a** $3 - 8 \div (-2)$
= 3 + ______
= ______

**b** $2 - 6 \div (-3)$
= 2 + ______
= ______

**c** $5 - 8 \div (-4)$
= 5 ______
= ______

**d** $6 - 10 \div (-2)$
= 6 ______
= ______

**e** $5 - 12 \div (-2)$
= ______
= ______

**f** $7 - 12 \div (-4)$
= ______
= ______

**g** $9 - 8 \div (-4)$
= ______
= ______

**h** $12 - 15 \div (-3)$
= ______
= ______

**i** $12 - 8 \div (-4)$
= ______
= ______

**j** $15 - 10 \div (-2)$
= ______
= ______

**k** $14 - 6 \div (-3)$
= ______
= ______

**l** $5 - 24 \div (-6)$
= ______
= ______

**m** $7 - 32 \div (-4)$
= ______
= ______

**n** $19 - 18 \div (-2)$
= ______
= ______

**2** Evaluate the following:

**a** $-5 + (-8) \div 4$
= −5 + (−______)
= ______

**b** $-10 + (-12) \div 3$
= −10 + (−______)
= ______

**c** $-6 + (-15) \div 5$
= −6 ______
= ______

**d** $-8 + (-20) \div 4$
= −8 ______
= ______

**e** $-9 + (-21) \div 7$
= ______
= ______

**f** $-5 + (-24) \div 6$
= ______
= ______

**g** $-12 + (-30) \div 5$
= ______
= ______

**h** $-4 + (-20) \div 5$
= ______
= ______

**i** $-6 + (-27) \div 3$
= ______
= ______

**j** $-5 + (-35) \div 5$
= ______
= ______

**k** $-8 + (-36) \div 6$
= ______
= ______

**l** $-11 + (-20) \div 10$
= ______
= ______

**m** $-7 + (-18) \div 6$
= ______
= ______

**n** $-18 + (-40) \div 5$
= ______
= ______

**3** Evaluate the following:

**a** $2 + 15 \div (-3)$
= 2 + (−______)
= ______

**b** $5 + 14 \div (-2)$
= 5 + (−______)
= ______

**c** $3 + 24 \div (-4)$
= 3 ______
= ______

**d** $7 + 18 \div (-2)$
= 7 ______
= ______

**e** $5 + 12 \div (-2)$
= ______
= ______

**f** $1 + 8 \div (-2)$
= ______
= ______

**g** $3 + 20 \div (-4)$
= ______
= ______

**h** $7 + 27 \div (-3)$
= ______
= ______

**i** $4 + 16 \div (-2)$
= ______
= ______

**j** $3 + 14 \div (-2)$
= ______
= ______

**k** $6 + 21 \div (-3)$
= ______
= ______

**l** $7 + 32 \div (-4)$
= ______
= ______

UNIT 16

# INTEGERS
## Order of operations

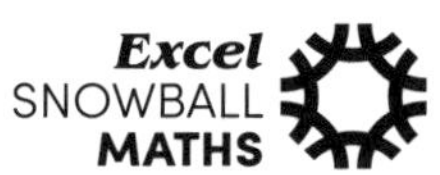

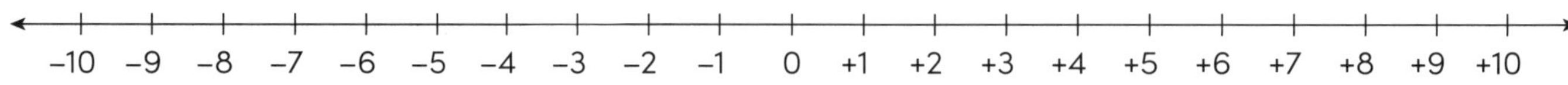


**1** Evaluate the following:

**a** $5 \times (2 - 5)$
$= 5 \times (-\_\_\_\_)$
$= \_\_\_\_$

**b** $2 \times (5 - 8)$
$= 2 \times (-\_\_\_\_)$
$= \_\_\_\_$

**c** $4 \times (4 - 7)$
$= 4 \times \_\_\_\_$
$= \_\_\_\_$

**d** $6 \times (3 - 5)$
$= 6 \times \_\_\_\_$
$= \_\_\_\_$

**e** $3 \times (2 - 5)$
$= 3 \_\_\_\_$
$= \_\_\_\_$

**f** $5 \times (2 - 4)$
$= 5 \_\_\_\_$
$= \_\_\_\_$

**g** $7 \times (3 - 4)$
$= \_\_\_\_$
$= \_\_\_\_$

**h** $7 \times (6 - 8)$
$= \_\_\_\_$
$= \_\_\_\_$

**i** $4 \times (5 - 8)$
$= \_\_\_\_$
$= \_\_\_\_$

**j** $5 \times (6 - 10)$
$= \_\_\_\_$
$= \_\_\_\_$

**k** $7 \times (6 - 10)$
$= \_\_\_\_$
$= \_\_\_\_$

**l** $4 \times (7 - 9)$
$= \_\_\_\_$
$= \_\_\_\_$

**m** $5 \times (10 - 15)$
$= \_\_\_\_$
$= \_\_\_\_$

**n** $10 \times (6 - 8)$
$= \_\_\_\_$
$= \_\_\_\_$

**2** Evaluate the following:

**a** $[3 + (-4)] \times (-2)$
$= -\_\_\_\_ \times (-2)$
$= \_\_\_\_$

**b** $[3 + (-5)] \times (-4)$
$= -\_\_\_\_ \times (-4)$
$= \_\_\_\_$

**c** $[6 + (-8)] \times (-3)$
$= \_\_\_\_ \times (-3)$
$= \_\_\_\_$

**d** $[7 + (-8)] \times (-5)$
$= \_\_\_\_ \times (-5)$
$= \_\_\_\_$

**e** $[6 + (-8)] \times (-4)$
$= \_\_\_\_ \times (\_\_\_\_)$
$= \_\_\_\_$

**f** $[6 + (-9)] \times (-5)$
$= \_\_\_\_ \times (\_\_\_\_)$
$= \_\_\_\_$

**g** $[7 + (-10)] \times (-2)$
$= \_\_\_\_ \times (\_\_\_\_)$
$= \_\_\_\_$

**h** $[8 + (-10)] \times (-3)$
$= \_\_\_\_ \times (\_\_\_\_)$
$= \_\_\_\_$

**i** $[5 + (-8)] \times (-3)$
$= \_\_\_\_$
$= \_\_\_\_$

**j** $[6 + (-8)] \times (-5)$
$= \_\_\_\_$
$= \_\_\_\_$

**k** $[9 + (-11)] \times (-6)$
$= \_\_\_\_$
$= \_\_\_\_$

**l** $[8 + (-11)] \times (-4)$
$= \_\_\_\_$
$= \_\_\_\_$

**m** $[9 + (-12)] \times (-3)$
$= \_\_\_\_$
$= \_\_\_\_$

**n** $[10 + (-15)] \times (-6)$
$= \_\_\_\_$
$= \_\_\_\_$

**3** Evaluate the following:

**a** $6 \div (-2 + 5)$
$= 6 \div \_\_\_\_$
$= \_\_\_\_$

**b** $8 \div (-2 + 4)$
$= 8 \div \_\_\_\_$
$= \_\_\_\_$

**c** $10 \div (-2 + 7)$
$= 10 \_\_\_\_$
$= \_\_\_\_$

**d** $12 \div (-2 + 6)$
$= 12 \_\_\_\_$
$= \_\_\_\_$

**e** $6 \div (-2 + 5)$
$= \_\_\_\_$
$= \_\_\_\_$

**f** $10 \div (-4 + 6)$
$= \_\_\_\_$
$= \_\_\_\_$

**g** $15 \div (-5 + 8)$
$= \_\_\_\_$
$= \_\_\_\_$

**h** $12 \div (-2 + 8)$
$= \_\_\_\_$
$= \_\_\_\_$

**i** $14 \div (-3 + 5)$
$= \_\_\_\_$
$= \_\_\_\_$

**j** $12 \div (-3 + 9)$
$= \_\_\_\_$
$= \_\_\_\_$

UNIT 17

## INTEGERS

# Order of operations

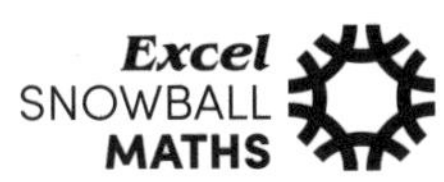


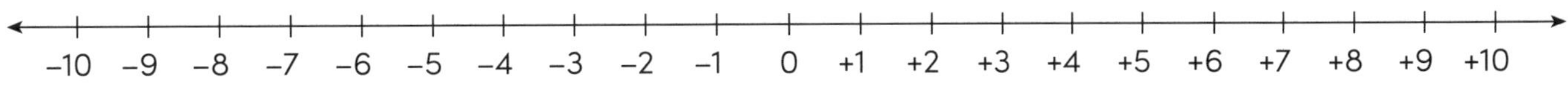


**1** Evaluate the following:

**a** $(-3 - 5) \div (-2)$
= −_____ ÷ (−2)
= _____

**b** $(-4 - 5) \div (-3)$
= −_____ ÷ (−3)
= _____

**c** $(-4 - 6) \div (-2)$
= _____ ÷ (−2)
= _____

**d** $(-5 - 7) \div (-4)$
= _____ ÷ (−4)
= _____

**e** $(-6 - 2) \div (-4)$
= _____ ÷ (_____)
= _____

**f** $(-9 - 1) \div (-5)$
= _____ ÷ (     )
= _____

**g** $(-10 - 2) \div (-3)$
= _____ ÷ (_____)
= _____

**h** $(-8 - 2) \div (-5)$
= _____ ÷ (_____)
= _____

**i** $(-8 - 4) \div (-4)$
= _____
= _____

**j** $(-7 - 5) \div (-2)$
= _____
= _____

**k** $(-9 - 3) \div (-2)$
= _____
= _____

**l** $(-7 - 7) \div (-2)$
= _____
= _____

**m** $(-8 - 6) \div (-7)$
= _____
= _____

**n** $(-12 - 4) \div (-8)$
= _____
= _____

**2** Evaluate the following:

**a** $(4 - 6) \times (-6 + 2)$
= −_____ × (−_____)
= _____

**b** $(3 - 5) \times (-7 + 2)$
= −_____ × (−_____)
= _____

**c** $(3 - 7) \times (-8 + 3)$
= _____ × _____
= _____

**d** $(4 - 7) \times (-9 + 4)$
= _____ × _____
= _____

**e** $(6 - 8) \times (-5 + 4)$
= _____
= _____

**f** $(4 - 6) \times (-3 + 1)$
= _____
= _____

**g** $(5 - 9) \times (-3 + 2)$
= _____
= _____

**h** $(5 - 6) \times (-4 + 2)$
= _____
= _____

**i** $(4 - 7) \times (-8 + 2)$
= _____
= _____

**j** $(2 - 7) \times (-9 + 4)$
= _____
= _____

**3** Evaluate the following:

**a** $-2 - 3 \times (-1 + 5)$
= −2 − 3 × _____
= −2 − _____
= _____

**b** $-3 - 4 \times (-2 + 5)$
= −3 − 4 × _____
= −3 − _____
= _____

**c** $-1 - 5 \times (-3 + 7)$
= −1 − 5 _____
= −1 − _____
= _____

**d** $-6 - 2 \times (-2 + 5)$
= −6 − 2 _____
= −6 − _____
= _____

**e** $-4 - 2 \times (-3 + 8)$
= −4 − 2 _____
= −4 _____
= _____

**f** $-2 - 3 \times (-2 + 5)$
= −2 − 3 _____
= −2 _____
= _____

**g** $-4 - 3 \times (-3 + 5)$
= −4 _____
= −4 _____
= _____

**h** $-5 - 2 \times (-1 + 6)$
= −5 _____
= −5 _____
= _____

**i** $-6 - 2 \times (-4 + 7)$
= _____
= _____
= _____

**j** $-7 - 2 \times (-3 + 5)$
= _____
= _____
= _____

UNIT 18

INTEGERS

# Order of operations

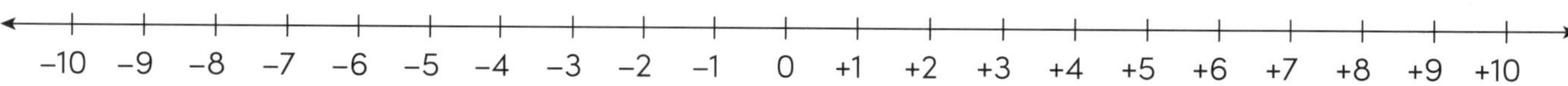


**1** Evaluate the following:

**a** $\frac{-9-3}{1+3} = \frac{-\square}{\square} =$ ______

**b** $\frac{-8-2}{2+3} = \frac{-\square}{\square} =$ ______

**c** $\frac{-7-8}{2+3} = \frac{-\square}{\square} =$ ______

**d** $\frac{-6-9}{1+2} = \frac{\square}{\square} =$ ______

**e** $\frac{-10-6}{1+3} = \frac{\square}{\square} =$ ______

**f** $\frac{-9-5}{4+3} =$ ______
= ______

**g** $\frac{-8-8}{1+3} =$ ______
= ______

**h** $\frac{-9-7}{5+3} =$ ______
= ______

**2** Evaluate the following:

**a** $-2 + 4 - 3$
= ______ $- 3$
= ______

**b** $-3 + 5 - 6$
= ______ $- 6$
= ______

**c** $-4 + 5 - 3$
= ______
= ______

**d** $-2 + 6 - 5$
= ______
= ______

**e** $-2 + 5 - 3$
= ______
= ______

**f** $-1 + 3 - 7$
= ______
= ______

**g** $-6 + 7 - 8$
= ______
= ______

**h** $-5 + 8 - 9$
= ______
= ______

**3** Evaluate the following:

**a** $-2 \times 6 \div 3$
= ______ $\div 3$
= ______

**b** $-3 \times 4 \div 6$
= ______ $\div 6$
= ______

**c** $-4 \times 5 \div 2$
= ______
= ______

**d** $-3 \times 6 \div 9$
= ______
= ______

**e** $-6 \times 4 \div 3$
= ______
= ______

**f** $-3 \times 10 \div 5$
= ______
= ______

**g** $-4 \times 5 \div 10$
= ______
= ______

**h** $-2 \times 8 \div 4$
= ______
= ______

**i** $-9 \times 4 \div 12$
= ______
= ______

**j** $-12 \times 5 \div 30$
= ______
= ______

**k** $-10 \times 4 \div 8$
= ______
= ______

**l** $-8 \times 6 \div 12$
= ______
= ______

UNIT 19

CHAPTER 1 REVIEW

# Integers

1 Write the correct symbol (either < or >) to show the larger number:

a −1 ☐ 3  b 0 ☐ −8  c −5 ☐ −2

2 Write the following sets of integers in descending order (largest to smallest):

a 3, −8, 6  b −4, 0, −8  c −9, 4, −5

3 Eve is currently 2 metres above sea level. What position will she be if she goes:

a 5 m down? ______

b 3 m up? ______

c 2 m up then 6 m down? ______

4 Evaluate the following:

a −8 + 2 = ______  b −10 + 3 = ______

c 12 + (−2) = ______  d 15 + (−4) = ______

e −5 − 15 = ______  f −1 − 20 = ______

g 4 − (−1) = ______  h 6 − (−2) = ______

i −8 − (−3) = ______  j −10 − (−2) = ______

5 The temperature increased by 12 °C from −3 °C. What was the final temperature?

______

6 Amara's bank balance is currently −$100. If she deposits $145, what will her balance be?

______

7 Solve:

a −3 × 6  b −6 × (−9)  c $(-2)^2$

= ______  = ______  = ______

8 Solve:

a −15 ÷ 3  b −42 ÷ (−6)  c $\frac{-16}{-4}$

= ______  = ______  = ______

9 Find the unknown:

a −4 × ? = −12  b ? × 6 = −24

______  ______

10 Evaluate:

a 3 − 5 × 2  b 5 − 3 × 3

= ______  = ______

11 Solve:

a −5 − 4 × (−2)  b −4 − 3 × (−2)

= ______  = ______

12 Evaluate:

a 5 × (6 − 8)  b 3 × (5 − 8)

= ______  = ______

13 Solve:

a 6 ÷ (−1 + 4)  b 9 ÷ (−2 + 5)

= ______  = ______

14 Evaluate:

a [6 + (−9)] × (−5)  b [2 + (−8)] × (−4)

= ______  = ______

15 Solve:

a (−2 − 10) ÷ (−4)  b (−4 − 10) ÷ (−2)

= ______  = ______

16 Evaluate:

a (4 − 6) × (−5 + 8)  b (2 − 4) × (−7 + 9)

= ______  = ______

17 Solve:

a $\frac{-12-4}{1+3}$  b $\frac{-18-2}{2+2}$

= ______  = ______

18 Evaluate:

a −2 + 4 − 3  b −3 + 5 − 6

= ______  = ______

19 Evaluate:

a −2 × 6 ÷ 3  b −3 × 4 ÷ 6

= ______  = ______

UNIT 1

# INDICES

## Indices

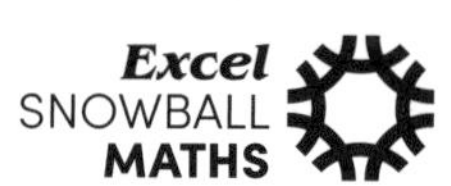

**1** Write the following in index form:

**a** $2 \times 2 \times 2 \times 2 = 2^{\square}$

**b** $3 \times 3 \times 3 \times 3 \times 3 = 3^{\square}$

**c** $5 \times 5 \times 5 =$ ______$^{\square}$

**d** $4 \times 4 \times 4 \times 4 =$ ______$^{\square}$

**e** $7 \times 7 =$ ______

**f** $8 \times 8 \times 8 \times 8 \times 8 \times 8 =$ ______

**g** $9 \times 9 \times 9 \times 9 \times 9 =$ ______

**h** $10 \times 10 \times 10 \times 10 =$ ______

**i** $6 \times 6 =$ ______

**j** $12 \times 12 \times 12 \times 12 \times 12 \times 12 =$ ______

**2** Write in expanded form and then evaluate without a calculator:

**a** $1^2 = 1 \times 1 =$ ______

**b** $2^2 =$ ______ $\times$ ______ $=$ ______

**c** $3^2 =$ ______ $\times$ ______ $=$ ______

**d** $4^2 =$ ______ $\times$ ______ $=$ ______

**e** $5^2 =$ ______ $\times$ ______ $=$ ______

**f** $6^2 =$ ____________________

**g** $7^2 =$ ____________________

**h** $8^2 =$ ____________________

**i** $9^2 =$ ____________________

**j** $10^2 =$ ____________________

**3** Evaluate with or without a calculator:

**a** $\sqrt{4} =$ ______ **b** $\sqrt{9} =$ ______

**c** $\sqrt{16} =$ ______ **d** $\sqrt{25} =$ ______

**e** $\sqrt{36} =$ ______ **f** $\sqrt{49} =$ ______

**g** $\sqrt{64} =$ ______ **h** $\sqrt{81} =$ ______

**i** $\sqrt{100} =$ ______ **j** $\sqrt{121} =$ ______

**4** Evaluate without a calculator:

**a** $2^2 =$ ______ **b** $5^2 =$ ______

**c** $4^2 =$ ______ **d** $6^2 =$ ______

**e** $3^2 =$ ______ **f** $7^2 =$ ______

**g** $\sqrt{4} =$ ______ **h** $\sqrt{16} =$ ______

**i** $\sqrt{25} =$ ______ **j** $\sqrt{9} =$ ______

**k** $\sqrt{100} =$ ______ **l** $\sqrt{36} =$ ______

**5** Write in expanded form and then evaluate with or without a calculator:

**a** $2^3 =$ ______ $\times$ ______ $\times$ ______ $=$ ______

**b** $3^3 =$ ______ $\times$ ______ $\times$ ______ $=$ ______

**c** $4^3 =$ ______ $\times$ ______ $\times$ ______ $=$ ______

**d** $5^3 =$ ______ $\times$ ______ $\times$ ______ $=$ ______

**e** $6^3 =$ ____________________

**f** $7^3 =$ ____________________

**g** $8^3 =$ ____________________

**h** $9^3 =$ ____________________

**i** $10^3 =$ ____________________

UNIT 2

# INDICES

## Indices

1 Evaluate with or without a calculator:

a $\sqrt[3]{8} =$ ______ b $\sqrt[3]{27} =$ ______

c $\sqrt[3]{64} =$ ______ d $\sqrt[3]{125} =$ ______

e $\sqrt[3]{216} =$ ______ f $\sqrt[3]{343} =$ ______

g $\sqrt[3]{512} =$ ______ h $\sqrt[3]{729} =$ ______

i $\sqrt[3]{1000} =$ ______ j $\sqrt[3]{1331} =$ ______

2 Evaluate with or without a calculator:

a $2^3 =$ ______ b $5^3 =$ ______

c $3^3 =$ ______ d $10^3 =$ ______

e $\sqrt[3]{27} =$ ______ f $\sqrt[3]{8} =$ ______

g $\sqrt[3]{64} =$ ______ h $\sqrt[3]{125} =$ ______

i $2^2 =$ ______ j $5^2 =$ ______

k $3^2 =$ ______ l $10^2 =$ ______

m $\sqrt{9} =$ ______ n $\sqrt{49} =$ ______

o $\sqrt{64} =$ ______ p $\sqrt{100} =$ ______

q $\sqrt[3]{1000} =$ ______ r $4^3 =$ ______

3 Write in expanded form and then evaluate with or without a calculator:

a $(-2)^2 = -2 \times -2 =$ ______

b $(-3)^2 =$ ______ $\times$ ______ $=$ ______

c $(-4)^2 =$ ______ $\times$ ______ $=$ ______

d $(-5)^2 =$ ______ $\times$ ______ $=$ ______

e $(-6)^2 =$ ______________________

f $(-7)^2 =$ ______________________

g $(-8)^2 =$ ______________________

h $(-2)^3 =$ ______ $\times$ ______ $\times$ ______ $=$ ______

i $(-3)^3 =$ ______ $\times$ ______ $\times$ ______ $=$ ______

j $(-4)^3 =$ ______ $\times$ ______ $\times$ ______ $=$ ______

k $(-5)^3 =$ ______ $\times$ ______ $\times$ ______ $=$ ______

l $(-6)^3 =$ ______________________

m $(-7)^3 =$ ______________________

n $(-8)^3 =$ ______________________

4 Evaluate the following without a calculator:

a $\sqrt{4} + 3$
$=$ ______ $+ 3$
$=$ ______

b $\sqrt{4} + 6$
$=$ ______ $+ 6$
$=$ ______

c $\sqrt{4} + 2$
$=$ ______ $+$ ______
$=$ ______

d $\sqrt{4} + 5$
$=$ ______ $+$ ______
$=$ ______

e $\sqrt{4} + 7$
$=$ ____________
$=$ ______

f $\sqrt{4} + 9$
$=$ ____________
$=$ ______

g $\sqrt{9} - 5$
$=$ ______ $- 5$
$=$ ______

h $\sqrt{9} - 4$
$=$ ______ $- 4$
$=$ ______

i $\sqrt{9} - 6$
$=$ ______ $-$ ______
$=$ ______

j $\sqrt{9} - 8$
$=$ ______ $-$ ______
$=$ ______

k $\sqrt{9} - 7$
$=$ ____________
$=$ ______

l $\sqrt{9} - 9$
$=$ ____________
$=$ ______

UNIT 3

**INDICES**

# Indices

**1** Evaluate the following without a calculator:

**a** $\sqrt{16} \times 2$

$= \_\_\_\_ \times 2$

$= \_\_\_\_$

**b** $\sqrt{16} \times 5$

$= \_\_\_\_ \times 5$

$= \_\_\_\_$

**c** $\sqrt{16} \times 4$

$= \_\_\_\_ \times \_\_\_\_$

$= \_\_\_\_$

**d** $\sqrt{16} \times 3$

$= \_\_\_\_ \times \_\_\_\_$

$= \_\_\_\_$

**e** $\sqrt{16} \times 6$

$= \_\_\_\_\_\_\_\_$

$= \_\_\_\_$

**f** $\sqrt{16} \times 7$

$= \_\_\_\_\_\_\_\_$

$= \_\_\_\_$

**g** $10 \div \sqrt{25}$

$= 10 \div \_\_\_\_$

$= \_\_\_\_$

**h** $20 \div \sqrt{25}$

$= 20 \div \_\_\_\_$

$= \_\_\_\_$

**i** $30 \div \sqrt{25}$

$= \_\_\_\_ \div \_\_\_\_$

$= \_\_\_\_$

**j** $15 \div \sqrt{25}$

$= \_\_\_\_ \div \_\_\_\_$

$= \_\_\_\_$

**k** $40 \div \sqrt{25}$

$= \_\_\_\_\_\_\_\_$

$= \_\_\_\_$

**l** $50 \div \sqrt{25}$

$= \_\_\_\_\_\_\_\_$

$= \_\_\_\_$

**2** Evaluate the following without a calculator:

**a** $\sqrt{4} + \sqrt{4}$

$= \_\_\_\_ + \_\_\_\_$

$= \_\_\_\_$

**b** $\sqrt{4} + \sqrt{9}$

$= \_\_\_\_ + \_\_\_\_$

$= \_\_\_\_$

**c** $\sqrt{9} + \sqrt{16}$

$= \_\_\_\_ + \_\_\_\_$

$= \_\_\_\_$

**d** $\sqrt{16} + \sqrt{25}$

$= \_\_\_\_ + \_\_\_\_$

$= \_\_\_\_$

**e** $\sqrt{25} + \sqrt{36}$

$= \_\_\_\_\_\_\_\_$

$= \_\_\_\_$

**f** $\sqrt{36} + \sqrt{49}$

$= \_\_\_\_\_\_\_\_$

$= \_\_\_\_$

**g** $\sqrt{9} - \sqrt{4}$

$= \_\_\_\_ - \_\_\_\_$

$= \_\_\_\_$

**h** $\sqrt{16} - \sqrt{9}$

$= \_\_\_\_ - \_\_\_\_$

$= \_\_\_\_$

**i** $\sqrt{25} - \sqrt{4}$

$= \_\_\_\_ - \_\_\_\_$

$= \_\_\_\_$

**j** $\sqrt{36} - \sqrt{9}$

$= \_\_\_\_ - \_\_\_\_$

$= \_\_\_\_$

**k** $\sqrt{49} - \sqrt{16}$

$= \_\_\_\_\_\_\_\_$

$= \_\_\_\_$

**l** $\sqrt{100} - \sqrt{81}$

$= \_\_\_\_\_\_\_\_$

$= \_\_\_\_$

**3** Evaluate the following without a calculator:

**a** $2^2 + 1$

$= \_\_\_\_ + 1$

$= \_\_\_\_$

**b** $2^2 + 3$

$= \_\_\_\_ + 3$

$= \_\_\_\_$

**c** $2^2 + 6$

$= \_\_\_\_ + \_\_\_\_$

$= \_\_\_\_$

**d** $2^2 + 5$

$= \_\_\_\_ + \_\_\_\_$

$= \_\_\_\_$

**e** $2^2 + 7$

$= \_\_\_\_\_\_\_\_$

$= \_\_\_\_$

**f** $2^2 + 10$

$= \_\_\_\_\_\_\_\_$

$= \_\_\_\_$

**g** $3^2 - 4$

$= \_\_\_\_ - 4$

$= \_\_\_\_$

**h** $3^2 - 5$

$= \_\_\_\_ - 5$

$= \_\_\_\_$

**i** $3^2 - 6$

$= \_\_\_\_ - \_\_\_\_$

$= \_\_\_\_$

**j** $3^2 - 3$

$= \_\_\_\_ - \_\_\_\_$

$= \_\_\_\_$

**k** $3^2 - 7$

$= \_\_\_\_\_\_\_\_$

$= \_\_\_\_$

**l** $3^2 - 8$

$= \_\_\_\_\_\_\_\_$

$= \_\_\_\_$

UNIT 4

# INDICES
## Indices

**1** Evaluate the following without a calculator:

**a** $4^2 \div 2$
= ________
= ________

**b** $4^2 \div 4$
= ________
= ________

**c** $4^2 \div 8$
= ________
= ________

**d** $4^2 \div 16$
= ________
= ________

**e** $5^2 - 10$
= ________
= ________

**f** $5^2 - 13$
= ________
= ________

**g** $5^2 - 12$
= ________
= ________

**h** $5^2 - 9$
= ________
= ________

**i** $2^2 + 3^2$
= ________
= ________

**j** $3^2 + 4^2$
= ________
= ________

**k** $4^2 + 5^2$
= ________
= ________

**l** $5^2 + 6^2$
= ________
= ________

**m** $10^2 - 3^2$
= ________
= ________

**n** $10^2 - 4^2$
= ________
= ________

**o** $2 \times 2^2$
= ________
= ________

**p** $2 \times 3^2$
= ________
= ________

**q** $4^2 \div 2^2$
= ________
= ________

**r** $2^3 \div 2^2$
= ________
= ________

**2** Evaluate:

**a** $2^2 - 3$
= ________
= ________

**b** $5^2 - 5$
= ________
= ________

**c** $6^2 + 4$
= ________
= ________

**d** $7^2 + 1$
= ________
= ________

**e** $2^2 \div 2$
= ________
= ________

**f** $5^2 \div 5$
= ________
= ________

**g** $3^2 \times 8$
= ________
= ________

**h** $2^2 \times 7$
= ________
= ________

**3** Solve:

**a** $6^2 \div 2^2$
= ________
= ________

**b** $10^2 \div 5^2$
= ________
= ________

**c** $5^2 + 2^2$
= ________
= ________

**d** $6^2 - 3^2$
= ________
= ________

**e** $5^2 + \sqrt{81}$
= ________
= ________

**f** $4^2 - \sqrt{36}$
= ________
= ________

**g** $5^2 \times \sqrt{9}$
= ________
= ________

**h** $6^2 \div \sqrt{9}$
= ________
= ________

UNIT 5

# INDICES

## Prime numbers and factor trees

**1** State whether each of the following is a **prime** or **composite** number:

a 5 ______ b 7 ______

c 8 ______ d 12 ______

e 17 ______ f 19 ______

g 20 ______ h 11 ______

i 24 ______ j 15 ______

k 23 ______ l 49 ______

**2** Write the following as a product of prime factors using index notation:

a

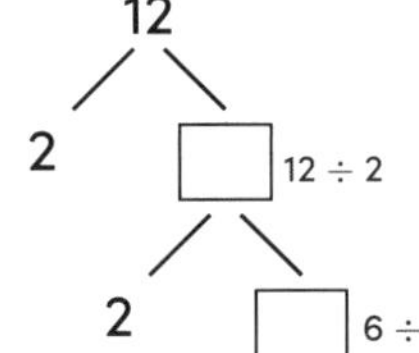


12 = 2 × 2 × ___

= 2□ × ___

b

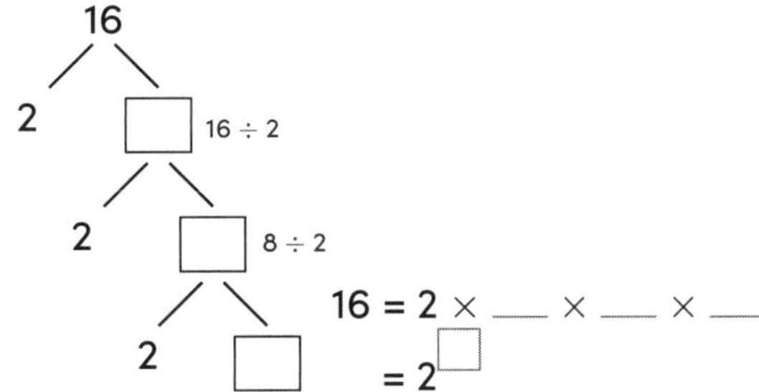


c

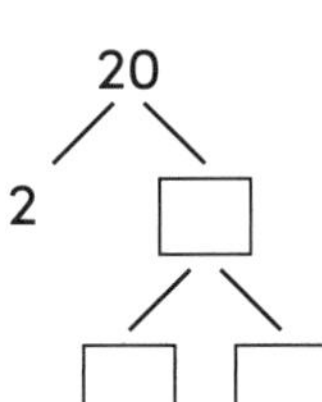


20 = ___ × ___ × ___

= ___□ × ___

d

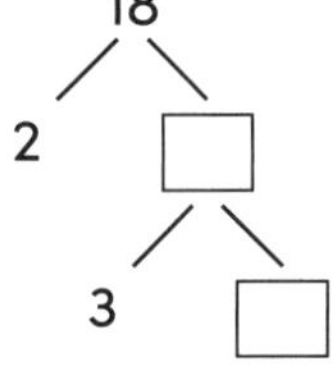


18 = ___ × ___ × ___

= ___ × ___□

e

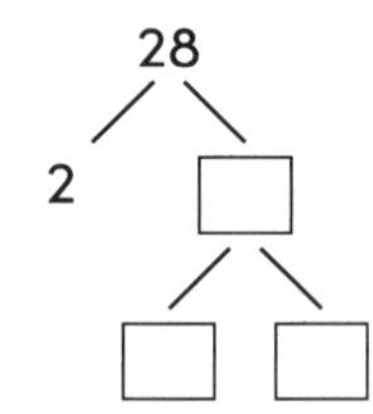


28 = ______

= ______

f

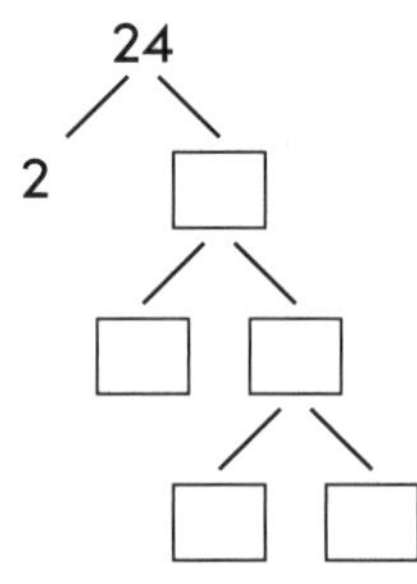


24 = ______

= ______

g

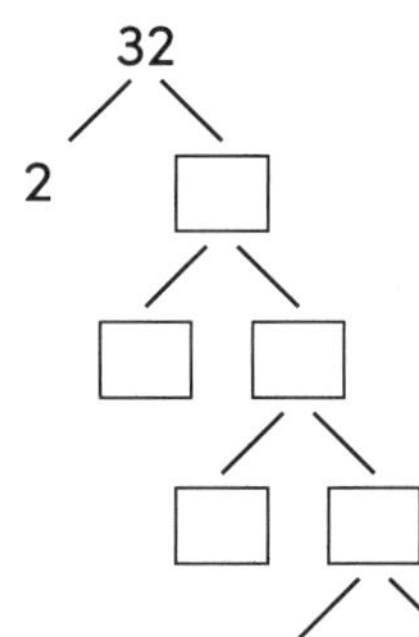


32 = ______

= ______

h

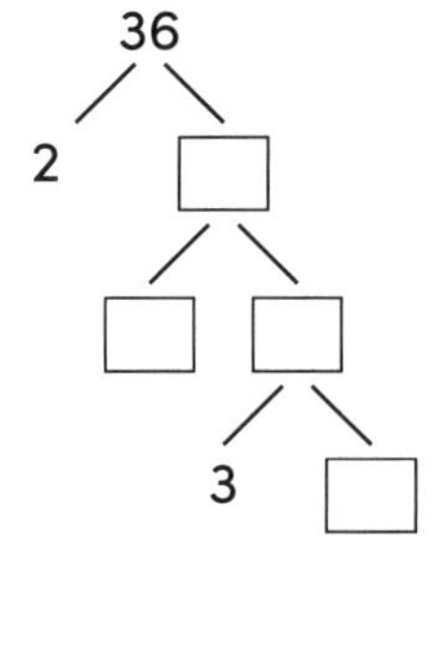


36 = ______

= ______

i

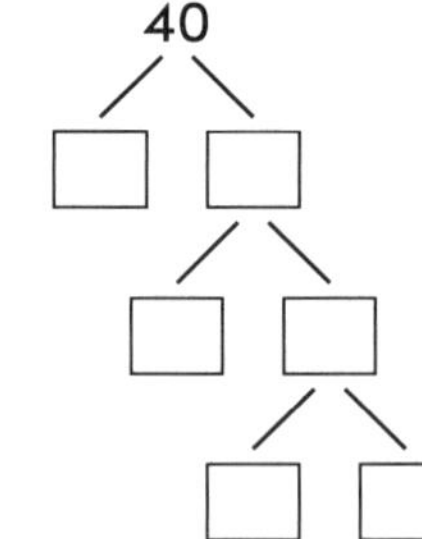


40 = ______

= ______

j

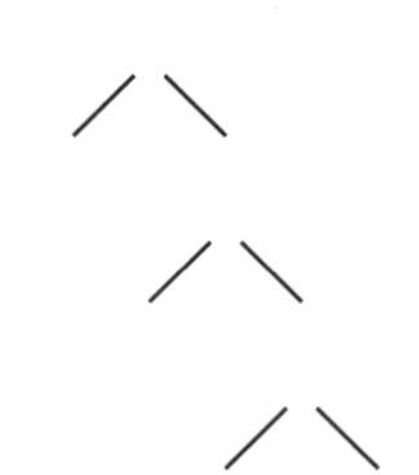


48 = ______

= ______

## INDICES

# Prime numbers and factor trees

1 Write the following as a product of prime factors using index notation:

a

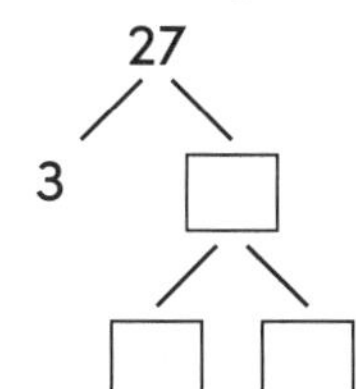


b

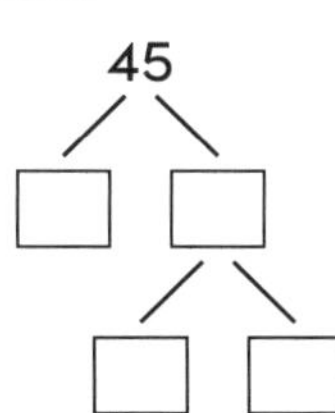


27 = ____________

= ____________

45 = ____________

= ____________

c

d

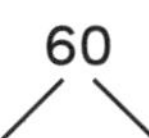


54 = ____________

= ____________

60 = ____________

= ____________

e

f

88 = ____________

= ____________

100 = ____________

= ____________

g 150

h 42

i 64

j 96

k 270

UNIT 7

CHAPTER 2 REVIEW

# Indices

**1** Write the following in index form:

a $8 \times 8 \times 8 \times 8 =$ ______

b $2 \times 2 \times 2 \times 2 \times 2 \times 2 =$ ______

c $7 \times 7 \times 7 \times 7 =$ ______

d $10 \times 10 \times 10 =$ ______

**2** Evaluate without a calculator:

a $4^2 =$ ______ b $7^2 =$ ______

c $8^2 =$ ______ d $5^2 =$ ______

e $9^2 =$ ______ f $10^2 =$ ______

**3** Evaluate without a calculator:

a $5^3 =$ ______ b $3^3 =$ ______

c $4^3 =$ ______ d $2^3 =$ ______

**4** Evaluate with or without a calculator:

a $\sqrt{16} =$ ______ b $\sqrt{49} =$ ______

c $\sqrt[3]{8} =$ ______ d $\sqrt[3]{64} =$ ______

e $3^2 =$ ______ f $6^2 =$ ______

g $6^3 =$ ______ h $10^3 =$ ______

i $\sqrt{81} =$ ______ j $\sqrt{100} =$ ______

k $\sqrt[3]{27} =$ ______ l $\sqrt{36} =$ ______

m $\sqrt{4} =$ ______ n $\sqrt[3]{125} =$ ______

**5** Evaluate without a calculator:

a $\sqrt{25} + 3$ ______ b $\sqrt{4} + 8$ ______

c $\sqrt{64} - 2$ ______ d $\sqrt{81} - 8$ ______

e $4 \times \sqrt{25}$ ______ f $8 \times \sqrt{16}$ ______

g $48 \div \sqrt{36}$ ______ h $42 \div \sqrt{49}$ ______

i $\sqrt{25} - \sqrt{9}$ ______ j $\sqrt{49} - \sqrt{36}$ ______

k $4^2 + 3^2$ ______ l $6^2 + 2^2$ ______

m $10^2 - 7^2$ ______ n $10 \times \sqrt{49}$ ______

o $3^2 \times 3$ ______ p $\sqrt[3]{27} - \sqrt{4}$ ______

**6** State whether each of these is a prime or composite number:

a 31 ______ b 48 ______

c 52 ______ d 17 ______

e 11 ______ f 44 ______

**7** Write the following as a product of prime factors using index notation:

a 44 ______

b 56 ______

c 80 ______

d 54 ______

e 136 ______

f 240 ______

g 396 ______

h 416 ______

i 342 ______

j 690 ______

UNIT 1

# FRACTIONS
## Equivalent fractions

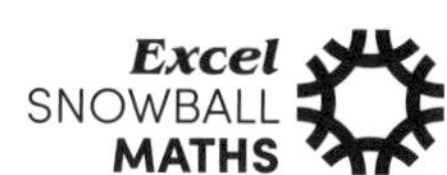

1 Write the fraction for the shaded parts in each diagram:

a 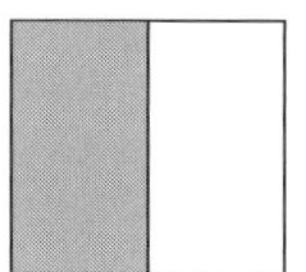

$\frac{\square}{\square}$

b 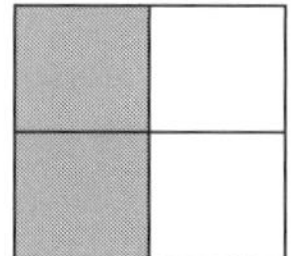

$\frac{\square}{\square} = \frac{1}{2}$

c 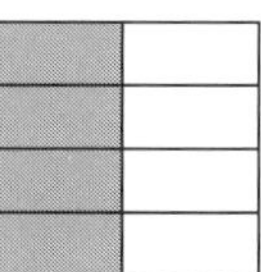

$\frac{\square}{\square} = \frac{1}{2}$

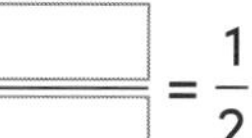

d 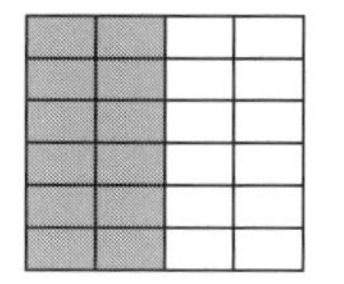

$\frac{\square}{\square} = \frac{1}{2}$

e 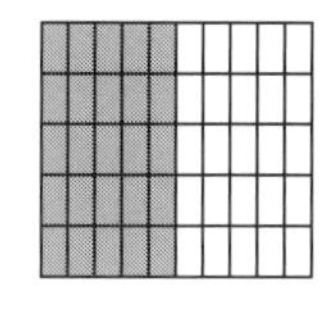

$\frac{\square}{\square} = \frac{1}{2}$

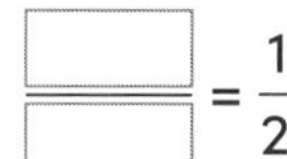

f 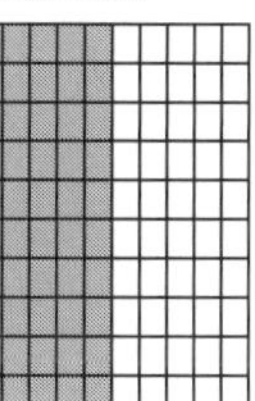

$\frac{\square}{\square} = \frac{1}{2}$

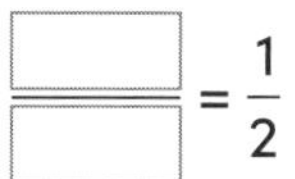

2 Write the fraction for the shaded parts in each diagram:

a 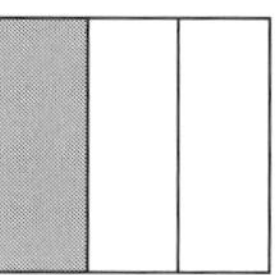

$\frac{\square}{\square}$

b 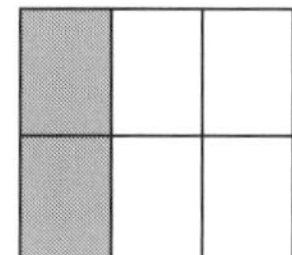

$\frac{\square}{\square} = \frac{1}{3}$

c 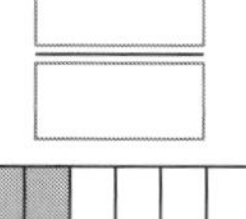

$\frac{\square}{\square} = \frac{1}{3}$

d 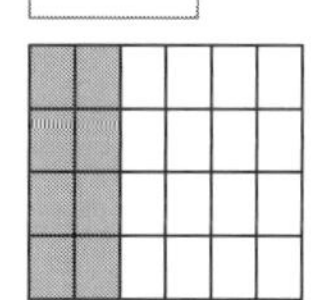

$\frac{\square}{\square} = \frac{1}{3}$

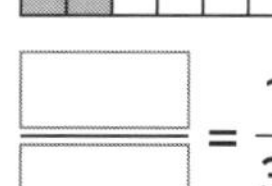

3 Solve:

a 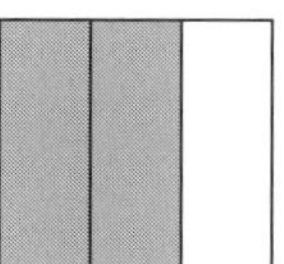

$\frac{\square}{\square}$

b 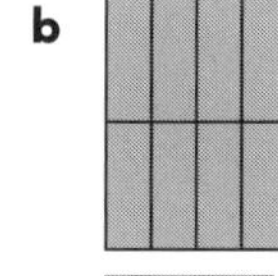

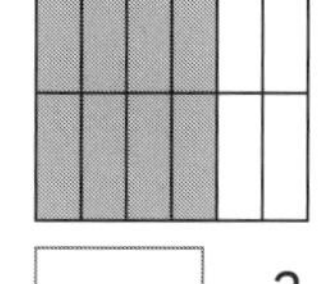

$\frac{\square}{\square} = \frac{2}{3}$

c 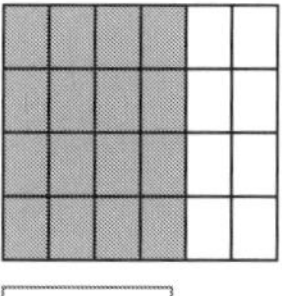

$\frac{\square}{\square} = \frac{2}{3}$

4 Solve:

a 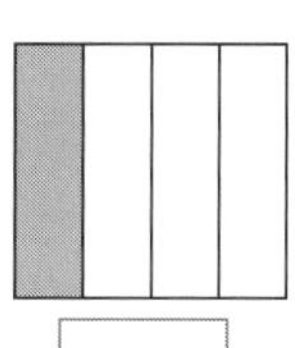

$\frac{\square}{\square}$

b 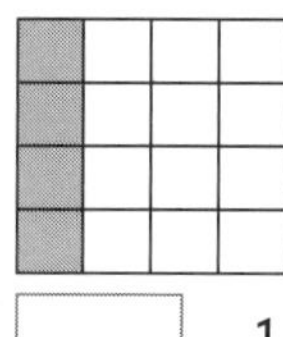

$\frac{\square}{\square} = \frac{1}{4}$

c 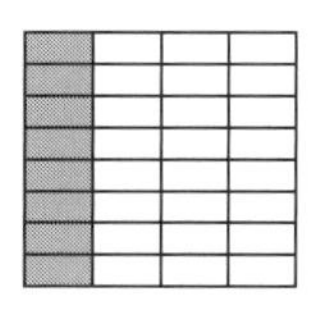

$\frac{\square}{\square} = \frac{1}{4}$

d 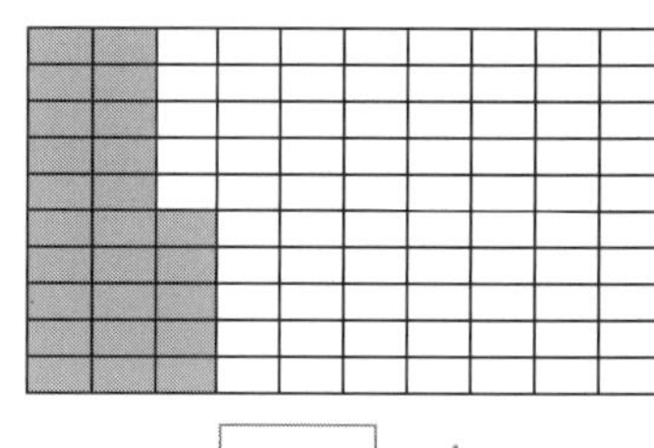

$\frac{\square}{\square} = \frac{1}{4}$

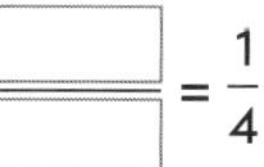

5 Solve:

a 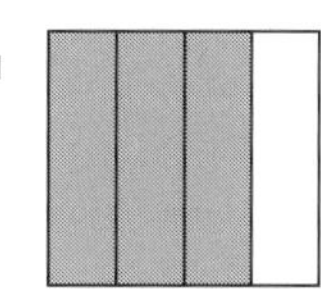

$\frac{\square}{\square}$

b 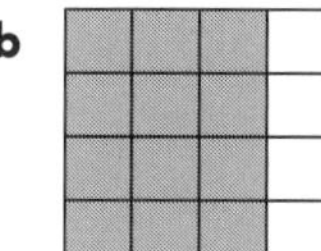

$\frac{\square}{\square} = \frac{3}{4}$

6 Complete each set of equivalent fractions by multiplying by 2:

a $\frac{1 \times 2}{2 \times 2} = \frac{\square}{4}$

b $\frac{2 \times 2}{3 \times 2} = \frac{\square}{6}$

c $\frac{5}{4} = \frac{\square}{8}$

d $\frac{6}{5} = \frac{\square}{10}$

e $\frac{2}{6} = \frac{4}{\square}$

f $\frac{5}{8} = \frac{10}{\square}$

g $\frac{7}{9} = \frac{14}{\square}$

h $\frac{3}{10} = \frac{6}{\square}$

i $\frac{4}{6} = \frac{\square}{12}$

j $\frac{3}{4} = \frac{\square}{8}$

k $\frac{8}{5} = \frac{16}{\square}$

l $\frac{9}{7} = \frac{18}{\square}$

UNIT 2

**FRACTIONS**

# Equivalent fractions

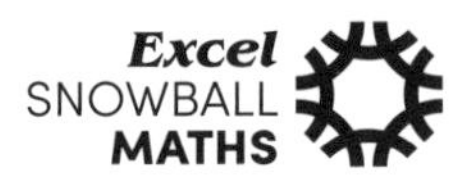


**1** Complete each set of equivalent fractions by multiplying by 3:

a $\frac{1 \times 3}{2 \times 3} = \frac{\square}{6}$ b $\frac{2 \times 3}{3 \times 3} = \frac{\square}{9}$

c $\frac{5}{4} = \frac{\square}{12}$ d $\frac{6}{5} = \frac{\square}{15}$

e $\frac{3}{6} = \frac{9}{\square}$ f $\frac{5}{8} = \frac{15}{\square}$

g $\frac{7}{9} = \frac{21}{\square}$ h $\frac{3}{10} = \frac{9}{\square}$

i $\frac{4}{6} = \frac{\square}{18}$ j $\frac{3}{4} = \frac{\square}{12}$

k $\frac{8}{5} = \frac{24}{\square}$ l $\frac{9}{7} = \frac{27}{\square}$

**2** Complete each set of equivalent fractions by multiplying by 4:

a $\frac{1 \times 4}{2 \times 4} = \frac{\square}{8}$ b $\frac{2 \times 4}{3 \times 4} = \frac{\square}{12}$

c $\frac{5}{4} = \frac{\square}{16}$ d $\frac{6}{5} = \frac{\square}{20}$

e $\frac{3}{6} = \frac{12}{\square}$ f $\frac{5}{8} = \frac{20}{\square}$

g $\frac{7}{9} = \frac{28}{\square}$ h $\frac{3}{10} = \frac{12}{\square}$

i $\frac{4}{6} = \frac{\square}{24}$ j $\frac{3}{4} = \frac{\square}{16}$

k $\frac{8}{5} = \frac{32}{\square}$ l $\frac{9}{7} = \frac{36}{\square}$

**3** Complete each set of equivalent fractions by multiplying by 5:

a $\frac{1 \times 5}{2 \times 5} = \frac{\square}{10}$ b $\frac{2 \times 5}{3 \times 5} = \frac{\square}{15}$

c $\frac{5}{4} = \frac{\square}{20}$ d $\frac{6}{5} = \frac{\square}{25}$

e $\frac{3}{6} = \frac{15}{\square}$ f $\frac{5}{8} = \frac{25}{\square}$

g $\frac{7}{9} = \frac{35}{\square}$ h $\frac{3}{10} = \frac{15}{\square}$

i $\frac{4}{6} = \frac{\square}{30}$ j $\frac{3}{4} = \frac{\square}{20}$

k $\frac{8}{5} = \frac{40}{\square}$ l $\frac{9}{7} = \frac{45}{\square}$

**4** Complete each set of equivalent fractions by dividing by 2:

a $\frac{4 \div 2}{8 \div 2} = \frac{\square}{4}$ b $\frac{4 \div 2}{6 \div 2} = \frac{\square}{3}$

c $\frac{10}{8} = \frac{\square}{4}$ d $\frac{12}{10} = \frac{\square}{5}$

e $\frac{4}{12} = \frac{2}{\square}$ f $\frac{10}{16} = \frac{5}{\square}$

UNIT 3

**FRACTIONS**

# Equivalent fractions

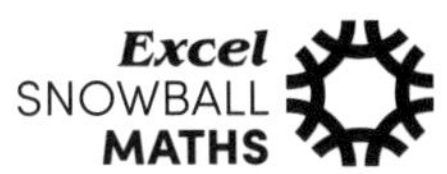


**1** Complete each set of equivalent fractions by dividing by 2:

**a** $\frac{14}{18} = \frac{7}{\square}$  **b** $\frac{6}{20} = \frac{3}{\square}$

**c** $\frac{8}{12} = \frac{\square}{6}$  **d** $\frac{6}{8} = \frac{\square}{4}$

**e** $\frac{16}{10} = \frac{8}{\square}$  **f** $\frac{18}{14} = \frac{9}{\square}$

**2** Complete each set of equivalent fractions by dividing by 3:

**a** $\frac{6 \div 3}{9 \div 3} = \frac{\square}{3}$  **b** $\frac{3 \div 3}{9 \div 3} = \frac{\square}{3}$

**c** $\frac{15}{12} = \frac{\square}{4}$  **d** $\frac{21}{15} = \frac{\square}{5}$

**e** $\frac{9}{24} = \frac{3}{\square}$  **f** $\frac{15}{27} = \frac{5}{\square}$

**g** $\frac{21}{30} = \frac{7}{\square}$  **h** $\frac{9}{33} = \frac{3}{\square}$

**i** $\frac{9}{18} = \frac{\square}{6}$  **j** $\frac{15}{12} = \frac{\square}{4}$

**k** $\frac{24}{18} = \frac{8}{\square}$  **l** $\frac{27}{24} = \frac{9}{\square}$

**3** Complete each set of equivalent fractions by dividing by 4:

**a** $\frac{8 \div 4}{16 \div 4} = \frac{\square}{4}$  **b** $\frac{12 \div 4}{24 \div 4} = \frac{\square}{6}$

**c** $\frac{24}{32} = \frac{\square}{8}$  **d** $\frac{28}{40} = \frac{\square}{10}$

**e** $\frac{16}{32} = \frac{4}{\square}$  **f** $\frac{24}{20} = \frac{6}{\square}$

**g** $\frac{32}{40} = \frac{8}{\square}$  **h** $\frac{16}{44} = \frac{4}{\square}$

**i** $\frac{12}{28} = \frac{\square}{7}$  **j** $\frac{24}{20} = \frac{\square}{5}$

**k** $\frac{36}{32} = \frac{9}{\square}$  **l** $\frac{28}{36} = \frac{7}{\square}$

**4** Complete each set of equivalent fractions by dividing by 5:

**a** $\frac{15 \div 5}{20 \div 5} = \frac{\square}{4}$  **b** $\frac{25 \div 5}{15 \div 5} = \frac{\square}{3}$

**c** $\frac{10}{20} = \frac{\square}{4}$  **d** $\frac{35}{25} = \frac{\square}{5}$

**e** $\frac{10}{30} = \frac{2}{\square}$  **f** $\frac{25}{40} = \frac{5}{\square}$

**g** $\frac{45}{30} = \frac{\square}{6}$  **h** $\frac{60}{20} = \frac{\square}{4}$

**i** $\frac{35}{50} = \frac{7}{\square}$  **j** $\frac{15}{35} = \frac{3}{\square}$

**k** $\frac{40}{5} = \frac{8}{\square}$  **l** $\frac{45}{15} = \frac{9}{\square}$

UNIT 4

FRACTIONS

# Ordering fractions

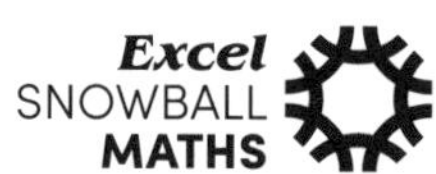

**1** Fill in each number line:

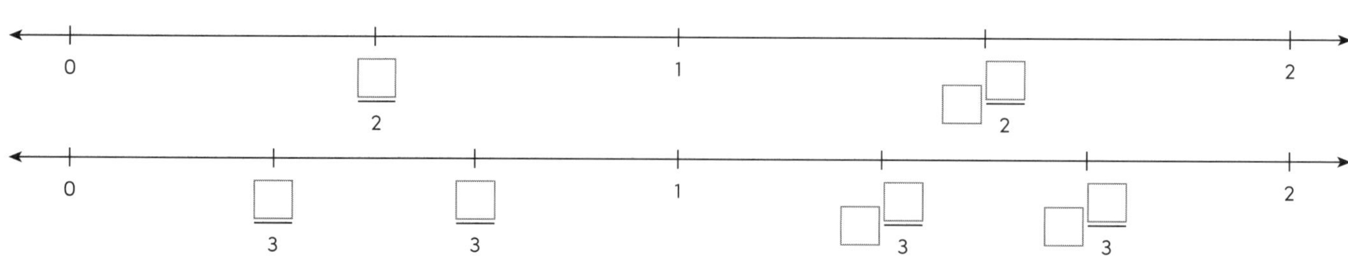

**2** Referring to the number lines above, write < or > to show the larger fraction:

**a** $\frac{1}{2}$ ☐ 1 **b** $1\frac{1}{2}$ ☐ $\frac{1}{2}$ **c** $\frac{2}{3}$ ☐ $\frac{1}{3}$ **d** $1\frac{1}{3}$ ☐ $1\frac{2}{3}$

**e** $\frac{1}{2}$ ☐ $\frac{1}{3}$ **f** $\frac{1}{2}$ ☐ $\frac{2}{3}$ **g** $1\frac{1}{3}$ ☐ $1\frac{1}{2}$ **h** $1\frac{2}{3}$ ☐ $1\frac{1}{2}$

**3** Fill in each number line:

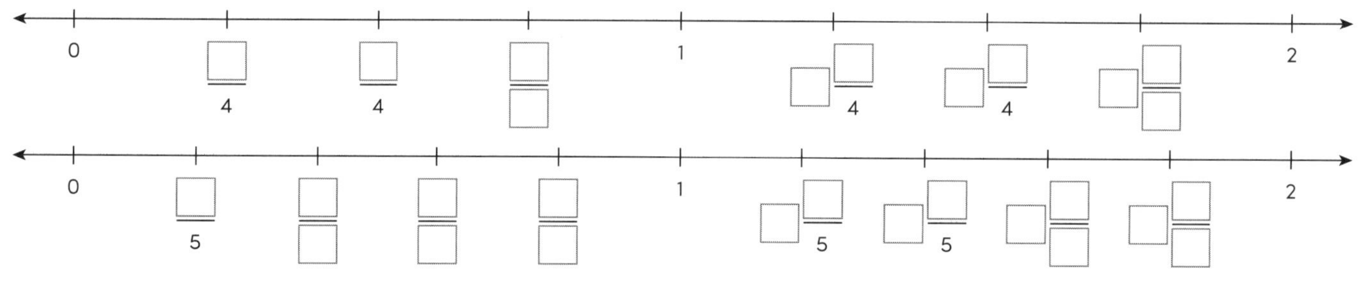

**4** Referring to the number lines in questions 1 and 3, write < , > or = to make each statement true:

**a** $\frac{1}{4}$ ☐ $\frac{1}{5}$ **b** $\frac{2}{4}$ ☐ $\frac{2}{5}$ **c** $\frac{3}{5}$ ☐ $\frac{3}{4}$ **d** $1\frac{1}{4}$ ☐ $1\frac{2}{5}$

**e** $1\frac{1}{5}$ ☐ $1\frac{1}{4}$ **f** $1\frac{3}{4}$ ☐ $1\frac{2}{5}$ **g** $1\frac{3}{4}$ ☐ $1\frac{3}{5}$ **h** $1\frac{4}{5}$ ☐ $1\frac{3}{4}$

**i** $\frac{1}{2}$ ☐ $\frac{1}{5}$ **j** $\frac{4}{5}$ ☐ $\frac{2}{3}$ **k** $\frac{1}{2}$ ☐ $\frac{2}{4}$ **l** $\frac{2}{3}$ ☐ $\frac{3}{4}$

**m** $1\frac{1}{5}$ ☐ $1\frac{1}{3}$ **n** $1\frac{1}{3}$ ☐ $1\frac{1}{4}$ **o** $1\frac{2}{5}$ ☐ $1\frac{1}{2}$ **p** $1\frac{2}{4}$ ☐ $1\frac{1}{2}$

UNIT 5

**FRACTIONS**

# Ordering fractions

**1** Fill in each number line:

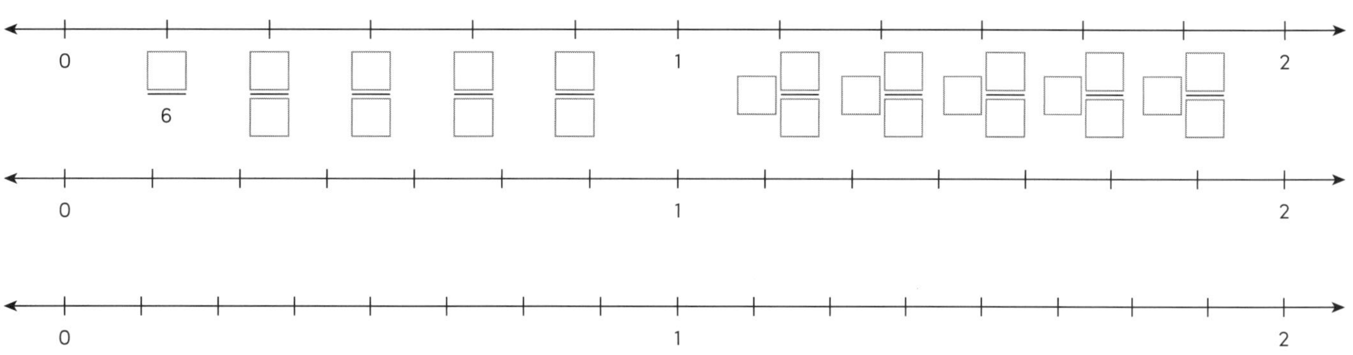

**2** Referring to the number lines above, write < , > or = to make each statement true:

a $\frac{4}{6}$ ☐ $\frac{4}{7}$  b $\frac{4}{8}$ ☐ $\frac{4}{7}$  c $\frac{3}{7}$ ☐ $\frac{3}{6}$  d $\frac{6}{7}$ ☐ $\frac{6}{8}$

e $1\frac{3}{6}$ ☐ $1\frac{4}{8}$  f $1\frac{2}{8}$ ☐ $1\frac{2}{7}$  g $1\frac{5}{6}$ ☐ $1\frac{7}{8}$  h $1\frac{4}{7}$ ☐ $1\frac{4}{8}$

i $\frac{5}{7}$ ☐ $\frac{4}{6}$  j $\frac{5}{7}$ ☐ $\frac{6}{8}$  k $1\frac{3}{7}$ ☐ $1\frac{3}{6}$  l $1\frac{7}{8}$ ☐ $1\frac{6}{7}$

**3** Fill in each number line:

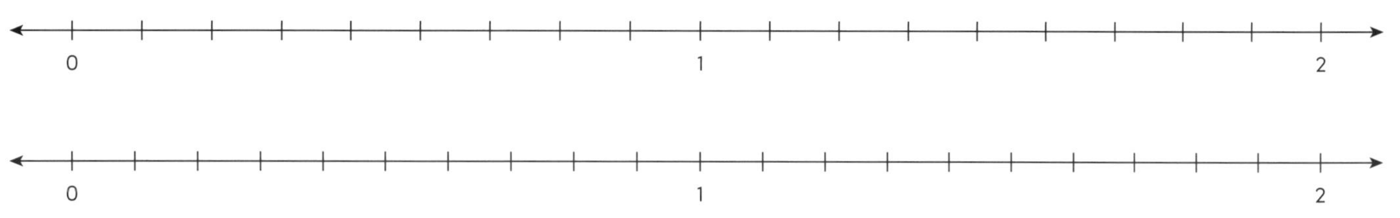

**4** Referring to the number lines above, write < , > or = to make each statement true:

a $\frac{1}{10}$ ☐ $\frac{1}{9}$  b $\frac{5}{10}$ ☐ $\frac{5}{9}$  c $\frac{3}{10}$ ☐ $\frac{3}{9}$  d $\frac{2}{9}$ ☐ $\frac{2}{10}$

e $1\frac{6}{10}$ ☐ $1\frac{6}{9}$  f $1\frac{5}{10}$ ☐ $1\frac{5}{9}$  g $1\frac{2}{9}$ ☐ $1\frac{2}{10}$  h $1\frac{8}{9}$ ☐ $1\frac{8}{10}$

i $\frac{4}{8}$ ☐ $\frac{4}{9}$  j $\frac{5}{10}$ ☐ $\frac{5}{8}$  k $\frac{3}{9}$ ☐ $\frac{3}{8}$  l $\frac{6}{7}$ ☐ $\frac{6}{9}$

m $\frac{2}{6}$ ☐ $\frac{3}{9}$  n $\frac{8}{10}$ ☐ $\frac{6}{8}$  o $1\frac{6}{9}$ ☐ $1\frac{4}{6}$  p $1\frac{7}{9}$ ☐ $1\frac{6}{8}$

# Ordering fractions

**1** Without the use of a number line, write < or > to show which fraction is larger:

**a** $\frac{1}{3} \square \frac{2}{3}$  **b** $\frac{3}{5} \square \frac{2}{5}$

**c** $\frac{4}{7} \square \frac{5}{7}$  **d** $\frac{2}{6} \square \frac{3}{6}$

**e** $\frac{5}{12} \square \frac{8}{12}$  **f** $\frac{4}{15} \square \frac{3}{15}$

**g** $\frac{11}{20} \square \frac{15}{20}$  **h** $\frac{12}{24} \square \frac{15}{24}$

**i** $\frac{19}{30} \square \frac{20}{30}$  **j** $\frac{24}{35} \square \frac{21}{35}$

**2** Make each statement true (the smaller denominator is larger):

**a** $\frac{2}{3} \square \frac{2}{4}$  **b** $\frac{3}{6} \square \frac{3}{5}$

**c** $\frac{5}{7} \square \frac{5}{9}$  **d** $\frac{2}{6} \square \frac{2}{7}$

**e** $\frac{8}{12} \square \frac{8}{10}$  **f** $\frac{9}{13} \square \frac{9}{15}$

**g** $\frac{11}{18} \square \frac{11}{20}$  **h** $\frac{12}{26} \square \frac{12}{24}$

**i** $\frac{19}{32} \square \frac{19}{30}$  **j** $\frac{20}{30} \square \frac{20}{35}$

**k** $\frac{10}{30} \square \frac{10}{40}$  **l** $\frac{15}{25} \square \frac{15}{20}$

**3** Which is larger?

**a** $\frac{1 \times 2}{2 \times 2}$ or $\frac{3}{4}$?

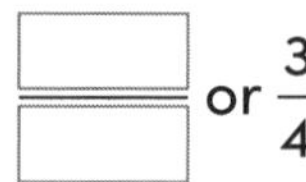

$\frac{\square}{\square}$ or $\frac{3}{4}$

Answer: $\frac{\square}{\square}$

**b** $\frac{2 \times 2}{5 \times 2}$ or $\frac{3}{10}$?

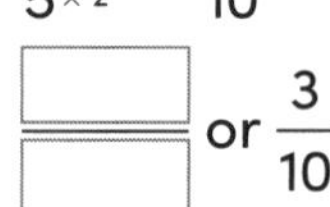

$\frac{\square}{\square}$ or $\frac{3}{10}$

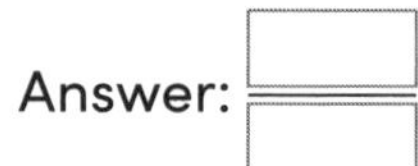

Answer: $\frac{\square}{\square}$

**c** $\frac{2}{3}$ or $\frac{5}{6}$?

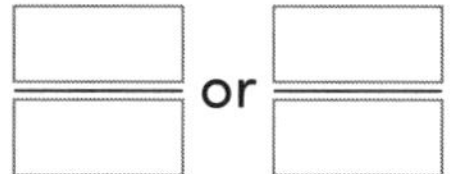

$\frac{\square}{\square}$ or $\frac{\square}{\square}$

Answer: $\frac{\square}{\square}$

**d** $\frac{3}{4}$ or $\frac{5}{8}$?

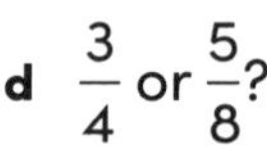

$\frac{\square}{\square}$ or $\frac{\square}{\square}$

Answer: $\frac{\square}{\square}$

**e** $\frac{5}{7}$ or $\frac{9}{14}$?

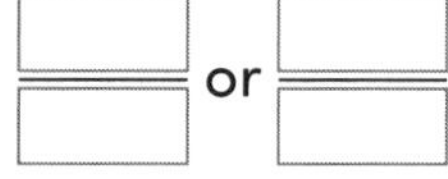

$\frac{\square}{\square}$ or $\frac{\square}{\square}$

Answer: $\frac{\square}{\square}$

**f** $\frac{5}{6}$ or $\frac{11}{12}$?

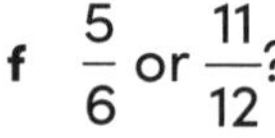

$\frac{\square}{\square}$ or $\frac{\square}{\square}$

Answer: $\frac{\square}{\square}$

**g** $\frac{8}{10}$ or $\frac{17}{20}$?

____________

Answer: ________

**h** $\frac{7}{16}$ or $\frac{3}{8}$?

____________

Answer: ________

**i** $\frac{3}{4}$ or $\frac{8}{12}$?

____________

Answer: ________

**j** $\frac{12}{15}$ or $\frac{3}{5}$?

____________

Answer: ________

UNIT 7

**FRACTIONS**

# Simplifying fractions

**1** Simplify the following fractions by dividing by 2:

a $\frac{2 \div 2}{4 \div 2} = \frac{\square}{\square}$

b $\frac{6 \div 2}{8 \div 2} = \frac{\square}{\square}$

c $\frac{10}{4} = \frac{\square}{\square}$

d $\frac{14}{6} = \frac{\square}{\square}$

e $\frac{18}{20} =$ ____

f $\frac{14}{24} =$ ____

g $\frac{16}{2} =$ ____

h $\frac{20}{2} =$ ____

**2** Simplify the following fractions by dividing by 3:

a $\frac{3 \div 3}{9 \div 3} = \frac{\square}{\square}$

b $\frac{12 \div 3}{15 \div 3} = \frac{\square}{\square}$

c $\frac{21}{12} = \frac{\square}{\square}$

d $\frac{18}{15} = \frac{\square}{\square}$

e $\frac{24}{27} =$ ____

f $\frac{9}{30} =$ ____

g $\frac{15}{3} =$ ____

h $\frac{24}{3} =$ ____

i $\frac{15}{21} =$ ____

j $\frac{12}{27} =$ ____

**3** Simplify the following fractions by dividing by 4:

a $\frac{4 \div 4}{8 \div 4} = \frac{\square}{\square}$

b $\frac{12 \div 4}{16 \div 4} = \frac{\square}{\square}$

c $\frac{24}{20} = \frac{\square}{\square}$

d $\frac{20}{12} = \frac{\square}{\square}$

e $\frac{28}{32} =$ ____

f $\frac{20}{32} =$ ____

g $\frac{16}{4} =$ ____

h $\frac{40}{4} =$ ____

**4** Simplify the following fractions by dividing by 5:

a $\frac{5 \div 5}{15 \div 5} = \frac{\square}{\square}$

b $\frac{10 \div 5}{15 \div 5} = \frac{\square}{\square}$

c $\frac{25}{20} = \frac{\square}{\square}$

d $\frac{35}{10} = \frac{\square}{\square}$

e $\frac{5}{45} =$ ____

f $\frac{40}{55} =$ ____

g $\frac{10}{5} =$ ____

h $\frac{50}{5} =$ ____

i $\frac{45}{50} =$ ____

j $\frac{25}{60} =$ ____

**5** Simplify the following fractions by dividing by 6:

a $\frac{6 \div 6}{12 \div 6} = \frac{\square}{\square}$

b $\frac{6 \div 6}{18 \div 6} = \frac{\square}{\square}$

c $\frac{18}{12} = \frac{\square}{\square}$

d $\frac{36}{30} = \frac{\square}{\square}$

e $\frac{24}{30} =$ ____

f $\frac{36}{42} =$ ____

g $\frac{12}{6} =$ ____

h $\frac{36}{6} =$ ____

**6** Simplify the following fractions by dividing by 7:

a $\frac{7 \div 7}{21 \div 7} = \frac{\square}{\square}$

b $\frac{21 \div 7}{28 \div 7} = \frac{\square}{\square}$

c $\frac{14}{21} = \frac{\square}{\square}$

d $\frac{28}{35} = \frac{\square}{\square}$

e $\frac{21}{49} =$ ____

f $\frac{14}{35} =$ ____

g $\frac{49}{7} =$ ____

h $\frac{7}{49} =$ ____

**FRACTIONS**

# Simplifying fractions

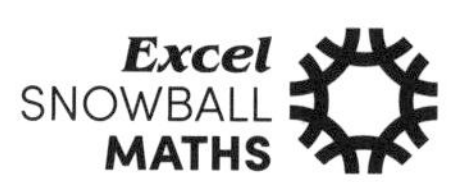


**1** Simplify the following fractions by dividing by 8:

a $\frac{8 \div 8}{16 \div 8} = \frac{\square}{\square}$ b $\frac{24 \div 8}{32 \div 8} = \frac{\square}{\square}$

c $\frac{40}{32} = \frac{\square}{\square}$ d $\frac{48}{40} = \frac{\square}{\square}$

e $\frac{56}{72} =$ ____ f $\frac{32}{56} =$ ____

g $\frac{24}{8} =$ ____ h $\frac{8}{8} =$ ____

**2** Simplify the following fractions by dividing by 9:

a $\frac{9 \div 9}{18 \div 9} = \frac{\square}{\square}$ b $\frac{27 \div 9}{45 \div 9} = \frac{\square}{\square}$

c $\frac{36}{45} = \frac{\square}{\square}$ d $\frac{54}{63} = \frac{\square}{\square}$

e $\frac{72}{81} =$ ____ f $\frac{63}{90} =$ ____

g $\frac{36}{9} =$ ____ h $\frac{18}{27} =$ ____

i $\frac{27}{36} =$ ____ j $\frac{63}{72} =$ ____

k $\frac{90}{63} =$ ____ l $\frac{81}{72} =$ ____

**3** Simplify the following fractions by dividing by 10:

a $\frac{10 \div 10}{30 \div 10} = \frac{\square}{\square}$ b $\frac{20 \div 10}{30 \div 10} = \frac{\square}{\square}$

c $\frac{40}{50} = \frac{\square}{\square}$ d $\frac{70}{100} = \frac{\square}{\square}$

e $\frac{90}{100} =$ ____ f $\frac{40}{50} =$ ____

g $\frac{50}{60} =$ ____ h $\frac{30}{100} =$ ____

**4** Simplify the following fractions by dividing by 11:

a $\frac{11}{22} =$ ____ b $\frac{33}{55} =$ ____

c $\frac{66}{77} =$ ____ d $\frac{44}{55} =$ ____

e $\frac{22}{99} =$ ____ f $\frac{88}{99} =$ ____

g $\frac{11}{110} =$ ____ h $\frac{121}{11} =$ ____

Write the simplified fraction for the shaded parts in each diagram:

**5** a 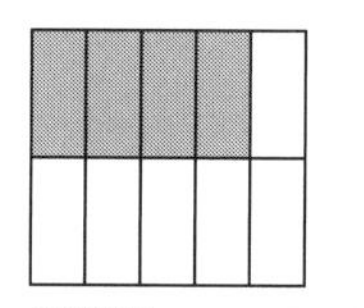 b 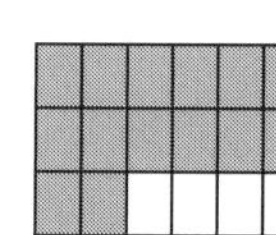 c 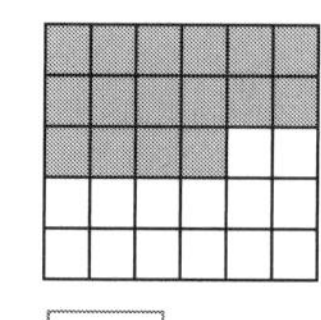

a $\frac{\square \div 2}{\square \div 2} = \frac{\square}{\square}$

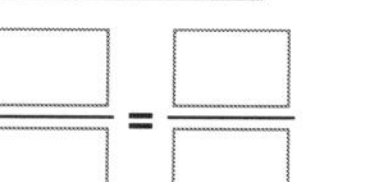

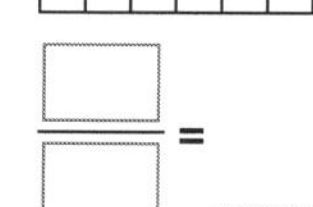

**6** a 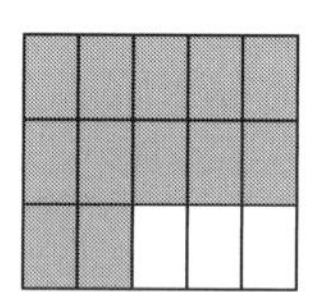 b 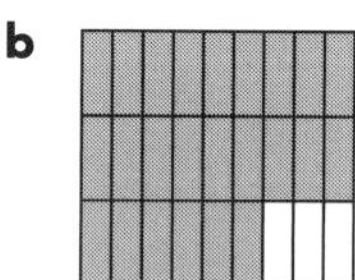 c 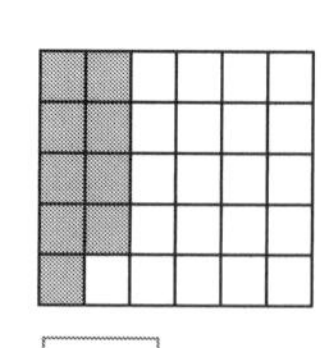

a $\frac{\square \div 3}{\square \div 3} = \frac{\square}{\square}$ b $\frac{\square}{\square} = \frac{\square}{\square}$ c $\frac{\square}{\square} =$ ____

**7** a 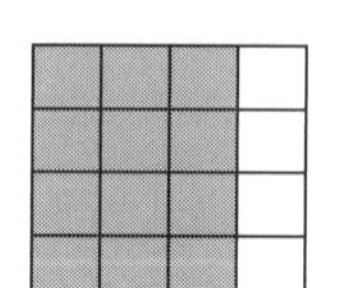 b 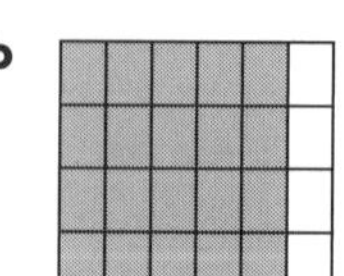 c 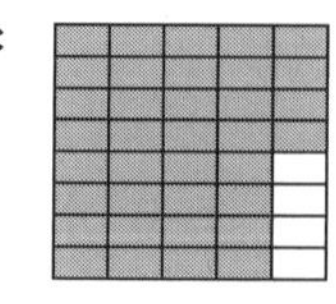

a $\frac{\square \div 4}{\square \div 4} = \frac{\square}{\square}$ b $\frac{\square}{\square} = \frac{\square}{\square}$

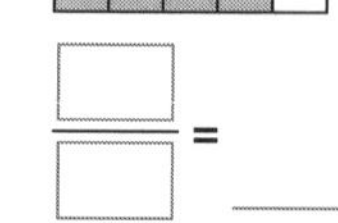

UNIT 9

**FRACTIONS**

# Adding and subtracting fractions

(with the same denominators)

**1** Evaluate the following:

a $\frac{1}{4}+\frac{2}{4}=\frac{\square}{4}$

b $\frac{1}{3}+\frac{1}{3}=\frac{\square}{3}$

c $\frac{2}{5}+\frac{2}{5}=\frac{\square}{\square}$

d $\frac{1}{5}+\frac{3}{5}=\frac{\square}{\square}$

e $\frac{3}{6}+\frac{2}{6}=$ ______

f $\frac{2}{7}+\frac{4}{7}=$ ______

g $\frac{1}{7}+\frac{5}{7}=$ ______

h $\frac{3}{8}+\frac{2}{8}=$ ______

i $\frac{4}{9}+\frac{1}{9}=$ ______

j $\frac{3}{10}+\frac{4}{10}=$ ______

**2** Add and simplify the following fractions:

a $\frac{1}{4}+\frac{1}{4}=\frac{\square}{4}\,\begin{smallmatrix}\div 2\\ \div 2\end{smallmatrix}=\frac{\square}{\square}$

b $\frac{1}{6}+\frac{3}{6}=\frac{\square}{6}\,\begin{smallmatrix}\div 2\\ \div 2\end{smallmatrix}=\frac{\square}{\square}$

c $\frac{1}{8}+\frac{1}{8}=\frac{\square}{\square}=\frac{\square}{\square}$

d $\frac{1}{8}+\frac{5}{8}=\frac{\square}{\square}=\frac{\square}{\square}$

e $\frac{1}{10}+\frac{5}{10}=$ ______

f $\frac{3}{10}+\frac{5}{10}=$ ______

g $\frac{5}{12}+\frac{5}{12}=$ ______

h $\frac{8}{14}+\frac{4}{14}=$ ______

**3** Subtract and simplify the following fractions:

a $\frac{4}{9}-\frac{1}{9}=\frac{\square}{9}\,\begin{smallmatrix}\div 3\\ \div 3\end{smallmatrix}=\frac{\square}{\square}$

b $\frac{8}{9}-\frac{2}{9}=\frac{\square}{9}\,\begin{smallmatrix}\div 3\\ \div 3\end{smallmatrix}=\frac{\square}{\square}$

c $\frac{9}{15}-\frac{3}{15}=\frac{\square}{\square}=\frac{\square}{\square}$

d $\frac{13}{15}-\frac{4}{15}=\frac{\square}{\square}=\frac{\square}{\square}$

e $\frac{11}{21}-\frac{2}{21}=$ ______

f $\frac{18}{27}-\frac{3}{27}=$ ______

g $\frac{20}{21}-\frac{2}{21}=$ ______

h $\frac{5}{30}-\frac{2}{30}=$ ______

**4** Add and simplify the following fractions:

a $\frac{3}{12}+\frac{1}{12}=\frac{\square}{12}\,\begin{smallmatrix}\div 4\\ \div 4\end{smallmatrix}=\frac{\square}{\square}$

b $\frac{6}{12}+\frac{2}{12}=\frac{\square}{12}\,\begin{smallmatrix}\div 4\\ \div 4\end{smallmatrix}=\frac{\square}{\square}$

c $\frac{9}{20}+\frac{3}{20}=\frac{\square}{\square}=\frac{\square}{\square}$

d $\frac{10}{20}+\frac{6}{20}=\frac{\square}{\square}=\frac{\square}{\square}$

e $\frac{15}{24}+\frac{5}{24}=$ ______

f $\frac{20}{32}+\frac{8}{32}=$ ______

g $\frac{12}{28}+\frac{4}{28}=$ ______

h $\frac{15}{36}+\frac{5}{36}=$ ______

## FRACTIONS
# Adding and subtracting fractions
(with the same denominators)

**1** Subtract and simplify the following fractions:

a $\frac{7}{10} - \frac{2}{10} = \frac{\square}{10} \,{}^{\div 5}_{\div 5} = \frac{\square}{\square}$

b $\frac{12}{15} - \frac{2}{15} = \frac{\square}{15} \,{}^{\div 5}_{\div 5} = \frac{\square}{\square}$

c $\frac{18}{20} - \frac{3}{20} = \frac{\square}{\square} = \frac{\square}{\square}$

d $\frac{24}{25} - \frac{4}{25} = \frac{\square}{\square} = \frac{\square}{\square}$

e $\frac{15}{25} - \frac{5}{25} =$ ______

f $\frac{29}{30} - \frac{4}{30} =$ ______

g $\frac{17}{35} - \frac{2}{35} =$ ______

h $\frac{15}{45} - \frac{5}{45} =$ ______

**2** Add and simplify the following fractions:

a $\frac{4}{12} + \frac{2}{12} = \frac{\square}{12} \,{}^{\div 6}_{\div 6} = \frac{\square}{\square}$

b $\frac{3}{18} + \frac{3}{18} = \frac{\square}{\square} \,{}^{\div 6}_{\div 6} = \frac{\square}{\square}$

c $\frac{10}{24} + \frac{8}{24} =$ ______

d $\frac{12}{30} + \frac{6}{30} =$ ______

e $\frac{18}{42} + \frac{6}{42} =$ ______

f $\frac{30}{42} + \frac{6}{42} =$ ______

g $\frac{15}{48} + \frac{3}{48} =$ ______

h $\frac{30}{54} + \frac{12}{54} =$ ______

**3** Subtract and simplify the following fractions:

a $\frac{9}{14} - \frac{2}{14} = \frac{\square}{\square} \,{}^{\div 7}_{\div 7} = \frac{\square}{\square}$

b $\frac{16}{21} - \frac{2}{21} = \frac{\square}{\square} \,{}^{\div 7}_{\div 7} = \frac{\square}{\square}$

c $\frac{20}{21} - \frac{13}{21} =$ ______

d $\frac{24}{35} - \frac{3}{35} =$ ______

e $\frac{40}{49} - \frac{5}{49} =$ ______

f $\frac{47}{49} - \frac{5}{49} =$ ______

**4** Subtract and simplify the following fractions:

a $\frac{10}{16} - \frac{2}{16} = \frac{\square}{\square} = \frac{\square}{\square}$

b $\frac{21}{24} - \frac{5}{24} = \frac{\square}{\square} = \frac{\square}{\square}$

c $\frac{28}{32} - \frac{4}{32} =$ ______

d $\frac{35}{40} - \frac{3}{40} =$ ______

e $\frac{18}{40} - \frac{2}{40} =$ ______

f $\frac{45}{56} - \frac{5}{56} =$ ______

**5** Add and simplify the following fractions:

a $\frac{7}{18} + \frac{2}{18} =$ ______

b $\frac{10}{27} + \frac{8}{27} =$ ______

c $\frac{30}{45} + \frac{6}{45} =$ ______

d $\frac{40}{54} + \frac{5}{54} =$ ______

# FRACTIONS
## Adding fractions
(with different denominators)

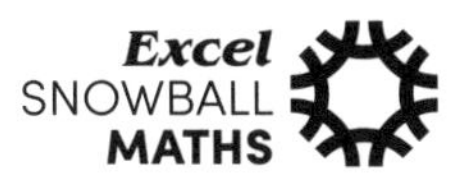

**1** Add the following fractions by multiplying by 2 first:

**a** $\frac{1 \times 2}{2 \times 2} + \frac{1}{4}$

$= \frac{\square}{\square} + \frac{1}{4}$

$= \frac{\square}{4}$

**b** $\frac{2 \times 2}{3 \times 2} + \frac{1}{6}$

$= \frac{\square}{\square} + \frac{1}{6}$

$= \frac{\square}{6}$

**c** $\frac{1}{4} + \frac{1}{8}$

$= \frac{\square}{\square} + \frac{1}{8}$

$= \frac{\square}{8}$

**d** $\frac{3}{4} + \frac{1}{8}$

$= \frac{\square}{\square} + \frac{1}{8}$

$= \frac{\square}{8}$

**e** $\frac{1}{5} + \frac{1}{10}$

$= \frac{\square}{\square} + \frac{1}{10}$

$= \frac{\square}{10}$

**f** $\frac{2}{5} + \frac{3}{10}$

$= \frac{\square}{\square} + \frac{3}{10}$

$= \frac{\square}{\square}$

**g** $\frac{3}{5} + \frac{3}{10}$

$= \frac{\square}{\square} + \frac{\square}{\square}$

$= \frac{\square}{\square}$

**h** $\frac{1}{6} + \frac{5}{12}$

$= \frac{\square}{\square} + \frac{\square}{\square}$

$= \frac{\square}{\square}$

**i** $\frac{1}{7} + \frac{1}{14}$

$= \frac{\square}{\square} + \frac{\square}{\square}$

$= \frac{\square}{\square}$

**j** $\frac{2}{7} + \frac{1}{14}$

$= \frac{\square}{\square} + \frac{\square}{\square}$

$= \frac{\square}{\square}$

**k** $\frac{3}{7} + \frac{3}{14}$

$= \frac{\square}{\square} + \frac{\square}{\square}$

$= \frac{\square}{\square}$

**l** $\frac{4}{7} + \frac{3}{14}$

$= \frac{\square}{\square} + \frac{\square}{\square}$

$= \frac{\square}{\square}$

**m** $\frac{5}{7} + \frac{1}{14}$

= ____________

= ____________

**n** $\frac{6}{7} + \frac{1}{14}$

= ____________

= ____________

**o** $\frac{3}{7} + \frac{5}{14}$

= ____________

= ____________

**p** $\frac{5}{8} + \frac{3}{16}$

= ____________

= ____________

**q** $\frac{4}{9} + \frac{5}{18}$

= ____________

= ____________

**r** $\frac{3}{10} + \frac{11}{20}$

= ____________

= ____________

**s** $\frac{7}{12} + \frac{5}{24}$

= ____________

= ____________

**t** $\frac{5}{18} + \frac{13}{36}$

= ____________

= ____________

UNIT 12

**FRACTIONS**

# Adding and subtracting fractions

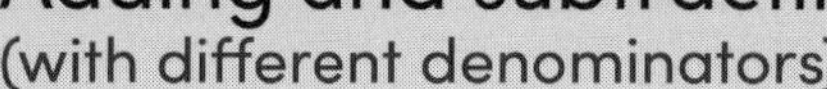
(with different denominators)

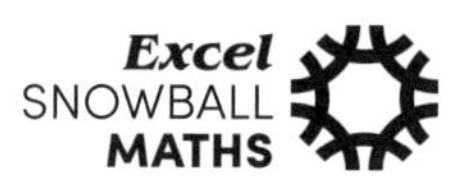


**1** Add the following fractions by multiplying by 3 first:

**a** $\frac{1^{\times 3}}{2_{\times 3}} + \frac{2}{6}$

$= \frac{\square}{\square} + \frac{2}{6}$

$= \frac{\square}{6}$

**b** $\frac{1^{\times 3}}{3_{\times 3}} + \frac{1}{9}$

$= \frac{\square}{\square} + \frac{1}{9}$

$= \frac{\square}{9}$

**c** $\frac{2}{3} + \frac{1}{9}$

$= \frac{\square}{\square} + \frac{1}{9}$

$= \frac{\square}{9}$

**d** $\frac{2}{3} + \frac{2}{9}$

$= \frac{\square}{\square} + \frac{2}{9}$

$= \frac{\square}{9}$

**e** $\frac{3}{4} + \frac{2}{12}$

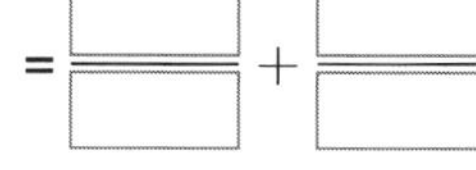

$= \frac{\square}{\square} + \frac{\square}{\square}$

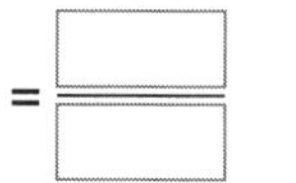

$= \frac{\square}{\square}$

**f** $\frac{1}{5} + \frac{1}{15}$

$= \frac{\square}{\square} + \frac{\square}{\square}$

$= \frac{\square}{\square}$

**g** $\frac{1}{6} + \frac{4}{18}$

= ______________

= ______________

**h** $\frac{3}{6} + \frac{4}{18}$

= ______________

= ______________

**i** $\frac{4}{6} + \frac{5}{18}$

= ______________

= ______________

**j** $\frac{2}{6} + \frac{5}{18}$

= ______________

= ______________

**2** Subtract the following fractions by multiplying by 2 first:

**a** $\frac{1^{\times 2}}{2_{\times 2}} - \frac{1}{4}$

$= \frac{\square}{\square} - \frac{1}{4}$

$= \frac{\square}{4}$

**b** $\frac{2^{\times 2}}{3_{\times 2}} - \frac{3}{6}$

$= \frac{\square}{\square} - \frac{3}{6}$

$= \frac{\square}{6}$

**c** $\frac{1}{4} - \frac{1}{8}$

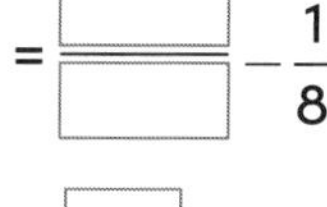

$= \frac{\square}{\square} - \frac{1}{8}$

$= \frac{\square}{8}$

**d** $\frac{3}{4} - \frac{3}{8}$

$= \frac{\square}{\square} - \frac{3}{8}$

$= \frac{\square}{8}$

**e** $\frac{3}{5} - \frac{3}{10}$

$= \frac{\square}{\square} - \frac{\square}{\square}$

$= \frac{\square}{\square}$

**f** $\frac{5}{6} - \frac{3}{12}$

$= \frac{\square}{\square} - \frac{\square}{\square}$

$= \frac{\square}{\square}$

**g** $\frac{5}{8} - \frac{3}{16}$

= ______________

= ______________

**h** $\frac{6}{8} - \frac{7}{16}$

= ______________

= ______________

**i** $\frac{8}{9} - \frac{5}{18}$

= ______________

= ______________

**j** $\frac{7}{10} - \frac{3}{20}$

= ______________

= ______________

# FRACTIONS
## Subtracting fractions
(with different denominators)

1 Subtract the following fractions by multiplying by 3 first:

a $\frac{1^{\times 3}}{2_{\times 3}} - \frac{2}{6}$
$= \frac{\square}{\square} - \frac{2}{6}$
$= \frac{\square}{6}$

b $\frac{1^{\times 3}}{3_{\times 3}} - \frac{1}{9}$
$= \frac{\square}{\square} - \frac{1}{9}$
$= \frac{\square}{9}$

c $\frac{2}{3} - \frac{1}{9}$
$= \frac{\square}{\square} - \frac{1}{9}$
$= \frac{\square}{9}$

d $\frac{2}{3} - \frac{2}{9}$
$= \frac{\square}{\square} - \frac{2}{9}$
$= \frac{\square}{9}$

e $\frac{3}{4} - \frac{2}{12}$
$= \frac{\square}{\square} - \frac{\square}{\square}$
$= \frac{\square}{\square}$

f $\frac{1}{5} - \frac{1}{15}$
$= \frac{\square}{\square} - \frac{\square}{\square}$
$= \frac{\square}{\square}$

g $\frac{3}{5} - \frac{2}{15}$
= ______
= ______

h $\frac{4}{5} - \frac{4}{15}$
= ______
= ______

i $\frac{3}{6} - \frac{4}{18}$
= ______
= ______

j $\frac{4}{6} - \frac{1}{18}$
= ______
= ______

k $\frac{4}{7} - \frac{4}{21}$
= ______
= ______

l $\frac{7}{10} - \frac{8}{30}$
= ______
= ______

m $\frac{6}{10} - \frac{7}{30}$
= ______
= ______

n $\frac{3}{8} - \frac{4}{24}$
= ______
= ______

o $\frac{7}{8} - \frac{14}{24}$
= ______
= ______

p $\frac{4}{9} - \frac{4}{27}$
= ______
= ______

q $\frac{7}{9} - \frac{7}{27}$
= ______
= ______

r $\frac{9}{11} - \frac{10}{33}$
= ______
= ______

s $\frac{5}{7} - \frac{4}{21}$
= ______
= ______

t $\frac{6}{10} - \frac{5}{30}$
= ______
= ______

u $\frac{3}{12} - \frac{4}{36}$
= ______
= ______

v $\frac{9}{12} - \frac{4}{36}$
= ______
= ______

UNIT 14

## FRACTIONS

# Adding fractions

(with different denominators)

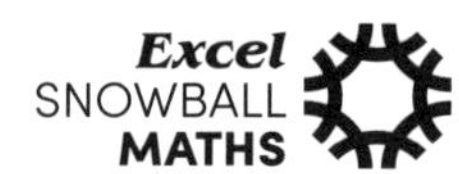

**1** Add the following fractions by multiplying by 4 first:

**a** $\frac{1}{8} + \frac{1^{\times 4}}{2_{\times 4}}$

$= \frac{1}{8} + \frac{\square}{\square}$

$= \frac{\square}{8}$

**b** $\frac{3}{8} + \frac{1^{\times 4}}{2_{\times 4}}$

$= \frac{3}{8} + \frac{\square}{\square}$

$= \frac{\square}{8}$

**c** $\frac{1}{12} + \frac{1}{3}$

$= \frac{1}{12} + \frac{\square}{\square}$

$= \frac{\square}{12}$

**d** $\frac{3}{12} + \frac{2}{3}$

$= \frac{3}{12} + \frac{\square}{\square}$

$= \frac{\square}{12}$

**e** $\frac{1}{16} + \frac{1}{4}$

$= \frac{1}{16} + \frac{\square}{\square}$

$= \frac{\square}{\square}$

**f** $\frac{1}{16} + \frac{3}{4}$

$= \frac{\square}{\square} + \frac{\square}{\square}$

$= \frac{\square}{\square}$

**g** $\frac{3}{16} + \frac{3}{4}$

$= \frac{\square}{\square} + \frac{\square}{\square}$

$= \frac{\square}{\square}$

**h** $\frac{1}{20} + \frac{2}{5}$

$= \frac{\square}{\square} + \frac{\square}{\square}$

$= \frac{\square}{\square}$

**i** $\frac{3}{20} + \frac{2}{5}$

= ________

= ________

**j** $\frac{5}{20} + \frac{3}{5}$

= ________

= ________

**k** $\frac{3}{20} + \frac{4}{5}$

= ________

= ________

**l** $\frac{1}{24} + \frac{1}{6}$

= ________

= ________

**m** $\frac{1}{24} + \frac{3}{6}$

= ________

= ________

**n** $\frac{3}{24} + \frac{5}{6}$

= ________

= ________

**o** $\frac{5}{32} + \frac{3}{8}$

= ________

= ________

**p** $\frac{13}{32} + \frac{3}{8}$

= ________

= ________

**q** $\frac{1}{40} + \frac{7}{10}$

= ________

= ________

**r** $\frac{11}{40} + \frac{5}{10}$

= ________

= ________

**s** $\frac{15}{48} + \frac{5}{12}$

= ________

= ________

**t** $\frac{17}{44} + \frac{6}{11}$

= ________

= ________

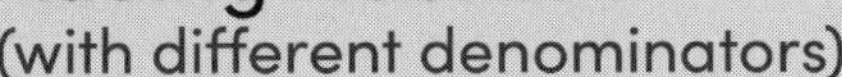

# FRACTIONS
## Adding fractions
(with different denominators)

**1** Add the following fractions by multiplying by 5 first:

**a** $\frac{2}{10}+\frac{1^{\times 5}}{2_{\times 5}}$

$=\frac{2}{10}+\frac{\square}{\square}$

$=\frac{\square}{10}$

**b** $\frac{1}{15}+\frac{2^{\times 5}}{3_{\times 5}}$

$=\frac{1}{15}+\frac{\square}{\square}$

$=\frac{\square}{15}$

**c** $\frac{4}{20}+\frac{1}{4}$

$=\frac{4}{20}+\frac{\square}{\square}$

$=\frac{\square}{\square}$

**d** $\frac{2}{20}+\frac{3}{4}$

$=\frac{2}{20}+\frac{\square}{\square}$

$=\frac{\square}{\square}$

**e** $\frac{8}{25}+\frac{2}{5}$

$=\frac{\square}{\square}+\frac{\square}{\square}$

$=\frac{\square}{\square}$

**f** $\frac{7}{25}+\frac{3}{5}$

$=\frac{\square}{\square}+\frac{\square}{\square}$

$=\frac{\square}{\square}$

**g** $\frac{6}{30}+\frac{1}{6}$

= ______

= ______

**h** $\frac{2}{30}+\frac{3}{6}$

= ______

= ______

**i** $\frac{9}{35}+\frac{2}{7}$

= ______

= ______

**j** $\frac{7}{35}+\frac{3}{7}$

= ______

= ______

**k** $\frac{11}{35}+\frac{4}{7}$

= ______

= ______

**l** $\frac{3}{40}+\frac{2}{8}$

= ______

= ______

**m** $\frac{12}{40}+\frac{3}{8}$

= ______

= ______

**n** $\frac{8}{40}+\frac{5}{8}$

= ______

= ______

**o** $\frac{18}{45}+\frac{2}{9}$

= ______

= ______

**p** $\frac{12}{45}+\frac{5}{9}$

= ______

= ______

**q** $\frac{5}{10}+\frac{4}{50}$

= ______

= ______

**r** $\frac{6}{10}+\frac{11}{50}$

= ______

= ______

**s** $\frac{9}{11}+\frac{6}{55}$

= ______

= ______

**t** $\frac{7}{12}+\frac{14}{60}$

= ______

= ______

**u** $\frac{5}{12}+\frac{18}{60}$

= ______

= ______

**v** $\frac{7}{11}+\frac{16}{55}$

= ______

= ______

# FRACTIONS

## Adding and subtracting fractions
(with different denominators)

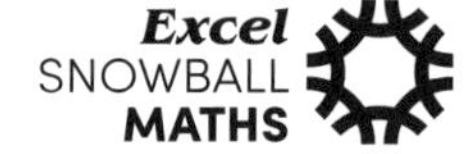


**1** Evaluate the following:

**a** 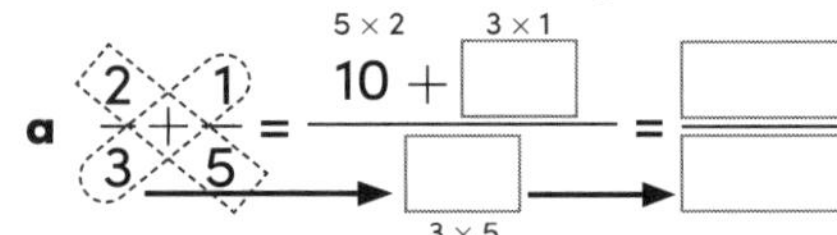


$\frac{2}{3}+\frac{1}{5} = \frac{10 + \square}{\square} = \frac{\square}{\square}$

**b** $\frac{1}{3}+\frac{1}{4} = \frac{4 + \square}{\square} = \frac{\square}{\square}$ (4 × 1, 3 × 1, 3 × 4)

**c** $\frac{1}{2}+\frac{1}{3} = \frac{\square + \square}{\square} = \frac{\square}{\square}$

**d** $\frac{2}{3}+\frac{1}{2} = \frac{\square + \square}{\square} = \frac{\square}{\square}$

**e** $\frac{2}{3}+\frac{1}{4} = \frac{\square + \square}{\square} = \frac{\square}{\square}$

**f** $\frac{2}{5}+\frac{1}{3} =$ ____________________

**g** $\frac{2}{7}+\frac{1}{4} =$ ____________________

**h** $\frac{3}{4}+\frac{1}{3} =$ ____________________

**i** $\frac{3}{7}+\frac{1}{4} =$ ____________________

**j** $\frac{3}{5}+\frac{1}{3} =$ ____________________

**k** $\frac{4}{7}+\frac{1}{2} =$ ____________________

**2** Evaluate the following:

**a** $\frac{2}{3}-\frac{1}{2} = \frac{\square - \square}{\square} = \frac{\square}{\square}$

**b** $\frac{2}{5}-\frac{1}{3} = \frac{\square - \square}{\square} = \frac{\square}{\square}$

**c** $\frac{2}{7}-\frac{1}{4} = \frac{\square - \square}{\square} = \frac{\square}{\square}$

**d** $\frac{2}{3}-\frac{1}{5} = \frac{\square - \square}{\square} = \frac{\square}{\square}$

**1** Evaluate the following:

**a** $\frac{2}{5}-\frac{1}{4} =$ ____________________

**b** $\frac{2}{5}-\frac{1}{6} =$ ____________________

**c** $\frac{4}{7}-\frac{1}{4} =$ ____________________

**d** $\frac{3}{5}-\frac{1}{4} =$ ____________________

**e** $\frac{3}{8}+\frac{1}{5} =$ ____________________

**f** $\frac{3}{8}+\frac{1}{7} =$ ____________________

**g** $\frac{4}{5}-\frac{1}{6} =$ ____________________

**h** $\frac{5}{6}-\frac{1}{5} =$ ____________________

**i** $\frac{5}{8}-\frac{1}{3} =$ ____________________

**j** $\frac{6}{7}+\frac{1}{4} =$ ____________________

**k** $\frac{7}{8}+\frac{1}{3} =$ ____________________

**l** $\frac{8}{9}-\frac{1}{2} =$ ____________________

**m** $\frac{2}{7}+\frac{2}{5} =$ ____________________

**n** $\frac{5}{6}-\frac{1}{9} =$ ____________________

**o** $\frac{1}{6}+\frac{3}{7} =$ ____________________

UNIT 17

## FRACTIONS

# Adding and subtracting fractions
(with different denominators)

**1** Evaluate the following:

a $\frac{3}{8} - \frac{2}{7} =$ ____________

b $\frac{4}{5} - \frac{4}{9} =$ ____________

c $\frac{5}{8} - \frac{3}{7} =$ ____________

d $\frac{7}{9} - \frac{3}{5} =$ ____________

e $\frac{8}{9} - \frac{4}{5} =$ ____________

**2** Evaluate the following:

a $\frac{5}{6} + \frac{2}{7} =$ ____________

b $\frac{5}{8} + \frac{4}{5} =$ ____________

c $\frac{4}{7} + \frac{5}{9} =$ ____________

d $\frac{3}{8} + \frac{2}{7} =$ ____________

e $\frac{5}{9} + \frac{3}{7} =$ ____________

**3** Evaluate the following:

a $\frac{5}{11} + \frac{3}{8} =$ ____________

b $\frac{8}{11} + \frac{2}{5} =$ ____________

c $\frac{2}{11} - \frac{1}{9} =$ ____________

d $\frac{4}{7} - \frac{3}{10} =$ ____________

e $\frac{7}{10} + \frac{2}{9} =$ ____________

f $\frac{3}{10} + \frac{2}{7} =$ ____________

g $\frac{9}{10} - \frac{5}{7} =$ ____________

h $\frac{2}{11} - \frac{1}{10} =$ ____________

i $\frac{7}{10} + \frac{3}{11} =$ ____________

j $\frac{8}{9} - \frac{2}{11} =$ ____________

k $\frac{3}{11} + \frac{4}{9} =$ ____________

l $\frac{6}{7} - \frac{7}{11} =$ ____________

**4** What is the sum of $\frac{1}{4}$ and $\frac{2}{5}$?

$\frac{1}{4} + \frac{2}{5} =$ ____________

**5** Increase $\frac{3}{5}$ by $\frac{2}{7}$.

____________

**6** What is the total of $\frac{9}{11}$ and $\frac{1}{6}$?

____________

**7** What is the difference between $\frac{2}{5}$ and $\frac{1}{6}$?

____________

**8** Decrease $\frac{8}{9}$ by $\frac{1}{7}$.

____________

UNIT 18

# FRACTIONS
## Types of fractions

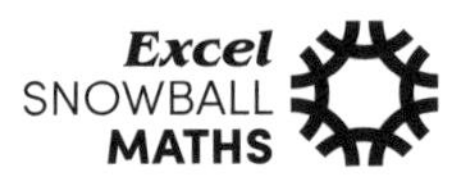

**1** State whether each fraction is improper, proper or a mixed numeral.

**a** $1\frac{3}{4}$ ______

**b** $3\frac{5}{6}$ ______

**c** $\frac{3}{5}$ ______

**d** $\frac{8}{5}$ ______

**e** $\frac{7}{6}$ ______

**f** $\frac{11}{20}$ ______

**2** Convert each mixed numeral to an improper fraction:

**a** $1\frac{+2}{\times 3} = \frac{\square}{3}$ (3 × 1 + 2)

**b** $1\frac{+4}{\times 7} = \frac{\square}{7}$ (7 × 1 + 4)

**c** $2\frac{2}{3} = \frac{\square}{3}$

**d** $2\frac{3}{4} = \frac{\square}{4}$

**e** $3\frac{1}{2} = \frac{\square}{\square}$

**f** $3\frac{2}{7} = \frac{\square}{\square}$

**g** $4\frac{3}{5} =$ ______

**h** $4\frac{1}{9} =$ ______

**i** $5\frac{2}{5} =$ ______

**j** $5\frac{6}{7} =$ ______

**k** $6\frac{1}{8} =$ ______

**l** $7\frac{4}{7} =$ ______

**m** $8\frac{5}{6} =$ ______

**n** $9\frac{1}{8} =$ ______

**3** Convert each improper fraction to a mixed numeral:

**a** $\frac{5}{3} = 1\frac{\square}{3}$ (5 − 3 × 1)

**b** $\frac{3}{2} = 1\frac{\square}{2}$ (3 − 2 × 1)

**c** $\frac{7}{4} = 1\frac{\square}{4}$

**d** $\frac{11}{8} = 1\frac{\square}{8}$

**e** $\frac{10}{7} = 1\frac{\square}{\square}$

**f** $\frac{10}{9} = 1\frac{\square}{\square}$

**g** $\frac{7}{6} = 1\frac{\square}{\square}$

**h** $\frac{11}{9} = 1\frac{\square}{\square}$

**i** $\frac{9}{7} = \square\frac{\square}{\square}$

**j** $\frac{4}{3} = \square\frac{\square}{\square}$

**k** $\frac{5}{4} =$ ______

**l** $\frac{13}{9} =$ ______

**4** Convert each improper fraction to a mixed numeral:

**a** $\frac{8}{3} = 2\frac{\square}{3}$ (8 − 3 × 2)

**b** $\frac{5}{2} = 2\frac{\square}{2}$ (5 − 2 × 2)

**c** $\frac{11}{4} = 2\frac{\square}{4}$

**d** $\frac{19}{8} = 2\frac{\square}{8}$

**e** $\frac{13}{6} = 2\frac{\square}{\square}$

**f** $\frac{13}{5} = 2\frac{\square}{\square}$

**g** $\frac{11}{5} = 2\frac{\square}{\square}$

**h** $\frac{20}{9} = 2\frac{\square}{\square}$

**i** $\frac{16}{7} = \square\frac{\square}{\square}$

**j** $\frac{14}{5} = \square\frac{\square}{\square}$

**k** $\frac{15}{7} =$ ______

**l** $\frac{18}{7} =$ ______

**m** $\frac{12}{5} =$ ______

**n** $\frac{17}{8} =$ ______

**o** $\frac{27}{10} =$ ______

**p** $\frac{22}{9} =$ ______

UNIT 19

FRACTIONS

# Types of fractions

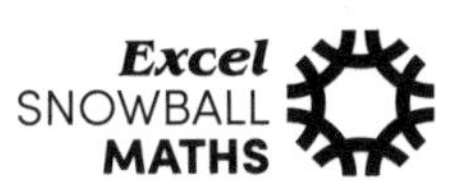


**1** Convert each improper fraction to a mixed numeral:

a $\frac{11}{3} = 3\frac{\square}{3}$ (11 − 3 × 3)

b $\frac{7}{2} = 3\frac{\square}{2}$ (7 − 2 × 3)

c $\frac{15}{4} = 3\frac{\square}{4}$

d $\frac{25}{8} = 3\frac{\square}{8}$

e $\frac{19}{6} = 3\frac{\square}{\square}$

f $\frac{17}{5} = 3\frac{\square}{\square}$

g $\frac{18}{5} = 3\frac{\square}{\square}$

h $\frac{13}{4} = 3\frac{\square}{\square}$

i $\frac{23}{7} = \square\frac{\square}{\square}$

j $\frac{27}{8} = \square\frac{\square}{\square}$

k $\frac{16}{5} =$ ________

l $\frac{19}{5} =$ ________

m $\frac{23}{6} =$ ________

n $\frac{25}{7} =$ ________

o $\frac{29}{9} =$ ________

p $\frac{29}{8} =$ ________

**2** Convert each improper fraction to a mixed numeral:

a $\frac{14}{3} = 4\frac{\square}{3}$ (14 − 3 × 4)

b $\frac{9}{2} = 4\frac{\square}{2}$ (9 − 2 × 4)

c $\frac{17}{4} = 4\frac{\square}{4}$

d $\frac{35}{8} = 4\frac{\square}{8}$

e $\frac{29}{6} = 4\frac{\square}{\square}$

f $\frac{22}{5} = 4\frac{\square}{\square}$

g $\frac{19}{4} =$ ________

h $\frac{38}{9} =$ ________

i $\frac{31}{7} =$ ________

j $\frac{13}{3} =$ ________

k $\frac{24}{5} =$ ________

l $\frac{47}{10} =$ ________

**3** Convert each improper fraction to a mixed numeral:

a $\frac{11}{2} = 5\frac{\square}{2}$ (11 − 2 × 5)

b $\frac{17}{3} = 5\frac{\square}{3}$ (17 − 3 × 5)

c $\frac{31}{6} = 5\frac{\square}{6}$

d $\frac{43}{8} = 5\frac{\square}{8}$

e $\frac{27}{5} = 5\frac{\square}{\square}$

f $\frac{49}{9} = 5\frac{\square}{\square}$

g $\frac{28}{5} =$ ________

h $\frac{41}{7} =$ ________

i $\frac{36}{7} =$ ________

j $\frac{53}{10} =$ ________

k $\frac{47}{9} =$ ________

l $\frac{38}{7} =$ ________

m $\frac{41}{8} =$ ________

n $\frac{51}{10} =$ ________

**4** Convert each improper fraction to a mixed numeral:

a $\frac{13}{2} = 6\frac{\square}{2}$

b $\frac{25}{4} = 6\frac{\square}{4}$

c $\frac{27}{4} = 6\frac{\square}{\square}$

d $\frac{31}{5} = 6\frac{\square}{\square}$

e $\frac{34}{5} =$ ________

f $\frac{20}{3} =$ ________

g $\frac{43}{7} =$ ________

h $\frac{56}{9} =$ ________

i $\frac{41}{6} =$ ________

j $\frac{51}{8} =$ ________

k $\frac{44}{7} =$ ________

l $\frac{45}{7} =$ ________

**FRACTIONS**

# Adding and subtracting mixed numerals

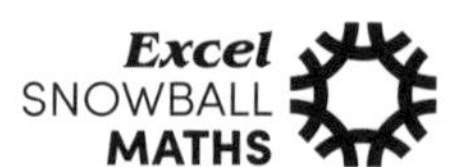


**1** Evaluate the following:

**a** $1\frac{1}{5} - \frac{2}{5}$ (+, ×)

$= \frac{\square}{\square} - \frac{2}{5}$

$= \frac{\square}{\square}$

**b** $1\frac{2}{5} - \frac{3}{5}$ (+, ×)

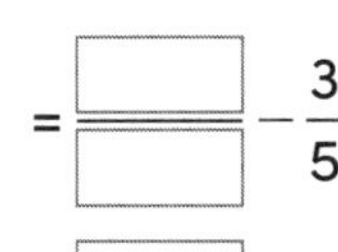

$= \frac{\square}{\square} - \frac{3}{5}$

$= \frac{\square}{\square}$

**c** $1\frac{1}{3} - \frac{2}{3}$

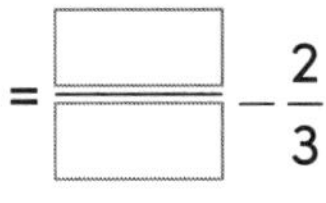

$= \frac{\square}{\square} - \frac{2}{3}$

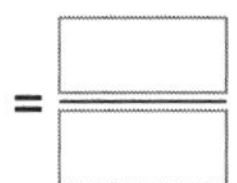

$= \frac{\square}{\square}$

**d** $1\frac{1}{4} - \frac{2}{4}$

$= \frac{\square}{\square} - \frac{2}{4}$

$= \frac{\square}{\square}$

**e** $1\frac{2}{5} - \frac{4}{5}$

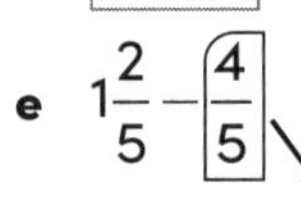

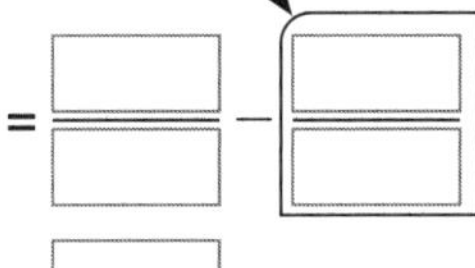

$= \frac{\square}{\square} - \frac{\square}{\square}$

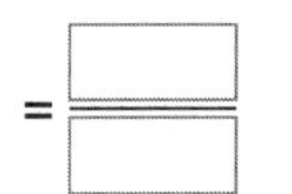

$= \frac{\square}{\square}$

**f** $1\frac{3}{5} - \frac{4}{5}$

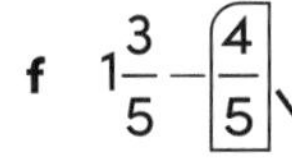

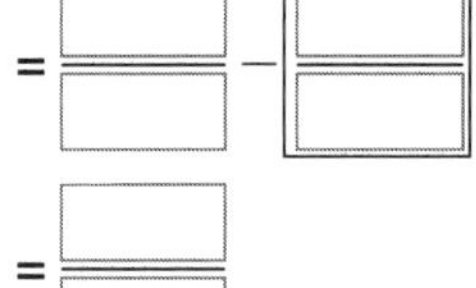

$= \frac{\square}{\square} - \frac{\square}{\square}$

$= \frac{\square}{\square}$

**g** $1\frac{1}{5} - \frac{3}{5}$

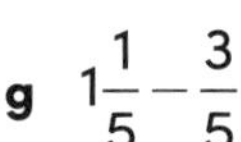

= __________

= __________

**h** $1\frac{1}{7} - \frac{4}{7}$

= __________

= __________

**i** $1\frac{3}{7} - \frac{4}{7}$

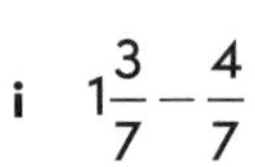

= __________

= __________

**j** $1\frac{5}{7} - \frac{6}{7}$

= __________

= __________

**k** $1\frac{2}{9} - \frac{4}{9}$

= __________

= __________

**l** $1\frac{4}{9} - \frac{5}{9}$

= __________

= __________

**2** Evaluate the following, leaving your answer as a mixed numeral:

**a** $1\frac{2}{5} + \frac{4}{5}$ (+, ×)

$= \frac{\square}{\square} + \frac{4}{5}$

$= \frac{\square}{\square}$

$= 2\frac{\square}{5}$

**b** $1\frac{2}{4} + \frac{3}{4}$ (+, ×)

$= \frac{\square}{\square} + \frac{3}{4}$

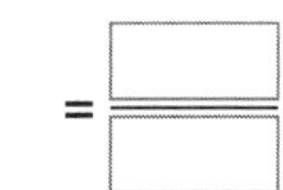

$= \frac{\square}{\square}$

$= 2\frac{\square}{4}$

**c** $1\frac{3}{5} + \frac{3}{5}$

$= \frac{\square}{\square} + \frac{\square}{\square}$

$= \frac{\square}{\square}$

$= 2\frac{\square}{5}$

**d** $1\frac{3}{5} + \frac{4}{5}$

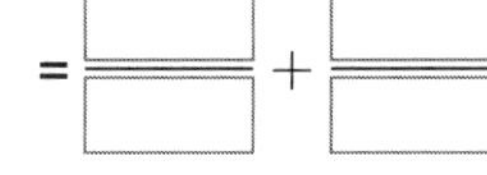

$= \frac{\square}{\square} + \frac{\square}{\square}$

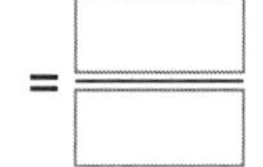

$= \frac{\square}{\square}$

$= 2\frac{\square}{5}$

**e** $1\frac{3}{4} + \frac{2}{4}$

$= \frac{\square}{\square} + \frac{\square}{\square}$

$= \frac{\square}{\square}$

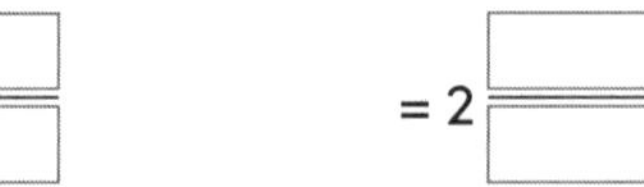

$= 2\frac{\square}{\square}$

**f** $1\frac{5}{6} + \frac{2}{6}$

$= \frac{\square}{\square} + \frac{\square}{\square}$

$= \frac{\square}{\square}$

$= 2\frac{\square}{\square}$

**g** $1\frac{2}{7} + \frac{6}{7}$

= __________

= __________

= __________

**h** $1\frac{4}{7} + \frac{5}{7}$

= __________

= __________

= __________

UNIT 21

**FRACTIONS**

# Adding and subtracting mixed numerals

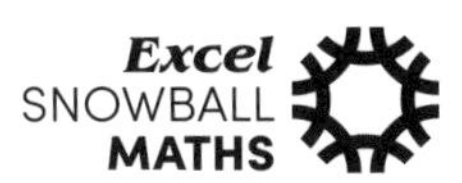

**1** Evaluate the following, leaving your answer as a mixed numeral:

**a** $3\frac{1}{3} - \frac{2}{3}$

$= \frac{\square}{\square} - \frac{2}{3}$

$= \frac{\square}{\square}$

$= 2\frac{\square}{\square}$

**b** $3\frac{2}{5} - \frac{4}{5}$

$= \frac{\square}{\square} - \frac{4}{5}$

$= \frac{\square}{\square}$

$= 2\frac{\square}{\square}$

**c** $3\frac{1}{5} - \frac{2}{5}$

$= \frac{\square}{\square} - \frac{\square}{\square}$

$= \frac{\square}{\square}$

$= \square\frac{\square}{\square}$

**d** $3\frac{1}{5} - \frac{3}{5}$

$= \frac{\square}{\square} - \frac{\square}{\square}$

$= \frac{\square}{\square}$

$= \square\frac{\square}{\square}$

**e** $3\frac{1}{4} - \frac{2}{4}$

$= \frac{\square}{\square} - \frac{\square}{\square}$

$= \frac{\square}{\square}$

$= \square\frac{\square}{\square}$

**f** $3\frac{2}{5} - \frac{3}{5}$

$= \frac{\square}{\square} - \frac{\square}{\square}$

$= \frac{\square}{\square}$

$= \square\frac{\square}{\square}$

**g** $3\frac{4}{6} - \frac{5}{6}$

= ____________

= ____________

= ____________

**h** $3\frac{1}{7} - \frac{4}{7}$

= ____________

= ____________

= ____________

**i** $3\frac{3}{5} - \frac{4}{5}$

= ____________

= ____________

= ____________

**j** $2\frac{1}{4} - \frac{2}{4}$

= ____________

= ____________

= ____________

**k** $2\frac{1}{3} - \frac{2}{3}$

= ____________

= ____________

= ____________

**l** $2\frac{1}{7} - \frac{2}{7}$

= ____________

= ____________

= ____________

**m** $4\frac{1}{3} - \frac{2}{3}$

= ____________

= ____________

= ____________

**n** $4\frac{1}{4} - \frac{2}{4}$

= ____________

= ____________

= ____________

**o** $2\frac{2}{5} + \frac{4}{5}$

= ____________

= ____________

= ____________

**p** $2\frac{5}{6} + \frac{2}{6}$

= ____________

= ____________

= ____________

**q** $2\frac{3}{4} + \frac{2}{4}$

= ____________

= ____________

= ____________

**r** $2\frac{4}{5} + \frac{3}{5}$

= ____________

= ____________

= ____________

UNIT 22

FRACTIONS

# Adding and subtracting mixed numerals

**1** Evaluate the following, leaving your answer as a mixed numeral:

**a** $1\frac{+2}{\times 5} + 1\frac{+4}{\times 5}$

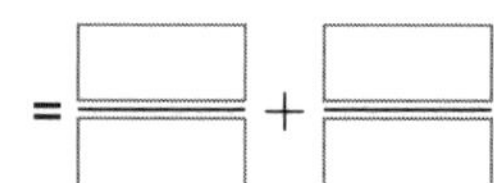

$= \frac{\square}{\square}$

$= 3\frac{\square}{\square}$

**b** $1\frac{+2}{\times 4} + 1\frac{+3}{\times 4}$

$= \frac{\square}{\square} + \frac{\square}{\square}$

$= \frac{\square}{\square}$

$= \square\frac{\square}{\square}$

**c** $2\frac{3}{5} + 1\frac{3}{5}$

=

=

=

**d** $2\frac{2}{4} + 2\frac{3}{4}$

=

=

=

**e** $3\frac{4}{5} + 1\frac{3}{5}$

=

=

=

**f** $2\frac{3}{5} + 3\frac{3}{5}$

=

=

=

**g** $3\frac{1}{3} - 1\frac{2}{3}$

=

=

=

**h** $3\frac{1}{5} - 1\frac{3}{5}$

=

=

=

**i** $4\frac{1}{3} - 2\frac{2}{3}$

=

=

=

**j** $3\frac{3}{7} - 2\frac{2}{7}$

=

=

=

**2** Simplify the following, leaving your answer as a mixed numeral:

**a** $1\frac{3}{4} + \frac{3}{4}$

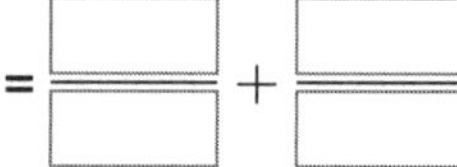

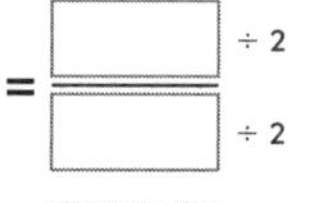


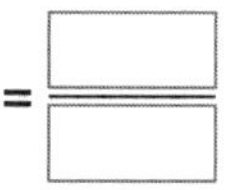

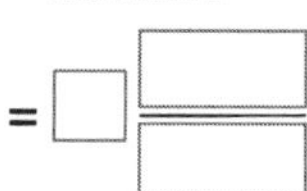

**b** $1\frac{5}{6} + \frac{5}{6}$

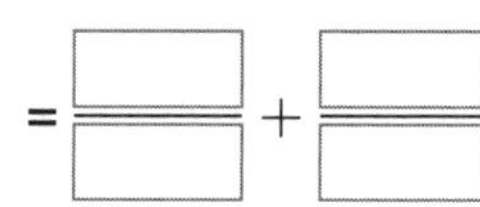

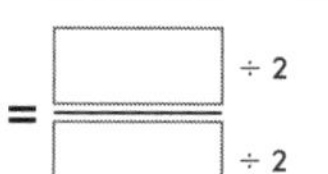


$= \frac{\square}{\square}$

$= \square\frac{\square}{\square}$

**c** $1\frac{5}{6} + \frac{3}{6}$

=

=

=

=

**d** $1\frac{5}{6} + \frac{4}{6}$

=

=

=

=

**e** $3\frac{1}{4} - \frac{3}{4}$

=

=

=

=

**f** $3\frac{1}{6} - \frac{5}{6}$

=

=

=

=

**FRACTIONS**

# Adding and subtracting mixed numerals

**1** Evaluate the following:

**a** $1\frac{2}{3} - \frac{5}{6}$ (+2, ×3)

$= \frac{\square}{\square}\,(\times 2 / \times 2) - \frac{5}{6}$

$= \frac{\square}{\square} - \frac{5}{6}$

$= \frac{\square}{\square}$

**b** $1\frac{1}{4} - \frac{3}{8}$ (+1, ×4)

$= \frac{\square}{\square}\,(\times 2 / \times 2) - \frac{3}{8}$

$= \frac{\square}{\square} - \frac{3}{8}$

$= \frac{\square}{\square}$

**c** $1\frac{1}{5} - \frac{9}{10}$

$= \frac{\square}{\square} - \frac{9}{10}$

$= \frac{\square}{\square} - \frac{9}{10}$

$= \frac{\square}{\square}$

**d** $1\frac{2}{5} - \frac{7}{10}$

$= \frac{\square}{\square} - \frac{7}{10}$

$= \frac{\square}{\square} - \frac{7}{10}$

$= \frac{\square}{\square}$

**e** $1\frac{2}{5} - \frac{5}{10}$

$= \frac{\square}{\square} - \frac{5}{10}$

$= \frac{\square}{\square} - \frac{5}{10}$

$= \frac{\square}{\square}$

**f** $1\frac{1}{4} - \frac{5}{8}$

$= \frac{\square}{\square} - \frac{5}{8}$

$= \frac{\square}{\square} - \frac{5}{8}$

$= \frac{\square}{\square}$

**g** $1\frac{1}{4} - \frac{7}{8}$

$= \frac{\square}{\square} - \frac{\square}{\square}$

$= \frac{\square}{\square} - \frac{\square}{\square}$

$= \frac{\square}{\square}$

**h** $1\frac{1}{5} - \frac{3}{10}$

$= \frac{\square}{\square} - \frac{\square}{\square}$

$= \frac{\square}{\square} - \frac{\square}{\square}$

$= \frac{\square}{\square}$

**i** $1\frac{3}{5} - \frac{7}{10}$

= ______

= ______

= ______

**j** $1\frac{3}{4} - \frac{7}{8}$

= ______

= ______

= ______

**k** $1\frac{4}{5} - \frac{9}{10}$

= ______

= ______

= ______

**l** $1\frac{3}{5} - \frac{9}{10}$

= ______

= ______

= ______

**m** $1\frac{1}{2} - \frac{3}{4}$

= ______

= ______

= ______

**n** $1\frac{1}{5} - \frac{3}{10}$

= ______

= ______

= ______

**o** $1\frac{1}{4} - \frac{3}{8}$

= ______

= ______

= ______

**p** $1\frac{2}{6} - \frac{5}{12}$

= ______

= ______

= ______

**q** $1\frac{3}{7} - \frac{7}{14}$

= ______

= ______

= ______

**r** $1\frac{1}{8} - \frac{5}{16}$

= ______

= ______

= ______

UNIT 24

**FRACTIONS**

# Adding and subtracting mixed numerals

**1** Evaluate the following, leaving your answer in mixed numeral form:

**a** $1\frac{1}{2}+\frac{3}{4}$

$= \frac{\square}{\square}\,{}^{\times 2}_{\times 2} + \frac{3}{4}$

$= \frac{\square}{\square} + \frac{3}{4}$

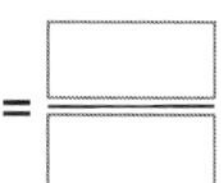

$= \frac{\square}{\square}$

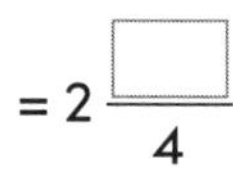

$= 2\frac{\square}{4}$

**b** $1\frac{1}{3}+\frac{5}{6}$

$= \frac{\square}{\square}\,{}^{\times 2}_{\times 2} + \frac{5}{6}$

$= \frac{\square}{\square} + \frac{5}{6}$

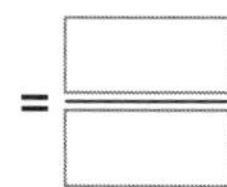

$= \frac{\square}{\square}$

$= 2\frac{\square}{6}$

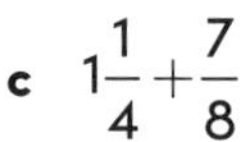

**c** $1\frac{1}{4}+\frac{7}{8}$

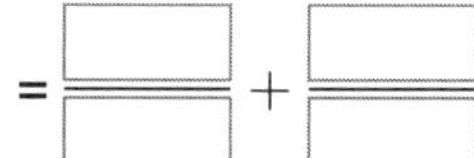

$= \frac{\square}{\square} + \frac{\square}{\square}$

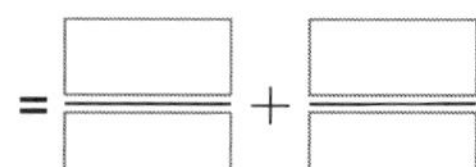

$= \frac{\square}{\square} + \frac{\square}{\square}$

$= \frac{\square}{\square}$

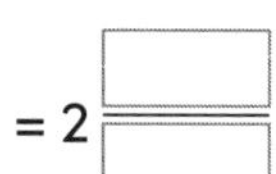

$= 2\frac{\square}{\square}$

**d** $1\frac{3}{4}+\frac{3}{8}$

$= \frac{\square}{\square} + \frac{\square}{\square}$

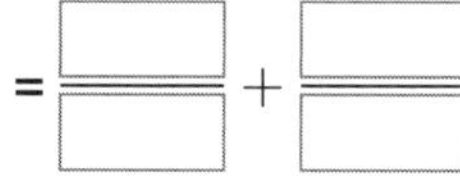

$= \frac{\square}{\square} + \frac{\square}{\square}$

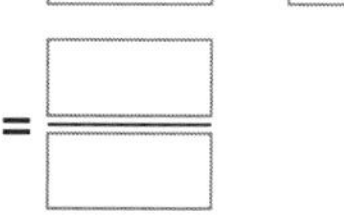

$= \frac{\square}{\square}$

$= 2\frac{\square}{\square}$

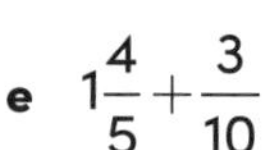

**e** $1\frac{4}{5}+\frac{3}{10}$

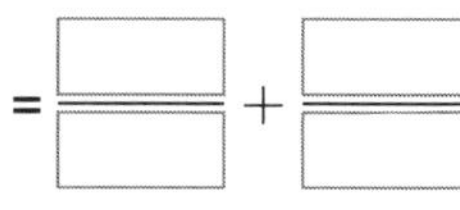

$= \frac{\square}{\square} + \frac{\square}{\square}$

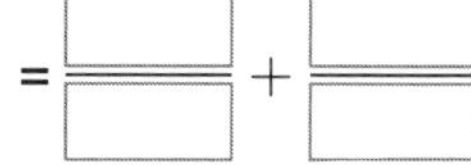

$= \frac{\square}{\square} + \frac{\square}{\square}$

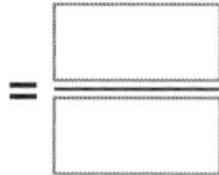

$= \frac{\square}{\square}$

$= \square\frac{\square}{\square}$

**f** $1\frac{3}{4}+\frac{5}{8}$

$= \frac{\square}{\square} + \frac{\square}{\square}$

$= \frac{\square}{\square} + \frac{\square}{\square}$

$= \frac{\square}{\square}$

$= \square\frac{\square}{\square}$

**g** $1\frac{3}{4}+\frac{7}{8}$

= ______

= ______

= ______

= ______

**h** $1\frac{2}{5}+\frac{7}{10}$

= ______

= ______

= ______

= ______

**i** $1\frac{4}{5}+\frac{9}{10}$

= ______

= ______

= ______

= ______

**j** $1\frac{2}{3}+\frac{7}{6}$

= ______

= ______

= ______

= ______

**k** $1\frac{3}{5}+\frac{7}{10}$

= ______

= ______

= ______

= ______

**l** $1\frac{3}{5}+\frac{11}{10}$

= ______

= ______

= ______

= ______

**m** $1\frac{1}{6}+\frac{11}{12}$

= ______

= ______

= ______

= ______

**n** $1\frac{5}{6}+\frac{7}{12}$

= ______

= ______

= ______

= ______

UNIT 25

## FRACTIONS

# Adding and subtracting mixed numerals

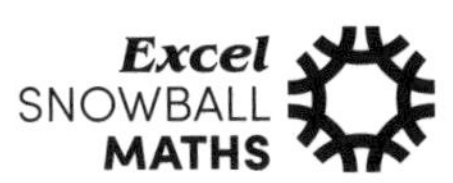



**1** Simplify the following, leaving your answer as a mixed numeral:

**a** $1\frac{1}{3}+\frac{1}{6}$

$=\frac{\square}{\square}\begin{smallmatrix}\times 2\\ \times 2\end{smallmatrix}+\frac{1}{6}$

$=\frac{\square}{\square}+\frac{\square}{\square}$

$=\frac{\square}{\square}\begin{smallmatrix}\div 3\\ \div 3\end{smallmatrix}$

$=\frac{\square}{\square}$

$=1\frac{\square}{\square}$

**b** $1\frac{2}{3}+\frac{5}{6}$

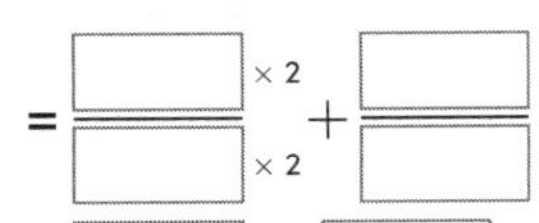

$=\frac{\square}{\square}\begin{smallmatrix}\times 2\\ \times 2\end{smallmatrix}+\frac{\square}{\square}$

$=\frac{\square}{\square}+\frac{\square}{\square}$

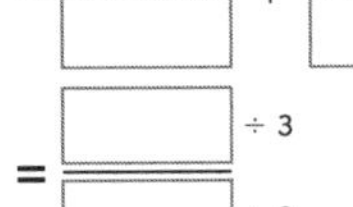

$=\frac{\square}{\square}\begin{smallmatrix}\div 3\\ \div 3\end{smallmatrix}$

$=\frac{\square}{\square}$

$=\square\frac{\square}{\square}$

**c** $2\frac{1}{3}-\frac{5}{6}$

$=\frac{\square}{\square}-\frac{\square}{\square}$

$=\frac{\square}{\square}-\frac{\square}{\square}$

$=\frac{\square}{\square}$

$=\frac{\square}{\square}$

$=\square\frac{\square}{\square}$

**d** $2\frac{2}{3}-\frac{1}{6}$

$=\frac{\square}{\square}-\frac{\square}{\square}$

$=\frac{\square}{\square}-\frac{\square}{\square}$

$=\frac{\square}{\square}$

$=\frac{\square}{\square}$

$=\square\frac{\square}{\square}$

**e** $3\frac{1}{3}-\frac{5}{6}$

=

=

=

=

=

**f** $3\frac{2}{3}+\frac{5}{6}$

=

=

=

=

=

**2** Simplify the following, leaving your answer as a mixed numeral:

**a** $2\frac{1}{5}-\frac{7}{10}$

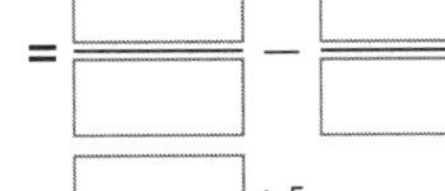

$=\frac{\square}{\square}\begin{smallmatrix}\times 2\\ \times 2\end{smallmatrix}-\frac{\square}{\square}$

$=\frac{\square}{\square}-\frac{\square}{\square}$

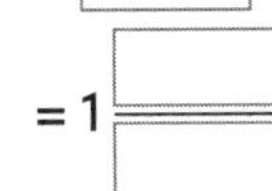

$=\frac{\square}{\square}\begin{smallmatrix}\div 5\\ \div 5\end{smallmatrix}$

$=\frac{\square}{\square}$

$=1\frac{\square}{\square}$

**b** $2\frac{4}{5}-\frac{13}{10}$

$=\frac{\square}{\square}\begin{smallmatrix}\times 2\\ \times 2\end{smallmatrix}-\frac{\square}{\square}$

$=\frac{\square}{\square}-\frac{\square}{\square}$

$=\frac{\square}{\square}\begin{smallmatrix}\div 5\\ \div 5\end{smallmatrix}$

$=\frac{\square}{\square}$

$=\square\frac{\square}{\square}$

**c** $1\frac{3}{5}+\frac{9}{10}$

=

=

=

=

=

**d** $1\frac{4}{5}+\frac{7}{10}$

=

=

=

=

=

**e** $2\frac{4}{5}+\frac{7}{10}$

=

=

=

=

=

**f** $3\frac{1}{5}-\frac{7}{10}$

=

=

=

=

=

# FRACTIONS
## Subtracting mixed numerals

**1** Evaluate the following:

**a** $1\frac{1}{4} - \frac{1}{3}$

$= \frac{5}{4} - \frac{1}{3}$

$= \frac{(3 \times 5) - (4 \times 1)}{4 \times 3}$

$= \frac{\square}{\square}$

**b** $1\frac{1}{5} - \frac{1}{2}$

$= \frac{6}{5} - \frac{1}{2}$

$= \frac{(2 \times 6) - (5 \times 1)}{5 \times 2}$

$= \frac{\square}{\square}$

**c** $1\frac{1}{4} - \frac{2}{3}$

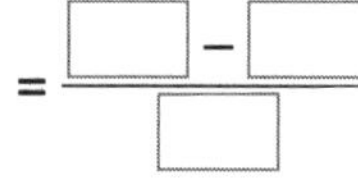

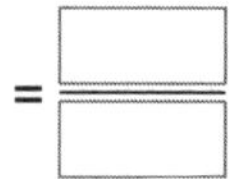

**d** $1\frac{1}{2} - \frac{2}{3}$

**e** $1\frac{1}{5} - \frac{1}{3}$

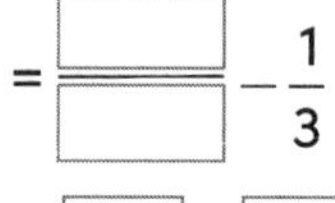

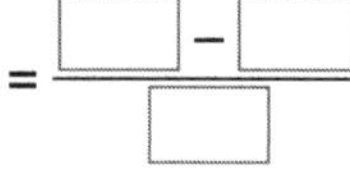

**f** $1\frac{1}{5} - \frac{1}{4}$

**g** $1\frac{1}{5} - \frac{2}{3}$

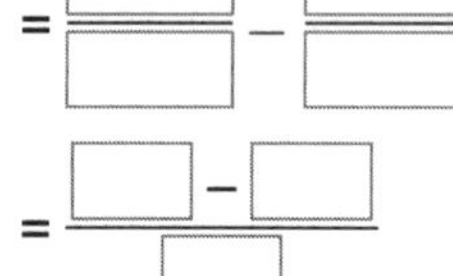

**h** $1\frac{2}{5} - \frac{2}{3}$

**i** $1\frac{3}{5} - \frac{2}{3}$

=

=

=

**j** $1\frac{1}{4} - \frac{2}{5}$

=

=

=

**k** $1\frac{1}{4} - \frac{3}{5}$

=

=

=

**l** $1\frac{1}{2} - \frac{4}{5}$

=

=

=

**2** Evaluate the following and leave your answer as a mixed numeral:

**a** $1\frac{3}{5} - \frac{1}{2}$

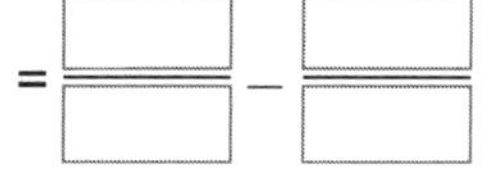

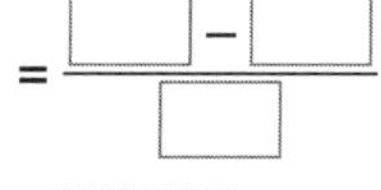

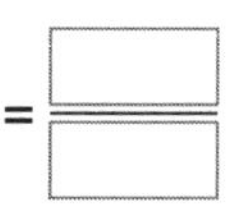

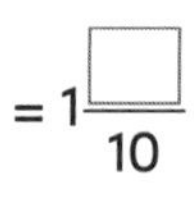

$= 1\frac{\square}{10}$

**b** $1\frac{4}{5} - \frac{1}{2}$

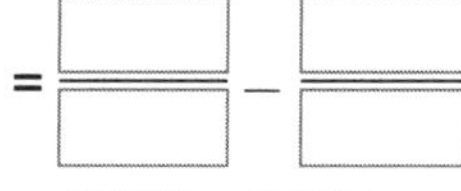

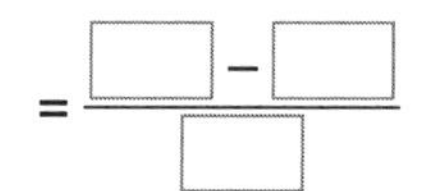

$= 1\frac{\square}{\square}$

**c** $1\frac{1}{4} - \frac{1}{5}$

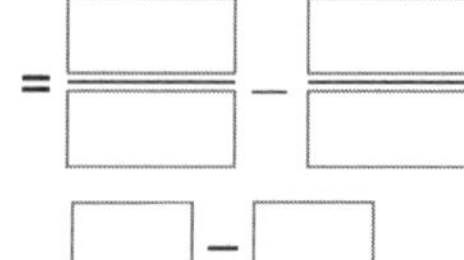

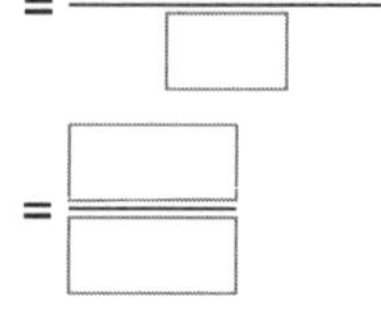

**d** $1\frac{2}{5} - \frac{1}{4}$

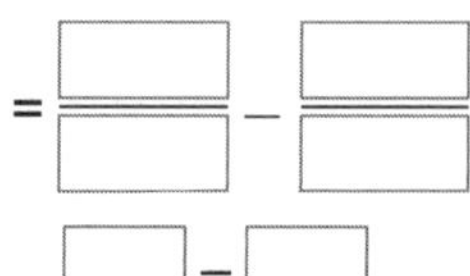

UNIT 27

# FRACTIONS
## Subtracting mixed numerals

**1** Evaluate the following and leave your answer as a mixed numeral:

**a** $1\frac{2}{5} - \frac{1}{3}$

=
=
=
=

**b** $1\frac{3}{4} - \frac{1}{3}$

=
=
=
=

**c** $1\frac{3}{4} - \frac{2}{3}$

=
=
=
=

**d** $1\frac{4}{5} - \frac{2}{3}$

=
=
=
=

**e** $1\frac{2}{7} - \frac{1}{5}$

=
=
=
=

**f** $1\frac{3}{7} - \frac{1}{5}$

=
=
=
=

**2** Evaluate the following:

**a** $1\frac{+3}{\times 5} - 1\frac{+1}{\times 2}$

=
=
=

**b** $1\frac{+4}{\times 5} - 1\frac{+1}{\times 2}$

=
=
=

**c** $1\frac{1}{3} - 1\frac{1}{4}$

=
=
=

**d** $1\frac{2}{3} - 1\frac{1}{4}$

=
=
=

**e** $1\frac{2}{5} - 1\frac{1}{4}$

=
=
=

**f** $1\frac{3}{5} - 1\frac{1}{4}$

=
=
=

**g** $1\frac{4}{5} - 1\frac{1}{4}$

=
=
=

**h** $1\frac{4}{5} - 1\frac{3}{4}$

=
=
=

**i** $1\frac{1}{2} - 1\frac{1}{5}$

=
=
=

**j** $1\frac{1}{3} - 1\frac{1}{5}$

=
=
=

**k** $1\frac{1}{4} - 1\frac{1}{5}$

=
=
=

**l** $1\frac{2}{3} - 1\frac{1}{5}$

=
=
=

# Multiplying fractions

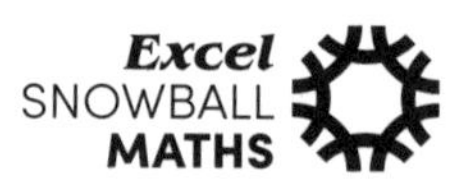


**1** Evaluate the following:

**a** $\frac{1}{5} \times \frac{2}{3} = \frac{\square}{\square}$ (top: $1 \times 2$; bottom: $5 \times 3$)

**b** $\frac{2}{3} \times \frac{1}{7} = \frac{\square}{\square}$ (top: $2 \times 1$; bottom: $3 \times 7$)

**c** $\frac{3}{4} \times \frac{5}{7} =$ ______

**d** $\frac{4}{7} \times \frac{3}{5} =$ ______

**e** $\frac{2}{3} \times \frac{2}{5} =$ ______

**f** $\frac{2}{3} \times \frac{4}{5} =$ ______

**g** $\frac{4}{5} \times \frac{1}{7} =$ ______

**h** $\frac{5}{8} \times \frac{1}{2} =$ ______

**i** $\frac{5}{4} \times \frac{3}{7} =$ ______

**j** $\frac{2}{5} \times \frac{4}{3} =$ ______

**2** Simplify the following:

**a** $\frac{1}{5} \times \frac{5}{3} = \frac{\square}{\square}$

**b** $\frac{2}{4} \times \frac{4}{7} = \frac{\square}{\square}$

**c** $\frac{1}{5} \times \frac{5}{3} =$ ______

**d** $\frac{4}{7} \times \frac{7}{5} =$ ______

**e** $\frac{6}{8} \times \frac{5}{6} =$ ______

**f** $\frac{8}{7} \times \frac{3}{8} =$ ______

**g** $\frac{6}{12} \times \frac{3}{6} = \frac{\square}{\square} \begin{smallmatrix}\div 3\\ \div 3\end{smallmatrix} = \frac{\square}{\square}$

**h** $\frac{7}{9} \times \frac{3}{7} = \frac{\square}{\square} \begin{smallmatrix}\div 3\\ \div 3\end{smallmatrix} = \frac{\square}{\square}$

**i** $\frac{7}{6} \times \frac{3}{7} =$ ______

**j** $\frac{6}{4} \times \frac{4}{9} =$ ______

**3** Evaluate the following, leaving your answer as a mixed numeral:

**a** $5 \times \frac{1}{4}$

$= \frac{5}{1} \times \frac{1}{4}$

$= \frac{\square}{\square}$

$= 1\frac{\square}{4}$

**b** $3 \times \frac{3}{5}$

$= \frac{3}{1} \times \frac{3}{5}$

$= \frac{\square}{\square}$

$= 1\frac{\square}{\square}$

**c** $4 \times \frac{2}{7}$

$= \frac{\square}{1} \times \frac{\square}{\square}$

$= \frac{\square}{\square}$

$= \square\frac{\square}{\square}$

**d** $5 \times \frac{1}{3}$

$= \frac{\square}{1} \times \frac{\square}{\square}$

$= \frac{\square}{\square}$

$= \square\frac{\square}{\square}$

**e** $4 \times \frac{2}{3}$

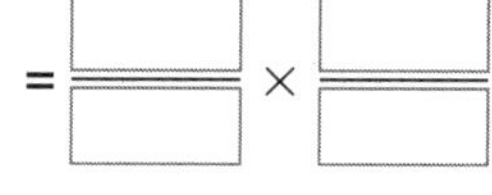

$= \frac{\square}{\square} \times \frac{\square}{\square}$

$= \frac{\square}{\square}$

$= 2\frac{\square}{\square}$

**f** $4 \times \frac{3}{5}$

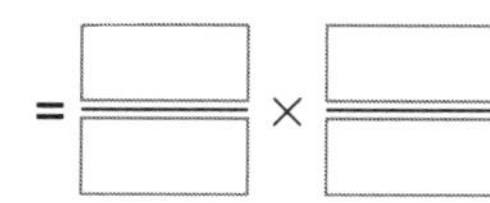

$= \frac{\square}{\square} \times \frac{\square}{\square}$

$= \frac{\square}{\square}$

$= \square\frac{\square}{\square}$

**g** $6 \times \frac{2}{5}$

= ______

= ______

= ______

**h** $7 \times \frac{2}{5}$

= ______

= ______

= ______

**i** $8 \times \frac{2}{7}$

= ______

= ______

= ______

**j** $9 \times \frac{2}{7}$

= ______

= ______

= ______

# Multiplying mixed numerals

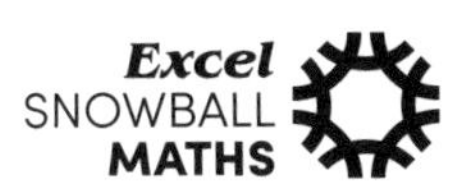

**1** Evaluate the following:

**a** $1\frac{2}{3} \times \frac{1}{2}$

$= \frac{\square}{\square} \times \frac{1}{2}$

$= \frac{\square}{\square}$

**b** $2\frac{1}{3} \times \frac{2}{5}$

$= \frac{\square}{\square} \times \frac{2}{5}$

$= \frac{\square}{\square}$

**c** $3\frac{2}{3} \times \frac{1}{5}$

$= \frac{\square}{\square} \times \frac{1}{5}$

$= \frac{\square}{\square}$

**d** $4\frac{2}{3} \times \frac{1}{5}$

$= \frac{\square}{\square} \times \frac{1}{5}$

$= \frac{\square}{\square}$

**e** $2\frac{1}{3} \times \frac{1}{5}$

$= \frac{\square}{\square} \times \frac{\square}{\square}$

$= \frac{\square}{\square}$

**f** $2\frac{1}{5} \times \frac{1}{3}$

$= \frac{\square}{\square} \times \frac{\square}{\square}$

$= \frac{\square}{\square}$

**g** $3\frac{1}{7} \times \frac{1}{5}$

= ______

= ______

**h** $3\frac{1}{2} \times \frac{1}{6}$

= ______

= ______

**c** $3\frac{1}{3} \times \frac{1}{3}$

= ______

= ______

= ______

**d** $4\frac{1}{2} \times \frac{1}{2}$

= ______

= ______

= ______

**e** $4\frac{1}{2} \times \frac{3}{5}$

= ______

= ______

= ______

**f** $5\frac{1}{2} \times \frac{3}{5}$

= ______

= ______

= ______

**2** Evaluate the following, leaving your answer as a mixed numeral:

**a** $1\frac{1}{3} \times \frac{4}{5}$

$= \frac{\square}{\square} \times \frac{\square}{\square}$

$= \frac{\square}{\square}$

$= 1\frac{\square}{\square}$

**b** $2\frac{1}{3} \times \frac{2}{3}$

$= \frac{\square}{\square} \times \frac{\square}{\square}$

$= \frac{\square}{\square}$

$= \square\frac{\square}{\square}$

**3** Simplify the following:

**a** $2\frac{1}{5} \times 1\frac{1}{2}$

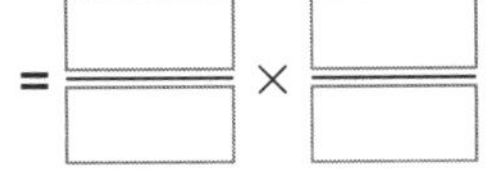

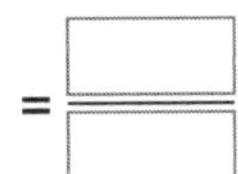

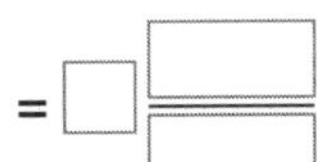

**b** $2\frac{1}{2} \times 1\frac{1}{4}$

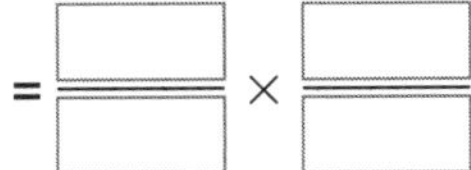

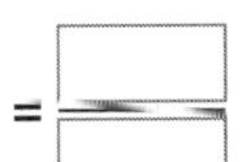

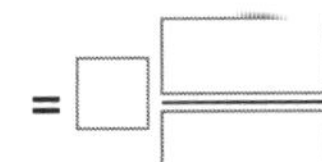

**c** $3\frac{1}{2} \times 1\frac{2}{3}$

= ______

= ______

= ______

**d** $4\frac{1}{2} \times 1\frac{1}{4}$

= ______

= ______

= ______

**FRACTIONS**

# Multiplying mixed numerals

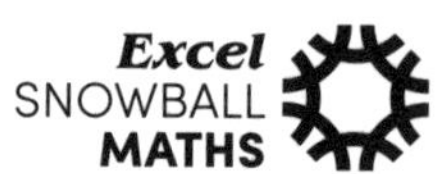


**1** Simplify the following:

**a** $1\frac{2}{5} \times 1\frac{2}{3}$

$= \frac{7}{\cancel{5}} \times \frac{\cancel{5}}{3}$

$= \frac{\square}{\square}$

$= \square\frac{\square}{\square}$

**b** $2\frac{1}{5} \times 1\frac{1}{4}$

$= \frac{11}{\cancel{5}} \times \frac{\cancel{5}}{4}$

$= \frac{\square}{\square}$

$= \square\frac{\square}{\square}$

**c** $1\frac{5}{7} \times 1\frac{2}{5}$

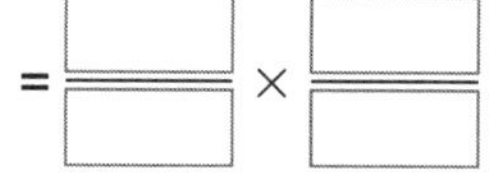

$= \frac{\square}{\square} \times \frac{\square}{\square}$

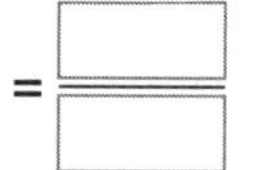

$= \frac{\square}{\square}$

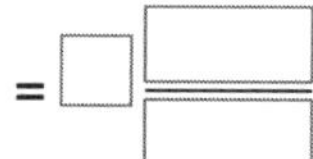

$= \square\frac{\square}{\square}$

**d** $1\frac{7}{8} \times 1\frac{1}{7}$

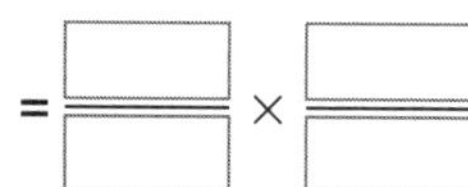

$= \frac{\square}{\square} \times \frac{\square}{\square}$

$= \frac{\square}{\square}$

$= \square\frac{\square}{\square}$

**e** $2\frac{3}{4} \times 1\frac{1}{3}$

= ______

= ______

= ______

**f** $2\frac{5}{6} \times 1\frac{1}{5}$

= ______

= ______

= ______

**g** $3\frac{1}{4} \times 1\frac{1}{3}$

= ______

= ______

= ______

**h** $3\frac{2}{5} \times 1\frac{1}{4}$

= ______

= ______

= ______

**i** $2\frac{2}{3} \times 1\frac{1}{2}$

= ______

= ______

= ______

**j** $2\frac{1}{2} \times 2\frac{2}{5}$

= ______

= ______

= ______

**k** $3\frac{1}{3} \times 1\frac{1}{2}$

= ______

= ______

= ______

**l** $4\frac{1}{2} \times 1\frac{1}{3}$

= ______

= ______

= ______

**m** $2\frac{1}{7} \times 2\frac{1}{3}$

= ______

= ______

= ______

**n** $2\frac{5}{8} \times 2\frac{2}{3}$

= ______

= ______

= ______

**o** $3\frac{3}{7} \times 2\frac{1}{3}$

= ______

= ______

= ______

**p** $7\frac{1}{5} \times 1\frac{1}{4}$

= ______

= ______

= ______

**q** $4\frac{2}{5} \times 2\frac{1}{2}$

= ______

= ______

= ______

**r** $5\frac{1}{7} \times 2\frac{1}{3}$

= ______

= ______

= ______

UNIT 31

**FRACTIONS**

# Dividing fractions

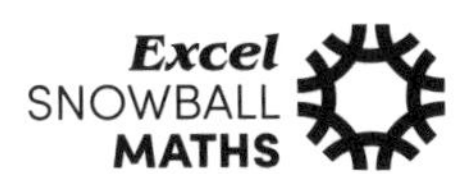

**1** Evaluate the following:

**a** $\frac{1}{5} \div \frac{3}{2}$ switch

$= \frac{1}{5} \times \frac{2}{3}$

$= \frac{\square}{\square}$

**b** $\frac{3}{4} \div \frac{7}{5}$ switch

$= \frac{3}{4} \times \frac{5}{7}$

$= \frac{\square}{\square}$

**c** $\frac{4}{7} \div \frac{5}{3}$

$= \frac{\square}{\square} \times \frac{\square}{\square}$

$= \frac{\square}{\square}$

**d** $\frac{2}{3} \div \frac{5}{2}$

$= \frac{\square}{\square} \times \frac{\square}{\square}$

$= \frac{\square}{\square}$

**e** $\frac{1}{5} \div \frac{8}{3}$

= ________

= ________

**f** $\frac{2}{5} \div \frac{3}{7}$

= ________

= ________

**2** Simplify the following:

**a** $\frac{1}{5} \div \frac{3}{5}$

$= \frac{1}{5} \times \frac{5}{3}$

$= \frac{\square}{\square}$

**b** $\frac{2}{4} \div \frac{7}{4}$

$= \frac{2}{4} \times \frac{4}{7}$

$= \frac{\square}{\square}$

**c** $\frac{2}{6} \div \frac{11}{6}$

$= \frac{\square}{\square} \times \frac{\square}{\square}$

$= \frac{\square}{\square}$

**d** $\frac{4}{7} \div \frac{5}{7}$

$= \frac{\square}{\square} \times \frac{\square}{\square}$

$= \frac{\square}{\square}$

**e** $\frac{5}{6} \div \frac{8}{6}$

= ________

= ________

**f** $\frac{3}{6} \div \frac{10}{6}$

= ________

= ________

**3** Simplify the following:

**a** $4 \div \frac{2}{3}$

$= \frac{4}{1} \times \frac{\square}{\square}$

$= \frac{\square}{\square}$

= ________

**b** $4 \div \frac{2}{5}$

$= \frac{4}{1} \times \frac{\square}{\square}$

$= \frac{\square}{\square}$

= ________

**c** $6 \div \frac{3}{2}$

$= \frac{\square}{\square} \times \frac{\square}{\square}$

$= \frac{\square}{\square}$

= ________

**d** $6 \div \frac{3}{5}$

$= \frac{\square}{\square} \times \frac{\square}{\square}$

$= \frac{\square}{\square}$

= ________

**e** $8 \div \frac{4}{5}$

= ________

= ________

**f** $8 \div \frac{2}{5}$

= ________

= ________

**g** $9 \div \frac{3}{2}$

= ________

= ________

**h** $9 \div \frac{3}{4}$

= ________

= ________

**i** $10 \div \frac{5}{3}$

= ________

= ________

**j** $10 \div \frac{5}{6}$

= ________

= ________

**FRACTIONS**

# Dividing mixed numerals and improper fractions

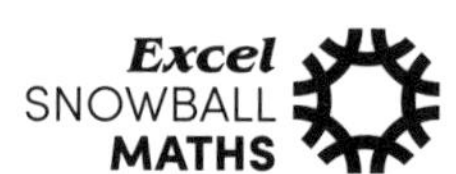


**1** Evaluate the following, leaving your answer as a mixed numeral:

**a** $2 \div \frac{5}{3}$

$= \frac{\square}{1} \times \frac{\square}{\square}$

$= \frac{\square}{\square}$

$= 1\frac{\square}{\square}$

**b** $3 \div \frac{5}{2}$

$= \frac{\square}{\square} \times \frac{\square}{\square}$

$= \frac{\square}{\square}$

$= 1\frac{\square}{\square}$

**c** $3 \div \frac{5}{4}$

$= \frac{\square}{\square} \times \frac{\square}{\square}$

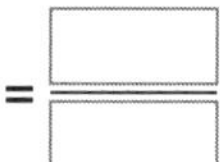

$= \frac{\square}{\square}$

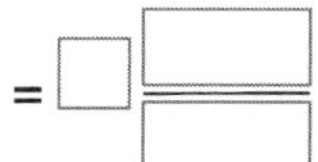

$= \square\frac{\square}{\square}$

**d** $4 \div \frac{3}{2}$

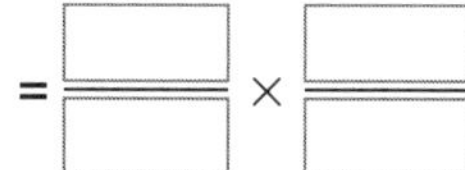

$= \frac{\square}{\square} \times \frac{\square}{\square}$

$= \frac{\square}{\square}$

$= \square\frac{\square}{\square}$

**e** $4 \div \frac{5}{3}$

= ____________

= ____________

= ____________

**f** $5 \div \frac{4}{3}$

= ____________

= ____________

= ____________

**2** Evaluate the following:

**a** $1\frac{1}{2} \div \frac{5}{3}$

$= \frac{\square}{\square} \times \frac{\square}{\square}$

$= \frac{\square}{\square}$

**b** $1\frac{1}{2} \div \frac{7}{3}$

$= \frac{\square}{\square} \times \frac{\square}{\square}$

$= \frac{\square}{\square}$

**c** $2\frac{1}{3} \div \frac{5}{2}$

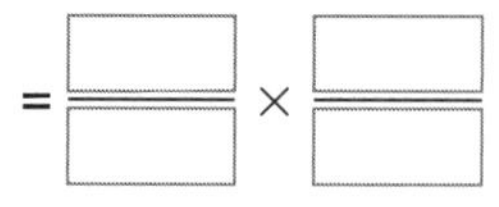

$= \frac{\square}{\square} \times \frac{\square}{\square}$

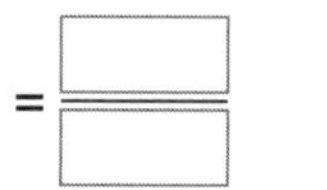

$= \frac{\square}{\square}$

**d** $2\frac{1}{5} \div \frac{5}{2}$

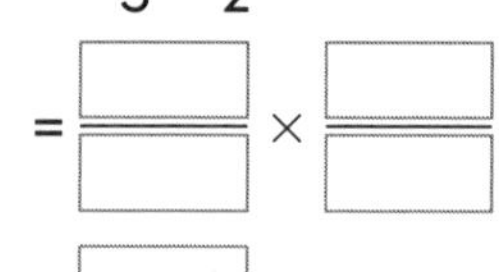

$= \frac{\square}{\square} \times \frac{\square}{\square}$

$= \frac{\square}{\square}$

**e** $3\frac{1}{2} \div \frac{11}{3}$

= ____________

= ____________

**f** $3\frac{1}{3} \div \frac{7}{2}$

= ____________

= ____________

**g** $2\frac{2}{3} \div \frac{7}{2}$

= ____________

= ____________

**h** $3\frac{2}{3} \div \frac{9}{2}$

= ____________

= ____________

**3** Evaluate the following, leaving your answer as a mixed numeral:

**a** $2\frac{1}{3} \div \frac{3}{2}$

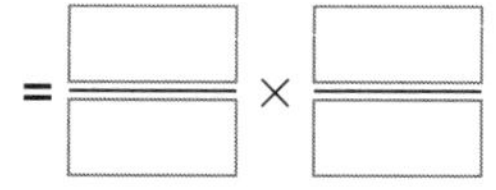

$= \frac{\square}{\square} \times \frac{\square}{\square}$

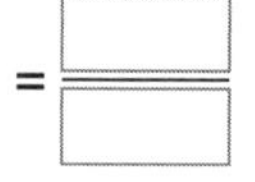

$= \frac{\square}{\square}$

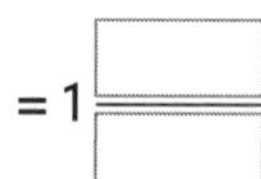

$= 1\frac{\square}{\square}$

**b** $2\frac{2}{3} \div \frac{5}{2}$

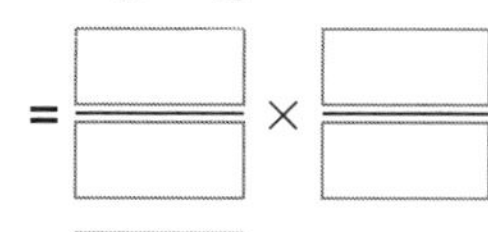

$= \frac{\square}{\square} \times \frac{\square}{\square}$

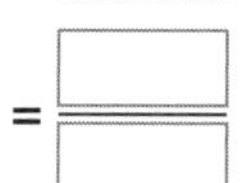

$= \frac{\square}{\square}$

$= \square\frac{\square}{\square}$

**c** $3\frac{2}{3} \div \frac{7}{2}$

= ____________

= ____________

= ____________

**d** $3\frac{1}{3} \div \frac{3}{2}$

= ____________

= ____________

= ____________

## FRACTIONS
# Dividing mixed numerals

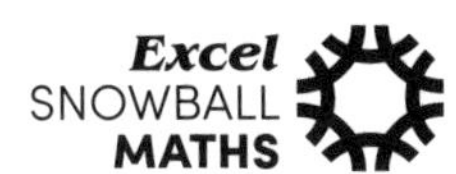

1 Evaluate the following, leaving your answer as a mixed numeral:

a $1\frac{2}{3} \div \frac{1}{2}$

=

=

=

b $2\frac{1}{3} \div \frac{1}{2}$

=

=

=

2 Simplify the following:

a $1\frac{{}^{+}1}{{}_{\times}2} \div 1\frac{{}^{+}1}{{}_{\times}3}$

$= \frac{\square}{\square} \div \frac{\square}{\square}$

$= \frac{\square}{\square} \times \frac{\square}{\square}$

$= \frac{\square}{\square}$

$= \square\frac{\square}{\square}$

b $3\frac{{}^{+}2}{{}_{\times}3} \div 3\frac{{}^{+}1}{{}_{\times}2}$

$= \frac{\square}{\square} \div \frac{\square}{\square}$

$= \frac{\square}{\square} \times \frac{\square}{\square}$

$= \frac{\square}{\square}$

$= \square\frac{\square}{\square}$

c $3\frac{1}{2} \div 1\frac{2}{3}$

$= \frac{\square}{\square} \div \frac{\square}{\square}$

$= \frac{\square}{\square} \times \frac{\square}{\square}$

$= \frac{\square}{\square}$

$= \square\frac{\square}{\square}$

d $4\frac{1}{2} \div 1\frac{2}{3}$

$= \frac{\square}{\square} \div \frac{\square}{\square}$

$= \frac{\square}{\square} \times \frac{\square}{\square}$

$= \frac{\square}{\square}$

$= \square\frac{\square}{\square}$

e $1\frac{1}{2} \div 1\frac{1}{4}$

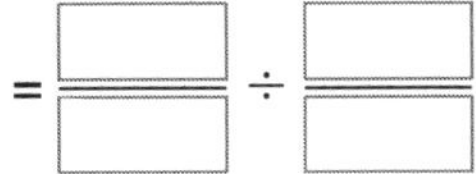

$= \frac{\square}{\square} \div \frac{\square}{\square}$

$= \frac{\square}{\square} \times \frac{\square}{\square}$

$= \frac{\square \div 2}{\square \div 2}$

$= \frac{\square}{\square}$

$= \square\frac{\square}{\square}$

f $2\frac{1}{2} \div 1\frac{3}{4}$

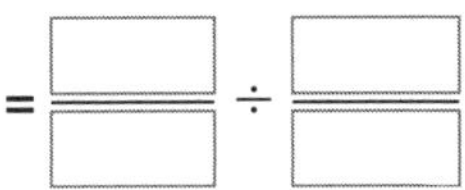

$= \frac{\square}{\square} \div \frac{\square}{\square}$

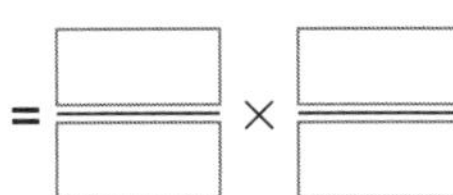

$= \frac{\square}{\square} \times \frac{\square}{\square}$

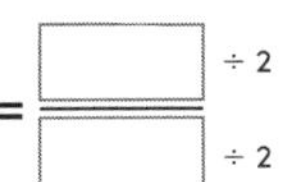

$= \frac{\square \div 2}{\square \div 2}$

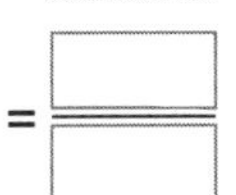

$= \frac{\square}{\square}$

$= \square\frac{\square}{\square}$

g $1\frac{1}{2} \div 1\frac{1}{6}$

h $3\frac{1}{3} \div 2\frac{1}{3}$

i $1\frac{3}{5} \div 1\frac{1}{3}$

j $2\frac{1}{4} \div 1\frac{1}{2}$

UNIT 34

**FRACTIONS**

# Fraction of a quantity

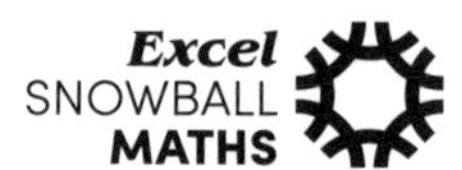


**1** Find each amount without using a calculator:

**a** $\frac{1}{2}$ of 12
= 12 ÷ ______
= ______

**b** $\frac{1}{5}$ of 10
= 10 ÷ ______
= ______

**c** $\frac{1}{4}$ of 24
= ______ ÷ ______
= ______

**d** $\frac{1}{8}$ of 16
= ______ ÷ ______
= ______

**e** $\frac{1}{3}$ of 15
= ______ ÷ ______
= ______

**f** $\frac{1}{4}$ of 20
= ______ ÷ ______
= ______

**g** $\frac{1}{6}$ of 36
= ______
= ______

**h** $\frac{1}{8}$ of 48
= ______
= ______

**i** $\frac{1}{10}$ of 30
= ______
= ______

**j** $\frac{1}{2}$ of 20
= ______
= ______

**k** $\frac{1}{5}$ of 45
= ______
= ______

**l** $\frac{1}{9}$ of 27
= ______
= ______

**2** Find each amount without using a calculator:

**a** $\frac{2}{3}$ of 12
= (12 ÷ ____) × ____
= ______

**b** $\frac{3}{5}$ of 10
= (10 ÷ ____) × ____
= ______

**c** $\frac{3}{4}$ of 20
= (____ ÷ ____) × ____
= ______

**d** $\frac{2}{5}$ of 15
= (____ ÷ ____) × ____
= ______

**e** $\frac{4}{5}$ of 30
= ______
= ______
= ______

**f** $\frac{3}{4}$ of 24
= ______
= ______
= ______

**g** $\frac{2}{3}$ of 18
= ______
= ______
= ______

**h** $\frac{5}{6}$ of 12
= ______
= ______
= ______

**i** $\frac{3}{8}$ of 40
= ______
= ______
= ______

**j** $\frac{5}{7}$ of 21
= ______
= ______
= ______

**3** Wally has 20 marbles and gave $\frac{1}{5}$ of them to his brother. How many marbles did his brother receive?

______ ÷ ______ = ______

**4** Lisa has 16 coins and gave $\frac{1}{2}$ to her brother. How many coins did her brother receive?

______ ÷ ______ = ______

**5** Lara gave $\frac{1}{6}$ of her colouring pencils to her sister. If Lara had 24 pencils, how many did her sister receive?

______

**6** Roland had 40 apples. If $\frac{1}{5}$ of his apples went missing, how many apples went missing?

______

**7** Michael had 42 nails. If $\frac{1}{6}$ of his nails went missing, how many nails went missing?

______

UNIT 35

# FRACTIONS
## Fraction of a quantity

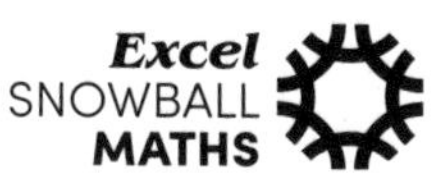

1 A classroom has 30 chairs. If $\frac{1}{10}$ of the chairs are taken, how many chairs are taken?

______

2 If $\frac{2}{3}$ of 24 slices of pizza were eaten, how many slices were eaten?

(24 ÷ ____) × ____ = ____

3 There were 36 posters at a shop and $\frac{3}{4}$ of them were sold. How many posters were sold?

(36 ÷ ____) × ____ = ____

4 Robert has $40 and decides to save $\frac{3}{5}$ of it. How much will Robert save?

______

______

______

5 Out of 30 students, $\frac{4}{5}$ passed the exam. How many students passed the exam?

______

______

______

6 Maggie has 32 loyalty points and redeems $\frac{3}{4}$ of them for a gift card. How many points did Maggie redeem?

______

______

______

7 Two-thirds of the balloons at a party are white. There are 18 balloons at the party.

a How many balloons are white?

(18 ÷ ____) × ____ = ____

b How many balloons are not white?

18 − ____ (a) = ____

8 Three-quarters of the stickers on a teacher's desk are blue. There are 20 stickers altogether.

a How many stickers are blue?

(20 ÷ ____) × ____ = ____

b How many stickers are not blue?

20 − ____ (a) = ____

9 Three-fifths of the students brought their calculators to class. There are 30 students in this class.

a How many students brought their calculators?

______

b How many students did not bring their calculators?

______

10 There are 20 coins. Two-fifths are gold.

a How many coins are gold?

______

b How many coins are not gold?

______

11 There are 30 cables. One-third are blue.

a How many cables are blue?

______

b How many cables are not blue?

______

UNIT 36

# FRACTIONS
## Fraction of a quantity

1 In a classroom, $\frac{3}{5}$ of the students are boys.

a What fraction of the students are girls?

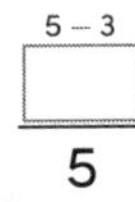

5 − 3

$\frac{\square}{5}$

b If there are 40 students, how many girls are there?

(40 ÷ ______) × ______ = ______

c How many boys are there?

40 − ______ = ______

2 In a bag full of marbles, $\frac{3}{7}$ of the marbles are green and the rest are blue.

a What fraction of the marbles are blue?

$\frac{\square}{7}$

b If there are 35 marbles, how many blue marbles are there?

(35 ÷ ______) × ______ = ______

c How many green marbles are there?

35 − ______ = ______

3 In a box full of capsules, $\frac{2}{5}$ of them are orange.

a What fraction of the capsules are not orange?

b If there are 45 capsules, how many capsules are not orange?

______

______

______

c How many orange capsules are there?

______

4 On a bookshelf, $\frac{4}{7}$ of the books are fiction.

a What fraction of the books are non-fiction?

______

b If there are 21 books, how many books are non-fiction?

______

______

______

c How many books are fiction?

______

5 At an auto repair shop, $\frac{4}{9}$ of the cars are white.

a What fraction of the cars are not white?

______

b If there are 27 cars, how many cars are not white?

______

______

______

c How many cars are white?

______

UNIT 37

FRACTIONS

# Expressing quantities as a fraction

**1** Express each quantity as a simplified fraction with or without a calculator:

**a** A test mark of 75 out of 100

$\frac{\square}{100} = \frac{\square}{\square}$

**b** A test mark of 30 out of 50

$\frac{\square}{50} = \frac{\square}{\square}$

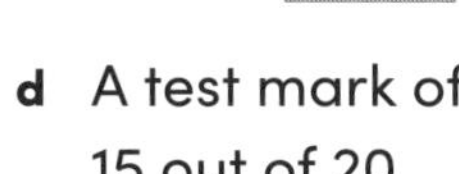

**c** A test mark of 25 out of 60 ________

**d** A test mark of 15 out of 20 ________

**e** A test mark of 35 out of 70 ________

**f** A test mark of 15 out of 40 ________

**g** \$10 out of \$50 ________

**h** \$20 out of \$80 ________

**i** \$30 out of \$95 ________

**j** \$90 out of \$500 ________

**k** 10 mL out of 120 mL ________

**l** 150 mL out of 200 mL ________

**m** 200 g out of 500 g ________

**n** 500 g out of 1000 g ________

**o** 2 cm out of 20 cm ________

**p** 10 m out of 25 m ________

**2** Express each quantity as a simplified fraction with or without a calculator:

**a** 50 cents out of \$1

$\frac{\square}{100} = \frac{\square}{\square}$

**b** 45 cents out of \$1

$\frac{\square}{100} = \frac{\square}{\square}$

**c** 20 cents out of \$1 ________

**d** 65 cents out of \$1 ________

**e** 85 cents out of \$1 ________

**f** 10 cents out of \$1 ________

**g** 80 cents out of \$2

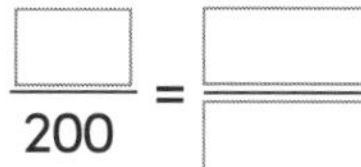

$\frac{\square}{200} = \frac{\square}{\square}$

**h** 40 cents out of \$2

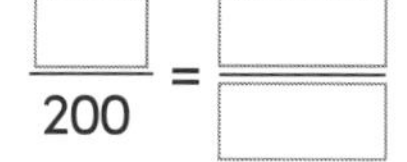

$\frac{\square}{200} = \frac{\square}{\square}$

**i** 55 cents out of \$2 ________

**j** 30 cents out of \$2 ________

**k** 70 cents out of \$2 ________

**l** 15 cents out of \$2 ________

**m** 90 cents out of \$3 ________

**n** 50 cents out of \$3 ________

**o** 85 cents out of \$5 ________

**p** 65 cents out of \$10 ________

**3** Express each quantity as a simplified fraction with or without a calculator:

**a** 15 min out of 1 h

$\frac{\square}{60} = \frac{\square}{\square}$

**b** 45 min out of 1 h

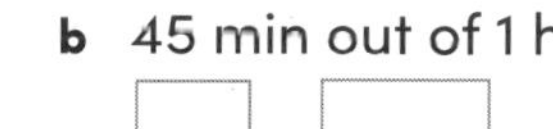

$\frac{\square}{60} = \frac{\square}{\square}$

**c** 30 min out of 1 h ________

**d** 20 min out of 1 h ________

**e** 40 min out of 1 h ________

**f** 50 min out of 1 h ________

**g** 60 min out of 2 h ________

**h** 40 min out of 2 h ________

## FRACTIONS

# Expressing quantities as a fraction

**1** Express each quantity as a simplified fraction with or without a calculator:

**a** 6 mm out of 4 cm

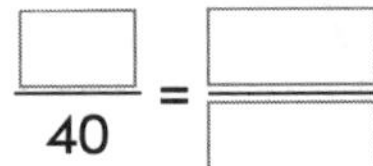

**b** 4 mm out of 3 cm

$\frac{\square}{30} = \frac{\square}{\square}$

**c** 5 mm out of 2 cm

**d** 6 mm out of 7 cm

**e** 10 mm out of 6 cm

**f** 8 mm out of 7 cm

**g** 5 mm out of 3 cm

**h** 12 mm out of 4 cm

**2** Express each quantity as a simplified fraction with or without a calculator:

**a** 120 g out of 4 kg

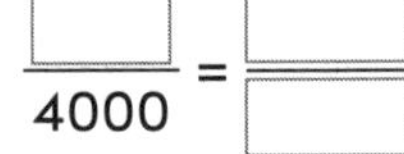

**b** 350 g out of 2 kg

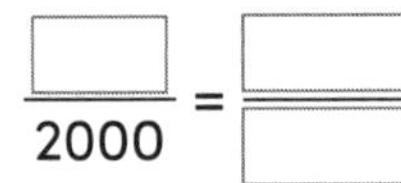

**c** 500 g out of 3 kg

**d** 720 g out of 5 kg

**e** 600 g out of 8 kg

**f** 470 g out of 3 kg

**g** 650 g out of 9 kg

**h** 850 g out of 10 kg

**3** Express each quantity as a simplified fraction with or without a calculator:

**a** 120 mL out of 3.6 L

$\frac{\square}{3600} = \frac{\square}{\square}$

**b** 500 mL out of 1.5 L

$\frac{\square}{1500} = \frac{\square}{\square}$

**c** 700 mL out of 2.1 L

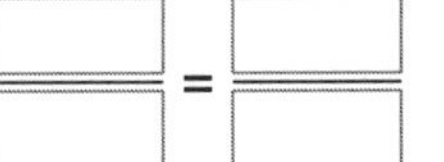

**d** 170 mL out of 3.4 L

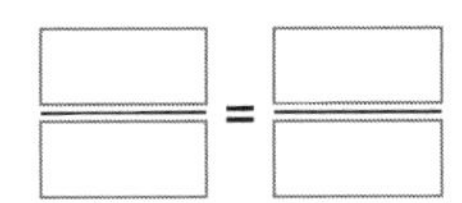

**e** 310 mL out of 6.2 L

**f** 900 mL out of 8.1 L

**g** 800 mL out of 9.6 L

**h** 250 mL out of 10.5 L

**4** Express each quantity as a simplified fraction with or without a calculator:

**a** 500 m out of 1 km

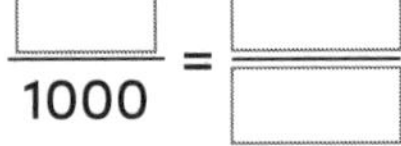

**b** 650 m out of 3 km

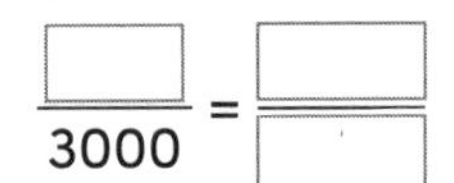

**c** 760 m out of 4 km

**d** 840 m out of 7 km

**e** 960 m out of 8 km

**f** 130 m out of 10 km

**g** 400 m out of 1.2 km

**h** 650 m out of 2.5 km

**5** Express each quantity as a simplified fraction with or without a calculator:

**a** 8 h out of 1 day

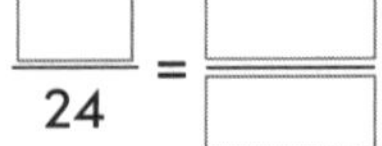

**b** 10 h out of 1 day

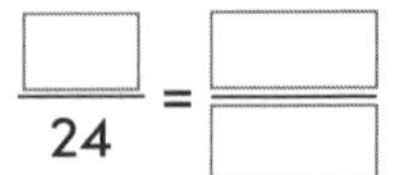

**c** 3 h out of 1 day

**d** 9 h out of 1 day

**e** 12 h out of 1 day

**f** 20 h out of 1 day

UNIT 39

CHAPTER 3 REVIEW

# Fractions

**1** Complete the following equivalent fractions:

**a** $\frac{2}{5} = \frac{\square}{10}$ **b** $\frac{4}{3} = \frac{\square}{12}$

**c** $\frac{15}{30} = \frac{\square}{6}$ **d** $\frac{50}{70} = \frac{5}{\square}$

**2** Simplify the following:

**a** $\frac{3}{10} + \frac{2}{10} =$ ______ **b** $\frac{5}{12} + \frac{3}{12} =$ ______

**c** $\frac{11}{20} - \frac{3}{20} =$ ______ **d** $\frac{20}{27} - \frac{2}{27} =$ ______

**3** Evaluate the following:

**a** $\frac{3}{8} + \frac{3}{16} =$ ______ **b** $\frac{4}{9} + \frac{7}{18} =$ ______

**4** Evaluate the following:

**a** $\frac{2}{7} + \frac{1}{3} =$ ______ **b** $\frac{3}{5} + \frac{1}{4} =$ ______

**c** $\frac{2}{5} - \frac{1}{7} =$ ______ **d** $\frac{5}{7} - \frac{1}{2} =$ ______

**5** Convert the following to mixed numerals:

**a** $\frac{7}{5} =$ ______ **b** $\frac{11}{4} =$ ______

**c** $\frac{25}{8} =$ ______ **d** $\frac{83}{10} =$ ______

**6** Convert the following to improper fractions:

**a** $1\frac{2}{5} =$ ______ **b** $2\frac{3}{4} =$ ______

**c** $3\frac{4}{7} =$ ______ **d** $4\frac{5}{6} =$ ______

**7** Evaluate the following, giving answers in simplified form:

**a** $1\frac{1}{3} + \frac{4}{6} =$ ______

**b** $2\frac{1}{2} + \frac{2}{4} =$ ______

**8** Evaluate the following, giving each as a mixed numeral:

**a** $2\frac{2}{3} - \frac{1}{2} =$ ______ **b** $3\frac{1}{2} - \frac{2}{5} =$ ______

**9** Simplify the following, giving your answer as a mixed numeral where possible:

**a** $\frac{2}{5} \times \frac{3}{7} =$ ______ **b** $\frac{5}{7} \times \frac{2}{7} =$ ______

**c** $\frac{2}{5} \div \frac{7}{4} =$ ______ **d** $\frac{4}{9} \div \frac{1}{3} =$ ______

**e** $3\frac{1}{2} \times \frac{2}{5} =$ ______ **f** $3\frac{1}{2} \div 1\frac{1}{2} =$ ______

**10** Find each amount without using a calculator:

**a** $\frac{1}{2}$ of 18 ______ **b** $\frac{1}{5}$ of 30 ______

**c** $\frac{4}{5}$ of 25 ______ **d** $\frac{3}{4}$ of 16 ______

**11** Peter has 15 colouring pens and gives $\frac{3}{5}$ to his sister. How many pens does his sister receive?

______

**12** In the fridge, $\frac{5}{7}$ of the apples are red and the rest are green.

**a** What fraction of the apples are green?

______

**b** If there are 14 apples, how many green apples are there?

______

**c** How many red apples are there?

______

**13** Express each quantity as a simplified fraction with or without a calculator:

**a** 35 cents out of \$1 ______ **b** 60 cents out of \$2 ______

**c** 30 min out of 1 h ______ **d** 90 min out of 2 h ______

**e** 50 mm out of 7 cm ______ **f** 25 mm out of 4 cm ______

UNIT 1

## FRACTIONS, DECIMALS AND PERCENTAGES

# Multiplying decimals by powers of ten

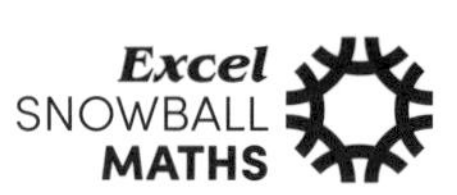

**1** Evaluate the following without using a calculator:

**a** 34.68 × 10 = ____________ **b** 23.49 × 10 = ____________

**c** 53.51 × 10 = ____________ **d** 16.23 × 10 = ____________

**e** 64.29 × 10 = ____________ **f** 83.63 × 10 = ____________

**g** 0.039 × 10 = ____________ **h** 0.062 × 10 = ____________

**i** 0.015 × 10 = ____________ **j** 0.027 × 10 = ____________

**k** 0.083 × 10 = ____________ **l** 0.035 × 10 = ____________

**m** 0.4 × 10 = ____________ **n** 0.2 × 10 = ____________

**o** 0.5 × 10 = ____________ **p** 0.8 × 10 = ____________

**q** 0.6 × 10 = ____________ **r** 0.9 × 10 = ____________

**2** Evaluate:

**a** 1.56 × 100 = ____________ **b** 3.42 × 100 = ____________

**c** 4.34 × 100 = ____________ **d** 9.87 × 100 = ____________

**e** 6.42 × 100 = ____________ **f** 7.99 × 100 = ____________

**3** Evaluate:

**a** 15.6 × 100 = ____________ **b** 34.2 × 100 = ____________

**c** 43.4 × 100 = ____________ **d** 98.7 × 100 = ____________

**e** 64.2 × 100 = ____________ **f** 79.9 × 100 = ____________

**g** 0.05 × 100 = ____________ **h** 0.07 × 100 = ____________

**i** 0.03 × 100 = ____________ **j** 0.09 × 100 = ____________

**k** 0.063 × 100 = ____________ **l** 0.078 × 100 = ____________

**m** 0.015 × 100 = ____________ **n** 0.084 × 100 = ____________

**4** Evaluate:

**a** 0.034 × 1000 = ____________ **b** 0.052 × 1000 = ____________

**c** 0.096 × 1000 = ____________ **d** 0.014 × 1000 = ____________

**e** 2.035 × 1000 = ____________ **f** 6.072 × 1000 = ____________

**g** 5.135 × 1000 = ____________ **h** 8.690 × 1000 = ____________

**i** 0.02 × 1000 = ____________ **j** 0.05 × 1000 = ____________

**k** 0.01 × 1000 = ____________ **l** 0.06 × 1000 = ____________

**m** 4.6 × 1000 = ____________ **n** 5.3 × 1000 = ____________

**o** 3.9 × 1000 = ____________ **p** 1.2 × 1000 = ____________

UNIT 2

# FRACTIONS, DECIMALS AND PERCENTAGES
## Dividing decimals by powers of ten

**1** Evaluate the following without using a calculator:

**a** 346.8 ÷ 10 = ______ **b** 234.9 ÷ 10 = ______

**c** 426.5 ÷ 10 = ______ **d** 756.4 ÷ 10 = ______

**e** 31.5 ÷ 10 = ______ **f** 42.7 ÷ 10 = ______

**g** 10.3 ÷ 10 = ______ **h** 20.4 ÷ 10 = ______

**i** 3.0 ÷ 10 = ______ **j** 5.0 ÷ 10 = ______

**k** 2 ÷ 10 = ______ **l** 7 ÷ 10 = ______

**m** 8 ÷ 10 = ______ **n** 1 ÷ 10 = ______

**o** 76 ÷ 10 = ______ **p** 89 ÷ 10 = ______

**q** 23 ÷ 10 = ______ **r** 65 ÷ 10 = ______

**2** Evaluate:

**a** 156 ÷ 100 = ______ **b** 234 : 100 = ______

**c** 426 ÷ 100 = ______ **d** 754 ÷ 100 = ______

**e** 3150 ÷ 100 = ______ **f** 5220 ÷ 100 = ______

**g** 1035 ÷ 100 = ______ **h** 2049 ÷ 100 = ______

**i** 3.0 ÷ 100 = ______ **j** 5.0 ÷ 100 = ______

**k** 2 ÷ 100 = ______ **l** 7 ÷ 100 = ______

**m** 8 ÷ 100 = ______ **n** 1 ÷ 100 = ______

**o** 23 ÷ 100 = ______ **p** 75 ÷ 100 = ______

**q** 46 ÷ 100 = ______ **r** 51 ÷ 100 = ______

**s** 60 ÷ 100 = ______ **t** 10 ÷ 100 = ______

**3** Evaluate:

**a** 23.9 ÷ 1000 = ______ **b** 16.7 ÷ 1000 = ______

**c** 46.5 ÷ 1000 = ______ **d** 76.4 ÷ 1000 = ______

**e** 35.6 ÷ 1000 = ______ **f** 20.7 ÷ 1000 = ______

**g** 203.4 ÷ 1000 = ______ **h** 563.7 ÷ 1000 = ______

**i** 192.8 ÷ 1000 = ______ **j** 745.2 ÷ 1000 = ______

**k** 100.3 ÷ 1000 = ______ **l** 203.4 ÷ 1000 = ______

**m** 4.9 ÷ 1000 = ______ **n** 5.8 ÷ 1000 = ______

**o** 2.3 ÷ 1000 = ______ **p** 7.5 ÷ 1000 = ______

**q** 1.1 ÷ 1000 = ______ **r** 6.6 ÷ 1000 = ______

# FRACTIONS, DECIMALS AND PERCENTAGES
## Multiplying decimals

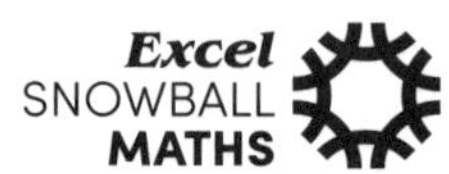


**1** Evaluate:

a
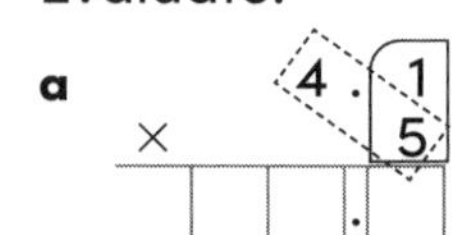

$\begin{array}{r} 4.1 \\ \times \quad 5 \\ \hline \end{array}$

b $\begin{array}{r} 3.2 \\ \times \quad 4 \\ \hline \end{array}$

c $\begin{array}{r} 7.3 \\ \times \quad 2 \\ \hline \end{array}$

d $\begin{array}{r} 8.4 \\ \times \quad 2 \\ \hline \end{array}$

e $\begin{array}{r} 5.4 \\ \times \quad 2 \\ \hline \end{array}$

f $\begin{array}{r} 6.2 \\ \times \quad 3 \\ \hline \end{array}$

g $\begin{array}{r} 7.2 \\ \times \quad 2 \\ \hline \end{array}$

h $\begin{array}{r} 9.2 \\ \times \quad 3 \\ \hline \end{array}$

i 5.3 × 2

j 6.2 × 4

k 7.4 × 2

l 4.3 × 3

m 0.3 × 2

n 0.4 × 2

o 0.2 × 3

p 0.3 × 3

q 0.5 × 7

r 0.6 × 7

**2** Evaluate:

a $\begin{array}{r} 3.2 \\ \times \quad 12 \\ \hline \end{array}$

b $\begin{array}{r} 4.1 \\ \times \quad 12 \\ \hline \end{array}$

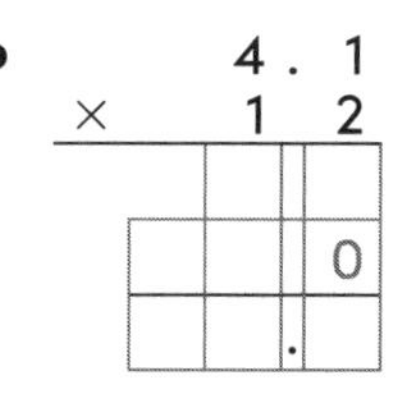

c $\begin{array}{r} 3.6 \\ \times \quad 11 \\ \hline \end{array}$

d $\begin{array}{r} 6.9 \\ \times \quad 11 \\ \hline \end{array}$

e $\begin{array}{r} 5.2 \\ \times \quad 21 \\ \hline \end{array}$

f $\begin{array}{r} 7.3 \\ \times \quad 21 \\ \hline \end{array}$

g $\begin{array}{r} 6.2 \\ \times \quad 21 \\ \hline \end{array}$

h $\begin{array}{r} 4.3 \\ \times \quad 32 \\ \hline \end{array}$

i $\begin{array}{r} 0.4 \\ \times \quad 12 \\ \hline \end{array}$

j $\begin{array}{r} 0.3 \\ \times \quad 23 \\ \hline \end{array}$

k $\begin{array}{r} 0.4 \\ \times \quad 21 \\ \hline \end{array}$

l $\begin{array}{r} 0.2 \\ \times \quad 41 \\ \hline \end{array}$

m $\begin{array}{r} 0.2 \\ \times \quad 34 \\ \hline \end{array}$

n $\begin{array}{r} 0.3 \\ \times \quad 33 \\ \hline \end{array}$

# Multiplying decimals

**1** Evaluate:

**a** 5 . 4
× 2 . 1

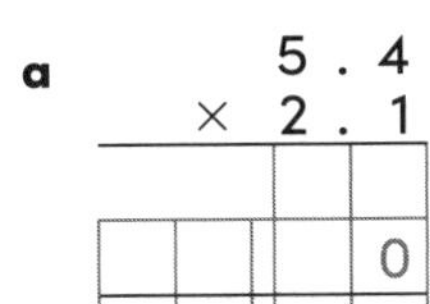

**b** 6 . 2
× 3 . 1

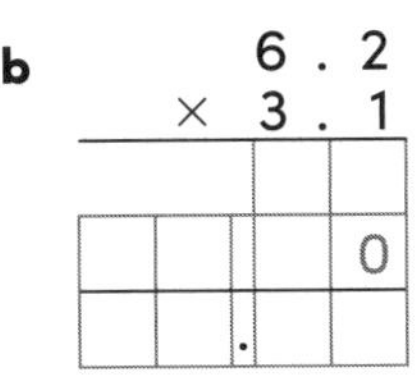

**c** 7.3 × 2.1

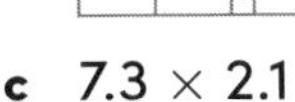

**d** 6.2 × 2.1

**e** 0.4 × 5.1

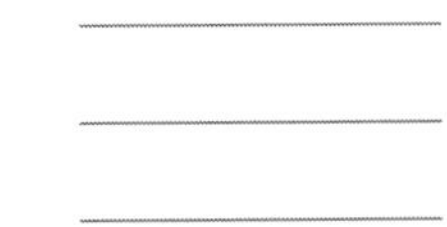

**f** 0.8 × 3.1

**g** 4.3 × 4.2

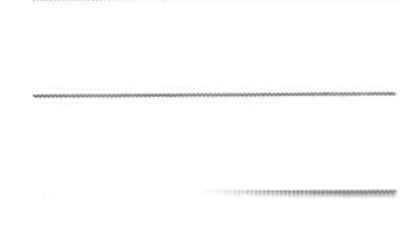

**h** 5.2 × 6.3

**i** 5.2 × 7.4

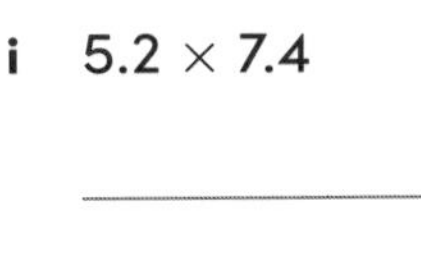

**j** 9.2 × 5.4

**2** Evaluate:

**a** 4.8 × 1.3

**b** 3.7 × 1.5

**c** 5.2 × 3.5

**d** 7.4 × 2.6

**e** 5.3 × 7.6

**f** 4.5 × 6.3

**g** 7.2 × 5.8

**h** 9.4 × 8.6

**i** 4.5 × 4.5

**j** 5.9 × 5.9

# Dividing decimals

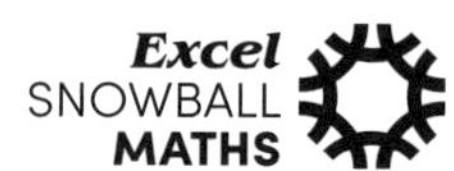

**1** Evaluate:

**a** 2.6 ÷ 2

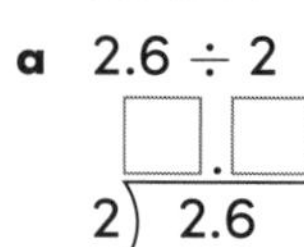

**b** 4.4 ÷ 2

2) 4.4

**c** 8.4 ÷ 4

4)

**d** 9.6 ÷ 3

3)

**e** 6.3 ÷ 3

**f** 5.5 ÷ 5

**g** 4.6 ÷ 2

**h** 8.2 ÷ 2

**i** 6.9 ÷ 3

**j** 9.9 ÷ 3

**2** Evaluate:

**a** 2 ÷ 5

0.
5)20

**b** 3 ÷ 5

0.
5)30

**c** 4 ÷ 5

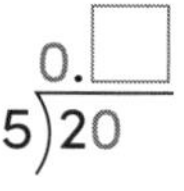

**d** 1 ÷ 2

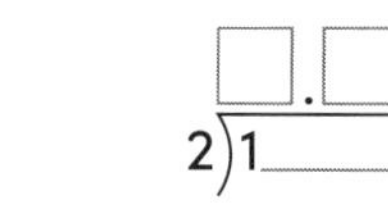

**e** 1 ÷ 5

5)1

**f** 4 ÷ 8

8)4

**g** 3 ÷ 6

6)

**h** 7 ÷ 10

10)

**i** 2 ÷ 10

**j** 3 ÷ 10

**k** 2 ÷ 4

**l** 5 ÷ 25

**3** Evaluate:

**a** 0.6 ÷ 0.2
= 6 ÷ 2
= ______

**b** 0.8 ÷ 0.4
= 8 ÷ 4
= ______

**c** 0.9 ÷ 0.3
= ____ ÷ ____
= ______

**d** 0.4 ÷ 0.2
= ____ ÷ ____
= ______

**e** 2.5 ÷ 0.5
= ______
= ______

**f** 3.6 ÷ 0.6
= ______
= ______

**g** 4.5 ÷ 0.9
= ______
= ______

**h** 7.2 ÷ 0.8
= ______
= ______

**i** 2 ÷ 0.5
= 20 ÷ 5
= ______

**j** 2 ÷ 0.4
= 20 ÷ 4
= ______

**k** 3 ÷ 0.5
= ______
= ______

**l** 4 ÷ 0.8
= ______
= ______

**m** 3 ÷ 0.6
= ______
= ______

**n** 1 ÷ 0.5
= ______
= ______

**o** 4 ÷ 0.2
= ______
= ______

**p** 5 ÷ 0.1
= ______
= ______

UNIT 6

# FRACTIONS, DECIMALS AND PERCENTAGES

## Rounding decimals

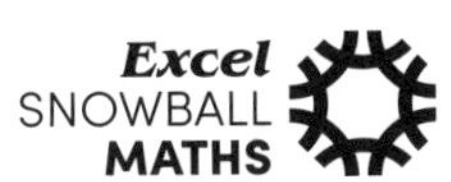


**1** Round each decimal to one decimal place. Circle the correct answer:

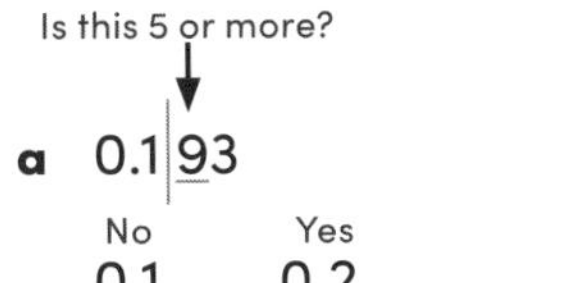


**b** 0.7|42 (Is this 5 or more?)
No 0.7 Yes 0.8

**c** 0.2|69
0.2 0.3

**d** 0.5|11
0.5 0.6

**2** Round each decimal to one decimal place:

**a** 0.3|54 = 0.______ **b** 0.8|72 = 0.______

**c** 0.5|31 = ______ **d** 0.7|15 = ______

**e** 0.496 = ______ **f** 0.369 = ______

**g** 0.216 = ______ **h** 0.321 = ______

**3** Round each decimal to the nearest tenth:

**a** 1.3|84 = 1.______ **b** 5.8|22 = 5.______

**c** 8.7|51 = ______ **d** 3.7|35 = ______

**e** 7.496 = ______ **f** 2.359 = ______

**g** 6.226 = ______ **h** 9.301 = ______

**i** 13.582 = 13.______ **j** 19.473 = ______

**k** 23.242 = ______ **l** 44.862 = ______

**m** 52.526 = ______ **n** 67.441 = ______

**o** 70.165 = ______ **p** 83.078 = ______

**q** 133.361 = ______ **r** 154.701 = ______

**s** 245.677 = ______ **t** 336.861 = ______

**u** 560.231 = ______ **v** 749.550 = ______

**4** Round each decimal to two decimal places. Circle the correct answer:

**a** 0.15|3
0.15 0.16

**b** 0.46|9
0.46 0.47

**c** 0.94|7
0.94 0.95

**d** 0.71|1
0.71 0.72

**5** Round each decimal to two decimal places:

**a** 0.25|4 = 0.2______ **b** 0.17|5 = 0.1______

**c** 0.43|8 = 0.______ **d** 0.81|4 = 0.______

**e** 0.75|6 = ______ **f** 0.46|3 = ______

**g** 0.516 = ______ **h** 0.712 = ______

**i** 0.925 = ______ **j** 0.567 = ______

**6** Round each decimal to the nearest hundredth:

**a** 6.82|5 = 6.______ **b** 2.35|4 = 2.______

**c** 8.53|7 = ______ **d** 5.73|4 = ______

**e** 9.45|6 = ______ **f** 1.35|9 = ______

**g** 6.211 = ______ **h** 7.315 = ______

**i** 10.582 = ______ **j** 34.877 = ______

**k** 23.218 = ______ **l** 62.526 = ______

**m** 73.488 = ______ **n** 82.169 = ______

**o** 59.645 = ______ **p** 94.273 = ______

**q** 154.361 = ______ **r** 365.118 = ______

**s** 745.607 = ______ **t** 649.344 = ______

**u** 926.843 = ______ **v** 852.309 = ______

UNIT 7

FRACTIONS, DECIMALS AND PERCENTAGES

# Rounding decimals

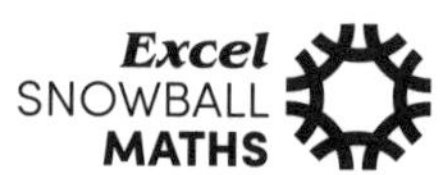


**1** Round each decimal to one decimal place. Circle the correct answer:

**a** 0.9|9
0.9 1.0

**b** 0.9|4
0.9 1.0

**c** 2.9|2
2.9 3.0

**d** 5.9|3
5.9 6.0

**2** Round each decimal to one decimal place:

**a** 7.9|8 = ________ **b** 8.9|1 = ________

**c** 3.9|3 = ________ **d** 5.9|5 = ________

**e** 6.95 = ________ **f** 2.96 = ________

**g** 1.91 = ________ **h** 5.94 = ________

**i** 4.97 = ________ **j** 7.92 = ________

**3** Round each decimal to one decimal place. Circle the correct answer:

**a** 23.9|12
23.9 24.0

**b** 52.9|86
52.9 53.0

**4** Round each decimal to one decimal place:

**a** 87.9|51 = ________ **b** 68.9|43 = ________

**c** 13.9|72 = ________ **d** 25.9|16 = ________

**e** 76.957 = ________ **f** 52.963 = ________

**g** 85.947 = ________ **h** 62.933 = ________

**i** 103.977 = ________ **j** 156.988 = ________

**k** 310.940 = ________ **l** 525.901 = ________

**5** Round the following to the nearest whole number:

**a** 5|.126 = ________ **b** 7|.817 = ________

**c** 16|.482 = ________ **d** 22|.632 = ________

**e** 11.955 = ________ **f** 53.389 = ________

**6** Determine whether each decimal is a recurring or a terminating decimal:

**a** 0.1333333... ________

**b** 1.526 ________

**c** 0.198 ________

**d** 0.1244444... ________

**e** 0.6534 ________

**f** $0.\dot{3}$ ________

**g** $0.\dot{2}3\dot{5}$ ________

**h** 1.531 ________

**7** Write each fraction as a recurring decimal using dot notation:

**a** $\frac{1}{3}$ = ________ **b** $\frac{1}{9}$ = ________

**c** $\frac{2}{3}$ = ________ **d** $\frac{4}{9}$ = ________

**e** $\frac{1}{6}$ = ________ **f** $\frac{5}{6}$ = ________

**g** $\frac{2}{11}$ = ________ **h** $\frac{9}{37}$ = ________

**8** Determine whether each fraction is a recurring or a terminating decimal:

**a** $\frac{4}{5}$ ________

**b** $\frac{2}{3}$ ________

**c** $\frac{3}{50}$ ________

**d** $\frac{5}{27}$ ________

**e** $1\frac{2}{5}$ ________

**f** $2\frac{1}{9}$ ________

**g** $5\frac{1}{3}$ ________

UNIT 8

# FRACTIONS, DECIMALS AND PERCENTAGES

## Fractions, decimals and percentages

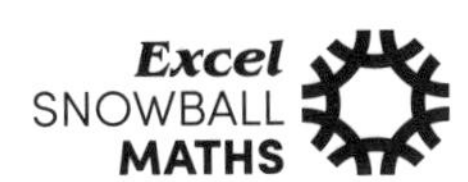

**1** Convert each percentage to a simplified fraction and a decimal, with or without a calculator:

**a** 50%

$= \frac{\square}{100}$

$= \frac{\square}{\square}$ Simplified fraction

= ______ Decimal

**b** 25%

$= \frac{\square}{100}$

$= \frac{\square}{\square}$ Simplified fraction

= ______ Decimal

**c** 70%

$= \frac{\square}{\square}$

$= \frac{\square}{\square}$

= ______

**d** 10%

$= \frac{\square}{\square}$

$= \frac{\square}{\square}$

= ______

**e** 5%

= ______

= ______

= ______

**f** 8%

= ______

= ______

= ______

**g** 16%

= ______

= ______

= ______

**h** 45%

= ______

= ______

= ______

**i** 120%

= ______

= ______

= ______

**j** 100%

= ______

= ______

= ______

**2** Using a calculator, convert each percentage to a simplified fraction and a decimal:

**a** $12\frac{1}{2}\%$

$= \frac{12\frac{1}{2}}{100}$

$= \frac{\square}{\square}$

= ______

**b** $6\frac{1}{4}\%$

$= \frac{6\frac{1}{4}}{100}$

$= \frac{\square}{\square}$

= ______

**c** $87\frac{1}{2}\%$

= ______

= ______

= ______

**d** $33\frac{1}{3}\%$

= ______

= ______

= ______

**e** 27.5%

= ______

= ______

= ______

**f** 0.4%

= ______

= ______

= ______

**3** Convert each fraction to a percentage, with or without a calculator:

**a** $\frac{1}{2}$

$= \frac{1}{2} \times$ 100

= ______ %

**b** $\frac{1}{4}$

$= \frac{1}{4} \times$ 100

= ______ %

**c** $\frac{3}{4}$

$= \frac{\square}{\square} \times$ ______

= ______ %

**d** $\frac{2}{5}$

$= \frac{\square}{\square} \times$ ______

= ______ %

UNIT 9

**FRACTIONS, DECIMALS AND PERCENTAGES**

# Fractions, decimals and percentages

Note: Answers should be correct to one decimal place.

**1** Convert each fraction to a percentage, with or without a calculator:

**a** $\frac{4}{9}$

$= \frac{\square}{\square} \times$ 100

= ______ %

**b** $\frac{3}{8}$

$= \frac{\square}{\square} \times$ 100

= ______ %

**c** $\frac{5}{2}$

= ______

= ______

**d** $\frac{11}{100}$

= ______

= ______

**e** $\frac{3}{25}$

= ______

= ______

**f** $2\frac{1}{4}$

= ______

= ______

**2** Convert each decimal to a percentage, with or without a calculator:

**a** 0.25

= ______ × 100

= ______ %

**b** 0.6

= ______ × 100

= ______ %

**c** 0.17

= ______ × ______

= ______ %

**d** 0.32

= ______ × ______

= ______ %

**e** 1.4

= ______

= ______

**f** 0.03

= ______

= ______

**g** 0.0405

= ______

= ______

**h** 3.65

= ______

= ______

**i** 5

= ______

= ______

**j** 1

= ______

= ______

**k** 2.08

= ______

= ______

**l** 7.03

= ______

= ______

**3** Complete the following tables leaving fractions in simplified form:

**a**

| Fraction | Decimal | Percentage |
|---|---|---|
| $\frac{1}{2}$ | | |
| $\frac{2}{5}$ | | |
| $\frac{3}{4}$ | | |
| $\frac{7}{5}$ | | |

**b**

| Fraction | Decimal | Percentage |
|---|---|---|
| | 0.15 | |
| | 0.2 | |
| | 0.05 | |
| | 0.86 | |

**c**

| Fraction | Decimal | Percentage |
|---|---|---|
| | | 12% |
| | | 75% |
| | | 125% |
| | | 12.5% |
| | 0.49 | |
| $\frac{5}{9}$ | | |

UNIT 10

**FRACTIONS, DECIMALS AND PERCENTAGES**

# Fractions, decimals and percentages

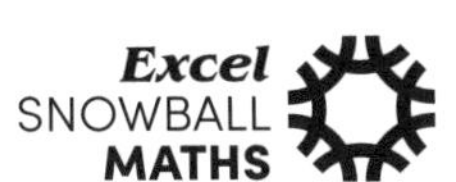

1 Determine which is larger:

a $\frac{14}{25}$ or 0.55?

$\frac{14}{25} \times$ 100 = ______%

0.55 × 100 = ______%

Answer: ________

b $\frac{6}{8}$ or 0.79?

$\frac{6}{8} \times$ 100 = ______%

0.79 × 100 = ______%

Answer: ________

c $\frac{7}{10}$ or 0.72?

= ________

= ________

= ________

d $\frac{9}{15}$ or 0.58?

= ________

= ________

= ________

e $\frac{7}{20}$ or 0.4?

= ________

= ________

= ________

f 0.17 or $\frac{9}{50}$?

= ________

= ________

= ________

2 Which is the largest: $\frac{8}{9}$, 0.84 or 90%?

3 Which is the largest: $\frac{19}{50}$, 0.35 or 34%?

4 Which is the largest: 0.11, $\frac{12}{110}$ or 12%?

5 Write 0.76, $\frac{7}{9}$ and 79% in descending order (largest to smallest):

6 Write 0.23, $\frac{3}{8}$ and 20% in descending order:

7 Write 42%, $\frac{5}{11}$ and 0.405 in ascending order:

8 Write 1.5%, $\frac{2}{100}$ and 0.021 in ascending order:

FRACTIONS, DECIMALS AND PERCENTAGES

# Expressing quantities as a percentage

Note: Answers should be correct to one decimal place.

**1** **a** What percentage of the balls are white?

$\frac{\square}{5}$ × ______ = ______ %

**b** What percentage of the balls are grey?

$\frac{\square}{5}$ × ______ = ______ %

**2** **a** What percentage of the shapes are square?

$\frac{\square}{\square}$ × ______ = ______ %

**b** What percentage of the shapes are crosses?

$\frac{\square}{\square}$ × ______ = ______ %

**3** **a** What percentage of the tiles are white?

$\frac{\square}{\square}$ × ______ = ______

**b** What percentage of the tiles are grey?

$\frac{\square}{\square}$ × ______ = ______

**4** Express each quantity as a percentage, with or without a calculator:

**a** 76 out of 100

$\frac{\square}{\square}$ × ______ = ______

**b** 82 out of 100

$\frac{\square}{\square}$ × ______ = ______

**c** 15 out of 100

______

**d** 34 out of 100

______

**e** $20 out of $50

______

**f** 30 mL out of 50 mL

______

**g** 40 min out of 60 min

______

**h** 6 m out of 10 m

______

**5** Express each quantity as a percentage, with or without a calculator:

**a** 50 cents out of $1

$\frac{\square}{100}$ × ______ = ______

**b** 35 cents out of $1

$\frac{\square}{100}$ × ______ = ______

**c** 40 cents out of $1

______

**d** 60 cents out of $1

______

**e** 90 cents out of $2

______

**f** 10 cents out of $5

______

**g** 70 cents out of $7

______

**h** 50 cents out of $10

______

UNIT 12

**FRACTIONS, DECIMALS AND PERCENTAGES**

# Expressing quantities as a percentage

Note: Answers should be correct to one decimal place.

**1** Express each quantity as a percentage, with or without a calculator:

**a** 25 min out of 1 h

$\frac{\square}{60} \times$ ______ = ______ %

**b** 10 min out of 1 h

$\frac{\square}{60} \times$ ______ = ______ %

**c** 12 min out of 1 h

______

**d** 55 min out of 1 h

______

**e** 40 min out of 2 h

$\frac{\square}{120} \times$ ______ = ______ %

**f** 15 min out of 2 h

______

**g** 30 min out of 2 h

______

**h** 15 min out of 3 h

______

**2** Express each quantity as a percentage, with or without a calculator:

**a** 250 m out of 1 km

$\frac{\square}{1000} \times$ ______ = ______ %

**b** 600 m out of 3 km

$\frac{\square}{3000} \times$ ______ = ______ %

**c** 470 m out of 2 km

______

**d** 300 m out of 5 km

______

**e** 840 m out of 7 km

______

**f** 960 m out of 8 km

______

**g** 800 m out of 6.4 km

______

**h** 230 m out of 9.2 km

______

**3** Express each quantity as a percentage, with or without a calculator:

**a** 6 h out of 1 day

$\frac{\square}{24} \times$ ______ = ______ %

**b** 12 h out of 1 day

$\frac{\square}{24} \times$ ______ = ______ %

**c** 15 h out of 1 day

______

**d** 21 h out of 1 day

______

**e** 24 h out of 2 days

$\frac{\square}{48} \times$ ______ = ______ %

**f** 36 h out of 2 days

______

**g** 30 h out of 2 days

______

**h** 42 h out of 2 days

______

UNIT 13

FRACTIONS, DECIMALS AND PERCENTAGES

# Expressing quantities as a percentage

Note: Answers should be correct to one decimal place.

**1** In a class of 20 students, 6 students like Maths, 10 students like English and the rest like other subjects.

**a** What percentage of the class likes Maths?

$\frac{\square}{20} \times$ ______ = ______ %

**b** What percentage of the class likes English?

$\frac{\square}{20} \times$ ______ = ______ %

**c** What percentage of the class likes other subjects?

____________________

**2** In a class survey of 30 students, 15 prefer online testing, 10 prefer paper tests and the rest have no preference.

**a** What percentage of the class prefers paper tests?

____________________

**b** What percentage of the class prefers online tests?

____________________

**c** What percentage of the class have no preference?

____________________

**3** There are 4 blue, 10 white and 2 grey shirts in a wardrobe.

**a** How many shirts are there in total?

____________________

**b** What percentage of the shirts are blue?

____________________

**c** What percentage of the shirts are grey?

____________________

**d** What percentage of the shirts are white?

____________________

**4** There are 8 green, 5 red and 6 blue watchbands in a store.

**a** How many watchbands are there in total?

____________________

**b** What percentage of the watchbands are green?

____________________

**c** What percentage of the watchbands are red?

____________________

**d** What percentage of the watchbands are red or blue?

____________________

**5** There are 10 white, 9 black and 6 brown clocks in an antique shop.

**a** What percentage of the clocks are black?

____________________

**b** What percentage of the clocks are brown?

____________________

**c** What percentage of the clocks are not brown?

____________________

**d** What percentage of the clocks are white or black?

____________________

UNIT 14

# FRACTIONS, DECIMALS AND PERCENTAGES

## Percentage of a quantity

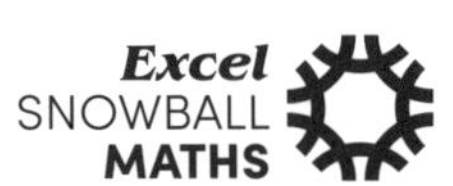


1 Evaluate without a calculator:

a What is **50**% of 10?

10 ÷ **2** = ______

b What is **50**% of 14?

14 ÷ **2** = ______

c What is 50% of $20?

______ ÷ ______ = ______

d What is 50% of 18 mL?

______ ÷ ______ = ______

e What is 50% of 16 grams?

______________________

f What is 50% of 24 hours?

______________________

2 Evaluate without a calculator:

a What is **25**% of 24?

24 ÷ **4** = ______

b What is **25**% of 32?

32 ÷ **4** = ______

c What is 25% of $40?

______________________

d What is 25% of 36 L?

______________________

e What is 25% of 20 cm?

______________________

f What is 25% of 28 min?

______________________

3 Evaluate without a calculator:

a What is **10**% of 50?

50 ÷ **10** = ______

b What is **10**% of 20?

20 ÷ **10** = ______

c What is 10% of 40 kg?

______________________

d What is 10% of 120 cm?

______________________

e What is 10% of 360°?

______________________

f What is 10% of $750?

______________________

4 Evaluate without a calculator:

a What is **20**% of 15?

15 ÷ **5** = ______

b What is **20**% of 45?

45 ÷ **5** = ______

c What is 20% of 30 min?

______________________

d What is 20% of 60 min?

______________________

e What is 20% of $100?

______________________

f What is 20% of 550 mL?

______________________

## FRACTIONS, DECIMALS AND PERCENTAGES
# Percentage of a quantity

**1** Evaluate using a calculator:

**a** 12% of $40

$\frac{\square}{100} \times 40 = \$$ ______

**b** 35% of $25

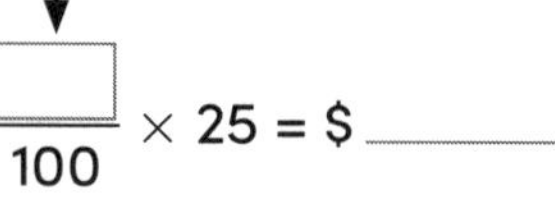

$\frac{\square}{100} \times 25 = \$$ ______

**c** 27% of $320

$\frac{\square}{100} \times$ ______ = ______

**d** 72% of $580

$\frac{\square}{100} \times$ ______ = ______

**e** 2% of 620 mL

$\frac{\square}{\square} \times$ ______ = ______

**f** 5% of 350 mL

$\frac{\square}{\square} \times$ ______ = ______

**g** 35% of 490 grams

______

**h** 21% of 830 grams

______

**i** 17.4% of 500 kg

______

**j** 43.2% of 110 kg

______

**k** $7\frac{1}{2}$% of 320 MB

______

**l** $12\frac{3}{4}$% of 800 GB

______

**2** There are 320 books in a bookstore and 60% of the books are fiction. How many books are fiction?

$\frac{\square}{\square} \times$ ______ = ______

**3** There were 800 students at the athletics carnival and 59% of the students participated in an event. How many students participated in an event?

______

**4** There are 420 computers in a warehouse and 35% of them need repairing. How many computers need repairing?

______

**5** In a survey of 400 television viewers, 65% of the viewers watch sport.

**a** How many viewers watch sport?

______

**b** How many viewers do not watch sport?

______

**6** At a car dealership, 5% of the cars are red. If there are 300 cars,

**a** how many cars are red?

______

**b** how many cars are not red?

______

UNIT 16

CHAPTER 4 REVIEW

# Fractions, decimals and percentages

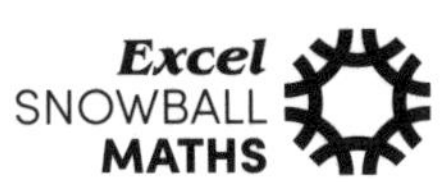


**1** Evaluate without a calculator:

a $24.67 \times 10 =$ ______ b $13.25 \times 10 =$ ______

c $1.35 \times 10 =$ ______ d $4.96 \times 10 =$ ______

e $0.04 \times 10 =$ ______ f $0.07 \times 10 =$ ______

**2** a $5.83 \times 100 =$ ______ b $6.49 \times 100 =$ ______

c $0.085 \times 100 =$ ______ d $0.094 \times 100 =$ ______

e $83.9 \times 100 =$ ______ f $72.6 \times 100 =$ ______

**3** a $15.8 \div 10 =$ ______ b $32.4 \div 10 =$ ______

c $6.4 \div 10 =$ ______ d $7.2 \div 10 =$ ______

e $13 \div 10 =$ ______ f $29 \div 10 =$ ______

**4** a $24 \div 100 =$ ______ b $52 \div 100 =$ ______

c $20.3 \div 100 =$ ______ d $19.6 \div 100 =$ ______

e $4.2 \div 100 =$ ______ f $8.9 \div 100 =$ ______

**5** a $4.2 \times 3 =$ ______ b $8.3 \times 2 =$ ______

c $0.4 \times 4 =$ ______ d $0.7 \times 3 =$ ______

e $6.2 \times 6.2 =$ ______ f $0.4 \times 7.1 =$ ______

**6** a $9.3 \div 3 =$ ______ b $8.6 \div 2 =$ ______

c $5 \div 10 =$ ______ d $6 \div 12 =$ ______

e $1.2 \div 0.4 =$ ______ f $2.1 \div 0.3 =$ ______

**7** Round the following to two decimal places:

a 0.738 ______ b 0.521 ______ c 7.266 ______

d 49.605 ______ e 37.234 ______ f 8.395 ______

**8** Round the following to one decimal place:

a 5.39 ______ b 4.23 ______ c 16.78 ______

d 62.41 ______ e 0.55 ______ f 0.99 ______

**9** Determine whether each fraction is a recurring or a terminating decimal:

a $\frac{3}{5}$ ______ b $\frac{7}{9}$ ______ c $1\frac{2}{3}$ ______

**10** Write 0.91, $\frac{7}{8}$ and 90% in ascending order.

______

**11** Complete the table:

| Fraction | Decimal | Percentage |
|---|---|---|
| $\frac{2}{5}$ | | |
| $1\frac{3}{4}$ | | |
| | 0.15 | |
| | | 25% |

**12** Express each quantity as a percentage:

a 16 out of 20 ______ b 65c out of $1 ______

c 40 cm out of 1 m ______

d 35 min out of 1 h ______

**13** In a bag there are 4 red, 7 blue and 5 white marbles.

a What percentage of the marbles are blue?

______

b What percentage of the marbles are not blue?

______

**14** Evaluate without a calculator:

a 50% of 12 cm ______ b 50% of $100 ______

c 25% of 32 kg ______ d 10% of $75 ______

**15** Diana would like to achieve 70% on her next maths test. If the test is out of 40 marks, how many marks does she need to achieve?

______

**16** Lina received 15% off a $140 dress.

a How much did Lina save?

______

b How much did she pay?

______

UNIT 1

## ALGEBRA

# Introduction to algebra

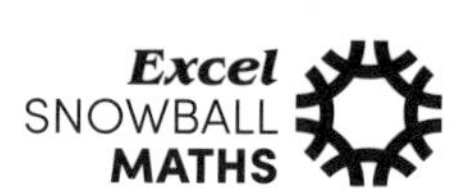


**1** Simplify each expression:

**a** $x + x + x + x =$ ____$x$

**b** $y + y + y =$ ____$y$

**c** $a + a + a + a + a + a =$ ________

**d** $m + m =$ ________

**e** $b + b + b + b =$ ________

**f** $c + c + c + c =$ ________

**g** $x^2 + x^2 + x^2 + x^2 + x^2 =$ ____$x^2$

**h** $y^2 + y^2 + y^2 =$ ____$y^2$

**i** $a^2 + a^2 =$ ________

**j** $h^2 + h^2 + h^2 + h^2 =$ ________

**k** $ab + ab + ba + ba =$ ____$ab$

**l** $xy + yx + xy =$ ____$xy$

**m** $pq + pq + pq + qp + qp =$ ________

**n** $vw + wv + wv + vw =$ ________

**2** Simplify each expression:

**a** $x + x + x + y + y =$ ____$x +$ ____$y$

**b** $a + a + b + b + b =$ ____$a +$ ____$b$

**c** $p + p + p + q + q =$ ____ $+$ ____

**d** $s + s + s + t =$ ____ $+$ ____

**e** $x + x + y + y + x =$ ____ $+$ ____

**f** $a + a + b + b + a + a =$ ____ $+$ ____

**g** $e + e + f + f + e =$ ________

**h** $n + m + m + n + n + m =$ ________

**i** $x + y + y + x + x + y =$ ________

**j** $v + w + w + v + w + v + v =$ ________

**k** $r + s + r + r + s + s + s =$ ________

**3** Simplify each expression:

**a** $x^2 + x^2 + y^2 + y^2 + y^2 =$ ____$x^2 +$ ____$y^2$

**b** $a^2 + a^2 + b^2 + b^2 + b^2 =$ ____$a^2 +$ ____$b^2$

**c** $p^2 + p^2 + p^2 + q^2 + q^2 =$ ____ $+$ ____

**d** $s^2 + s^2 + t^2 + t^2 =$ ____ $+$ ____

**e** $x^2 + x^2 + y^2 + y^2 + x^2 =$ ________

**f** $a^2 + a^2 + b^2 + b^2 + a^2 + a^2 =$ ________

**g** $r^2 + r^2 + x^2 + x^2 + r^2 =$ ________

**h** $y^2 + w + w + y^2 + y^2 + w =$ ________

**i** $k^2 + m + m + k^2 + k^2 + m =$ ________

**4** Simplify each expression:

**a** $xy + xy + y + y =$ ____$xy +$ ____$y$

**b** $wd + wd + wd + d + d + d + d$

$=$ ____$wd +$ ____$d$

**c** $ab + ab + a + a + a =$ ____ $+$ ____

**d** $rm + rm + rm + m + m =$ ____ $+$ ____

**e** $wt + wt + t + t + t + wt =$ ________

**f** $ab + ab + b + b + ab + ab =$ ________

**g** $pq + pq + q + q + qp + pq =$ ________

**h** $xy + y + y + xy + yx + y =$ ________

**i** $t + kt + kt + kt + t + tk =$ ________

UNIT 2

## ALGEBRA

# Introduction to algebra

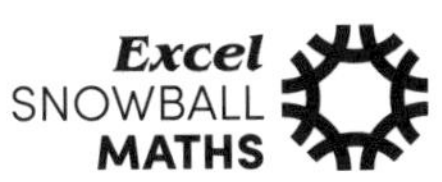

**1** Simplify each expression:

**a** $2x + 5x =$ ______ $x$

**b** $4y + 6y =$ ______ $y$

**c** $3a + 8a =$ ______

**d** $4m + 7m =$ ______

**e** $2x^2 + 7x^2 =$ ______

**f** $5a^2 + 3a^2 =$ ______

**g** $8e^2 + 5e^2 =$ ______

**h** $7w^2 + 6w^2 =$ ______

**i** $2xy + 6xy =$ ______

**j** $3ab + 5ba =$ ______

**k** $4yd + 8dy =$ ______

**l** $10cf + 4cf =$ ______

**m** $2x + 5x + 2y =$ ______ $x +$ ______ $y$

**n** $4a + 3a + 7b =$ ______ $a +$ ______ $b$

**o** $3p + 5p + 6r =$ ______ $+$ ______

**p** $5r + 4r + 5t =$ ______ $+$ ______

**q** $6x + 3x + 4t =$ ______

**r** $5k + k + 3g =$ ______

**s** $7v + v + 4d =$ ______

**t** $y + 7y + 8p =$ ______

**2** Simplify each expression:

**a** $6x - 2x + 3 =$ ______ $x +$ ______

**b** $10y - 6y + 5 =$ ______ $y +$ ______

**c** $8a - 3a + 4 =$ ______ $+$ ______

**d** $10m - 7m + 2 =$ ______ $+$ ______

**e** $12z^2 - 4z^2 + 3 =$ ______

**f** $15h^2 - 3h^2 + 4 =$ ______

**g** $18e^2 - 5e^2 + 8 =$ ______

**h** $17w^2 - 6w^2 + 9 =$ ______

**i** $12xy - 6xy + 3 =$ ______

**j** $8ab - 5ba + 4 =$ ______

**k** $10x - 5x + 6y - 2y =$ ______

**l** $9p - 4p + 6r - 3r =$ ______

**m** $5r - 4r + 3t - t =$ ______

**n** $9x - 3x + 4t - 2t =$ ______

**o** $5k - k + 3g - g =$ ______

**p** $12h - 4h + 6p - 5p$ ______

**q** $15s - 4s + 8q - q$ ______

**r** $7y - 3y + 5x - 3x$ ______

**s** $11k - 2k + 13r - r$ ______

**3** Simplify each expression:

**a** $5x - 7x - 6y + 2y =$ ______ $x -$ ______ $y$

**b** $4p - 9p - 7r + 3r =$ ______ $p -$ ______ $r$

**c** $4r - 5r - 4t + 2t =$ ______

**d** $4x - 7x - 5t + 3t =$ ______

**e** $2k - 5k - 8g + 5g =$ ______

**f** $4h - 8h - 6p + 5p =$ ______

**g** $5s - 7s - 5q + 2q =$ ______

**h** $3y - 7y - 8x + 4x =$ ______

**i** $2k - 9k - 5r + 3r =$ ______

**4** From the term $7a$ state:

**a** the pronumeral ______

**b** the coefficient ______

**5** From the term $12x^2$ state:

**a** the pronumeral ______

**b** the coefficient ______

**6** From the term $5xy$ state:

**a** the pronumeral(s) ______

**b** the coefficient ______

**7** From the term $-10a^2b$ state:

**a** the pronumeral(s) ______

**b** the coefficient ______

UNIT 3

**ALGEBRA**

# Adding like terms

**1** Simplify the following expressions:

**a** $\boxed{5x + 2x} + 3 =$ ______$x +$ ______

**b** $\boxed{8x + 3x} + 5 =$ ______$x +$ ______

**c** $2x + 6x + 4 =$ ____________

**d** $4x + 6x + 8 =$ ____________

**e** $5x + x + 9 =$ ____________

**f** $7x + 2x + 3 =$ ____________

**2** **a** $\boxed{7m} + 3n \boxed{+ 4m} =$ ______$m +$ ______$n$

**b** $\boxed{4m} + 7n \boxed{+ 3m} =$ ______$m +$ ______$n$

**c** $10m + 6n + 8m =$ ____________

**d** $7m + n + 4m =$ ____________

**e** $2m + 6n + 3m =$ ____________

**f** $m + 3n + 5m =$ ____________

**3** **a** $\boxed{5xy} + 8 \boxed{+ 2xy} =$ ______$xy +$ ______

**b** $\boxed{10xy} + 9 \boxed{+ 4xy} =$ ______$xy +$ ______

**c** $2xy + 5 + 9xy =$ ____________

**d** $6xy + 3 + 5xy =$ ____________

**e** $8xy + 6 + 2xy =$ ____________

**f** $6xy + 4 + 3xy =$ ____________

**g** $5xy + 6 + 3yx =$ ____________

**h** $7xy + 2 + yx =$ ____________

**i** $xy + 5 + xy =$ ____________

**4** **a** $\boxed{3ab} + 5 \boxed{+ 2ab} + 1 =$ ______$ab +$ ______

**b** $\boxed{4ab} + 2 \boxed{+ 3ab} + 7 =$ ______$ab +$ ______

**c** $6ab + 4 + 2ab + 3 =$ ____________

**d** $5ab + 7 + 3ab + 2 =$ ____________

**e** $2ab + 3 + 4ba + 5 =$ ____________

**f** $4ab + 6 + ab + 2 =$ ____________

**5** **a** $\boxed{9a} + 2b \boxed{+ 2a} + 5b =$ ______$a +$ ______$b$

**b** $\boxed{10a} + 3b \boxed{+ 5a} + 6b =$ ______$a +$ ______$b$

**c** $8a + 3b + 2a + 4b =$ ____________

**d** $4a + 4b + 3a + 2b =$ ____________

**e** $7a + 3b + a + 2b =$ ____________

**f** $2a + 5b + 4a + b =$ ____________

**6** **a** $\boxed{2w} + 5vw + 3vw \boxed{+ 4w} =$ ______$w +$ ______$vw$

**b** $\boxed{3w} + 4vw + 7vw \boxed{+ 6w} =$ ______$w +$ ______$vw$

**c** $7w + 5vw + 2vw + 3w =$ ____________

**d** $10w + 6vw + 3vw + 2w =$ ____________

**e** $9w + 3vw + vw + 2w =$ ____________

**f** $4w + 5vw + 2wv + 6w=$ ____________

**7** **a** $\boxed{2x^2} + 3x \boxed{+ 5x^2} =$ ______$x^2 +$ ______$x$

**b** $\boxed{8x^2} + 3x \boxed{+ 2x^2} =$ ______$x^2 +$ ______$x$

**c** $4x^2 + 6x + 3x^2 =$ ____________

**d** $3x^2 + 2x + x^2 =$ ____________

**e** $5x^2 + 3x + x^2 =$ ____________

**f** $x^2 + x + x^2 =$ ____________

**8** **a** $\boxed{6a^2} + 2a \boxed{+ 2a^2} + 4a =$ ______$a^2 +$ ______$a$

**b** $\boxed{4a^2} + 3a \boxed{+ 5a^2} + 7a =$ ______$a^2 +$ ______$a$

**c** $8a^2 + 4a + 3a^2 + 2a =$ ____________

**d** $4a^2 + 5a + 2a^2 + a =$ ____________

**e** $5a^2 + 3a + 4a^2 + a =$ ____________

**f** $3a^2 + a + 3a^2 + 5a =$ ____________

UNIT 4

**ALGEBRA**

# Adding and subtracting terms

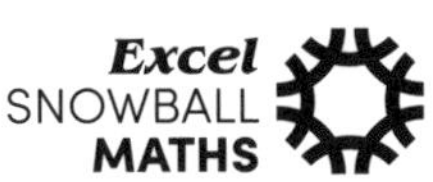

**1** Simplify the following expressions:

**a** $3x + 2y + 2x + 5y + 1$

$= \_\_\_x + \_\_\_y + \_\_\_$

**b** $4x + 3y + 2x + y + 2$

$= \_\_\_x + \_\_\_y + \_\_\_$

**c** $7x + 4y + 2x + 3y + 5$

$=$ __________

**d** $6x + 5y + x + 2y + 4$

$=$ __________

**e** $5x + 4y + 3x + y + 7$

$=$ __________

**f** $7x + 5y + x + 2y + 9$

$=$ __________

**2** Simplify the following expressions:

**a** $5a - 3a + 2b = \_\_\_a + \_\_\_b$

**b** $8a - 2a + 6b = \_\_\_a + \_\_\_b$

**c** $7a - 3a + 5b =$ __________

**d** $9a - a + 2b =$ __________

**e** $8a - 3a + b =$ __________

**f** $2a - a + 2b =$ __________

**3** Simplify the following expressions:

**a** $3x + 2y - 5x = \_\_\_x + \_\_\_y$

**b** $4x + 5y - 7x = \_\_\_x + \_\_\_y$

**c** $2x + 7y - 6x =$ __________

**d** $3x + 5y - 8x =$ __________

**e** $4x + y - 5x =$ __________

**f** $x + 2y - 6x =$ __________

**4** Simplify the following expressions:

**a** $5a - 4b - 2a + 2b = \_\_\_a - \_\_\_b$

**b** $6a - 5b - 4a + 3b = \_\_\_a - \_\_\_b$

**c** $9a - 8b - 5a + 4b =$ __________

**d** $6a - 9b - 5a + 2b =$ __________

**e** $4a - 7b - a + 3b =$ __________

**f** $3a - 5b - a + 2b =$ __________

**g** $8a - 6b - 2a + 3b =$ __________

**h** $10a - 4b - 6a + 2b =$ __________

**i** $12a - 8b - 7a + b =$ __________

**j** $13a - 2b - 10a + b =$ __________

**5** Simplify the following expressions:

**a** $-4h + 5j - 2h - 2j = \_\_\_h + \_\_\_j$

**b** $-6h + 7j - 3h - 4j = \_\_\_h + \_\_\_j$

**c** $-2h + 8j - 3h - 5j =$ __________

**d** $-4h + 9j - 6h - 3j =$ __________

**e** $-3h + 10j - h - 6j =$ __________

**f** $-7h + 3j - 2h - j =$ __________

UNIT 5

# ALGEBRA
## Adding and subtracting terms

**1** Simplify the following expressions:

a $\boxed{5ab} - 3a \boxed{-2ab} - 7a =$ ______ $ab -$ ______ $a$

b $\boxed{6ab} - 5a \boxed{-4ab} - 2a =$ ______ $ab -$ ______ $a$

c $8ab - 5a - 3ab - 4a =$ ____________

d $7ab - 2a - 3ab - 5a =$ ____________

e $9ab - 4a - ab - 8a =$ ____________

f $10ab - 5a - 5ba - 3a =$ ____________

g $9ab - 6a - ba - a =$ ____________

**2** Simplify the following expressions:

a $\boxed{-7pq} + 4p - 2p \boxed{+2pq} =$ ______ $pq +$ ______ $p$

b $\boxed{-9pq} + 5p - 3p \boxed{+2pq} =$ ______ $pq +$ ______ $p$

c $-8pq + 7p - 2p + 5pq =$ ____________

d $-7pq + 6p - 3p + 3pq =$ ____________

e $-8pq + 9p - 5p + 2qp =$ ____________

f $-9pq + 9p - 7p + 4qp =$ ____________

g $-6pq + 2p - p + 3pq =$ ____________

**3** Simplify the following expressions:

a $\boxed{6p^2} - 7pq + 3pq \boxed{-4p^2} =$ ______ $p^2 -$ ______ $pq$

b $\boxed{7p^2} - 6pq + 4pq \boxed{-5p^2} =$ ______ $p^2 -$ ______ $pq$

c $9p^2 - 6pq + 3pq - 5p^2 =$ ____________

d $7p^2 - 5pq + 3pq - 2p^2 =$ ____________

e $10p^2 - 4pq + 3qp - 6p^2$

$=$ ____________

f $12p^2 - 8pq + 2qp - 7p^2$

$=$ ____________

g $11p^2 - 7pq + pq - 3p^2$

$=$ ____________

**4** Simplify the following expressions:

a $\boxed{8a^2} - 3a \boxed{-4a^2} + 5a$

$=$ ______ $a^2 +$ ______ $a$

b $\boxed{9a^2} - 2a \boxed{-4a^2} + 6a$

$=$ ______ $a^2 +$ ______ $a$

c $6a^2 - 4a - 5a^2 + 7a =$ ____________

d $9a^2 - 7a - 3a^2 + 9a =$ ____________

e $10a^2 - 2a - 5a^2 + 6a =$ ____________

f $12a^2 - 6a - 4a^2 + 7a =$ ____________

g $11a^2 - a - 5a^2 + 5a =$ ____________

**5** Simplify the following expressions:

a $\boxed{8x^2} - 7x \boxed{-3x^2} - 2x \,(+4)$

$=$ ______ $x^2 -$ ______ $x +$ ______

b $\boxed{6x^2} - 5x \boxed{-4x^2} - 6x \,(+9)$

$=$ ______ $x^2 -$ ______ $x +$ ______

c $8x^2 - 4x - 5x^2 - 3x + 4 =$ ____________

d $6x^2 - 8x - 2x^2 - 5x + 3 =$ ____________

e $9x^2 - 6x - 3x^2 - 2x + 7 =$ ____________

f $3x^2 - 4x - 2x^2 - x + 4 =$ ____________

g $7x^2 - 8x - 4x^2 - 2x + 8 =$ ____________

h $10x^2 - 8x - 6x^2 - 5x + 9$

$=$ ____________

UNIT 6

**ALGEBRA**

# Multiplying terms

1 Simplify the following:

a $3a \times 2 =$ ______ $a$

b $7y \times 5 =$ ______ $y$

c $4x \times 3 =$ ______

d $5w \times 7 =$ ______

e $4 \times 2e =$ ______

f $9 \times 2f =$ ______

g $3a \times (-8) =$ ______

h $5w \times (-6) =$ ______

i $(-7h) \times 8 =$ ______

j $-4b \times 5 =$ ______

k $-2a \times (-6) =$ ______

l $-5b \times (-4) =$ ______

2 Simplify the following:

a $6a \times 2b =$ ______ $ab$

b $7y \times 4w =$ ______ $yw$

c $4x \times y =$ ______

d $5w \times 9k =$ ______

e $3a \times (-2t) =$ ______

f $5w \times (-8g) =$ ______

g $(-7h) \times 3t =$ ______

h $-6t \times 2p =$ ______

i $-2a \times (-6b) =$ ______

j $-4x \times (-2y) =$ ______

3 Simplify the following:

a $5a \times 2a =$ ______ $a^{\square}$

b $7y \times 3y =$ ______ $y^{\square}$

c $4x \times 3x =$ ______

d $7w \times w =$ ______

e $4m \times 3m =$ ______

f $6h \times 4h =$ ______

g $(-5h) \times 3h =$ ______

h $-6t \times 3t =$ ______

i $-5p \times (-5p) =$ ______

j $-4q \times (-7q) =$ ______

4 Simplify the following:

a $9e \times 2em =$ ______ $e^{\square}m$

b $3f \times 2fg =$ ______ $f^{\square}g$

c $2s \times 10st =$ ______ $^{\square}$ ______

d $6g \times 6gh =$ ______ $^{\square}$ ______

e $7x \times 4xy =$ ______

f $5w \times 4wt =$ ______

g $2at \times (-8a) =$ ______

h $8wv \times (-6w) =$ ______

i $-3ph \times 8p =$ ______

j $-3ad \times a =$ ______

k $-2wv \times 6v =$ ______

l $-4dh \times -6h =$ ______

UNIT 7

**ALGEBRA**

# Multiplying terms

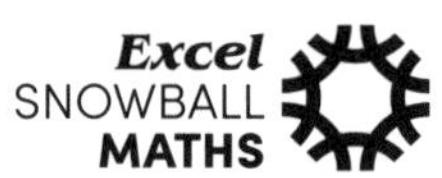

**1** Simplify the following:

**a** $5a \times 4ab \times 2b =$ ______ $a^{\square}b^{\square}$

**b** $2y \times 6yt \times 4t =$ ______ $y^{\square}t^{\square}$

**c** $5x \times 6xw \times 3w =$ ______ $^{\square}$ ______ $^{\square}$

**d** $7x \times xt \times 2t =$ ______ $^{\square}$ ______ $^{\square}$

**e** $3y \times 6ym \times 2m =$ ______

**f** $4u \times uv \times 5v =$ ______

**g** $2nt \times (-9n) \times 5t =$ ______

**h** $8wv \times (-3w) \times 6v =$ ______

**i** $(-3ph) \times 2p \times 6h =$ ______

**j** $-5pq \times 3p \times 6q =$ ______

**k** $-4xy \times 3y \times (-2x) =$ ______

**l** $-5ab \times 2b \times (-3a) =$ ______

**m** $4zk \times (-8k) \times 6z =$ ______

**n** $9ap \times (-2p) \times 5 =$ ______

**o** $8v \times 5w \times (-6v) =$ ______

**2** Simplify the following:

**a** $4a^2 \times 2a =$ ______ $a^{\square}$

**b** $2y^2 \times 8y =$ ______ $y^{\square}$

**c** $5x \times 6x^2 =$ ______

**d** $w \times 7w^2 =$ ______

**e** $4b \times 4b^2 =$ ______

**f** $3e^2 \times 2em =$ ______ $e^{\square}m$

**g** $5f^2 \times 2fg =$ ______ $f^{\square}g$

**h** $2z^2 \times 7zt =$ ______

**i** $6g^2 \times gh =$ ______

**j** $6at \times (-8a^2) =$ ______

**k** $2wv \times (-6w^2) =$ ______

**l** $-2hp \times (-8p^2) =$ ______

**m** $-4ab \times (-6b^2) =$ ______

**n** $3yp \times (-2y^2) \times 4p =$ ______

**o** $5kd \times (-4d^2) \times 2k =$ ______

**p** $6xy^2 \times y \times (-3x) =$ ______

**3** Simplify:

**a** $8abc \times 2ab =$ ______ $^{\square}$ ______ $^{\square}$ ______

**b** $6xyz \times 4xy =$ ______ $^{\square}$ ______ $^{\square}$ ______

**c** $5srt \times 3sr =$ ______

**d** $7acd \times 2ca =$ ______

**e** $10khj \times (-3hk) =$ ______

**f** $-9cde \times 4de =$ ______

**g** $3mnp \times (-6np) =$ ______

**h** $-7dgh \times (-3hg) =$ ______

**i** $4cd \times 2ce \times 3e =$ ______

**j** $5kg \times 4kw \times 6w =$ ______

**k** $-2wp \times 6pz \times 4z =$ ______

**l** $4rs \times (-6st) \times 2t =$ ______

**m** $6tmv \times 5t \times (-9m) =$ ______

**n** $8rpt \times 3p \times 7t =$ ______

**o** $-4zl \times 5kl \times 7z =$ ______

**p** $-10pq \times 5p \times (-8qr) =$ ______

UNIT 8

ALGEBRA

# Dividing terms

**1** Simplify the following:

a $6a \div 2 =$ ______$a$

b $35y \div 7 =$ ______$y$

c $12x \div 3 =$ ______

d $35w \div 5 =$ ______

e $10e \div 2 =$ ______

f $40z \div 10 =$ ______

g $(-24a) \div 3 =$ ______

h $-30w \div 6 =$ ______

i $(-56w) \div (-7) =$ ______

j $-72a \div -6 =$ ______

**2** Simplify the following:

a $\frac{20a}{4} =$ ______$a$ b $\frac{12j}{4} =$ ______$j$

c $\frac{15y}{3} =$ ______ d $\frac{16t}{4} =$ ______

e $\frac{8m}{2} =$ ______ f $\frac{18a}{9} =$ ______

g $\frac{36h}{-4} =$ ______ h $\frac{-20t}{5} =$ ______

**3** Simplify the following:

a $\frac{20}{5b} = \frac{\square}{b}$ b $\frac{10}{5n} = \frac{\square}{n}$

c $\frac{24}{3c} = \frac{\square}{\square}$ d $\frac{9}{3t} = \frac{\square}{\square}$

e $\frac{8}{2m} =$ ______ f $\frac{12}{3a} =$ ______

g $\frac{12}{-6a} =$ ______ h $\frac{-18}{6c} =$ ______

i $\frac{-36}{4h} =$ ______ j $\frac{-25}{5b} =$ ______

**4** Simplify the following:

a $\frac{4a}{20} = \frac{a}{\square}$ b $\frac{2j}{12} = \frac{j}{\square}$

c $\frac{3y}{15} = \frac{\square}{\square}$ d $\frac{5t}{30} = \frac{\square}{\square}$

e $\frac{2m}{8} =$ ______ f $\frac{9a}{18} =$ ______

g $\frac{12x}{18} = \frac{\square\, x}{\square}$ h $\frac{15y}{40} = \frac{\square\, y}{\square}$

i $\frac{14b}{8} =$ ______ j $\frac{20w}{15} =$ ______

**5** Simplify the following:

a $\frac{20\not{a}}{5\not{a}} =$ ______ b $\frac{16\not{b}}{4\not{b}} =$ ______

c $\frac{30x}{10x} =$ ______ d $\frac{24a}{8a} =$ ______

e $\frac{2\not{a}}{14\not{a}} = \frac{1}{\square}$ f $\frac{2\not{b}}{16\not{b}} = \frac{1}{\square}$

g $\frac{5x}{20x} =$ ______ h $\frac{7a}{35a} =$ ______

i $\frac{6a}{9a} =$ ______ j $\frac{16n}{20n} =$ ______

k $\frac{-20p}{12p} =$ ______ l $\frac{25w}{-15w} =$ ______

**6** Simplify the following:

a $\frac{35a\not{t}}{7\not{t}} =$ ______$a$ b $\frac{15h\not{w}}{5\not{w}} =$ ______$h$

c $\frac{25bg}{5g} =$ ______ d $\frac{30ym}{6m} =$ ______

e $\frac{24pq}{6q} =$ ______ f $\frac{40rm}{8m} =$ ______

g $\frac{-18cd}{2c} =$ ______ h $\frac{-40nh}{5n} =$ ______

**ALGEBRA**

# Dividing terms

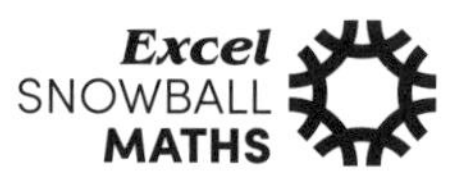


**1** Simplify the following:

a $\frac{7pq}{49p} = \frac{q}{\square}$ b $\frac{7at}{35t} = \frac{a}{\square}$

c $\frac{5hw}{15w} = \frac{\square}{\square}$ d $\frac{10bg}{20b} = \frac{\square}{\square}$

e $\frac{6ym}{18y} =$ ______ f $\frac{5pq}{30q} =$ ______

g $\frac{8rs}{32s} =$ ______ h $\frac{7cd}{35c} =$ ______

i $\frac{15vw}{20v} =$ ______ j $\frac{14kl}{16k} =$ ______

k $\frac{24st}{-20t} =$ ______ l $\frac{32xy}{-24y} =$ ______

**2** Simplify the following:

a $\frac{5ab}{20abc} = \frac{1}{\square}$ b $\frac{4xy}{16xyz} = \frac{1}{\square}$

c $\frac{5fg}{25fgh} =$ ______ d $\frac{7pq}{28qpr} =$ ______

e $\frac{3mn}{21nmp} =$ ______ f $\frac{4tx}{8xvt} =$ ______

g $\frac{2uv}{20wvu} =$ ______ h $\frac{8cb}{24cwb} =$ ______

i $\frac{-4ap}{28ahp} =$ ______ j $\frac{-6gb}{36bgh} =$ ______

**3** Simplify the following:

a $\frac{20a^2}{15a} = \frac{\square\, a}{\square}$ b $\frac{14b^2}{16b} = \frac{\square\, b}{\square}$

c $\frac{10x^2}{18x} =$ ______ d $\frac{35a^2}{25a} =$ ______

e $\frac{24e^2}{16e} =$ ______ f $\frac{20a^2}{45a} =$ ______

g $\frac{4b^2}{20b} =$ ______ h $\frac{6x^2}{36x} =$ ______

i $\frac{7w^2}{28w} =$ ______ j $\frac{3b^2}{21b} =$ ______

k $\frac{8p^2}{-24p} =$ ______ l $\frac{12k^2}{-60k} =$ ______

**4** Simplify the following:

a $\frac{15x^2}{20xy} = \frac{\square\, x}{\square\, y}$ b $\frac{8d^2}{12de} = \frac{\square\, d}{\square\, e}$

c $\frac{12v^2}{16vw} =$ ______ d $\frac{6z^2}{15yz} =$ ______

e $\frac{25p^2}{15qp} =$ ______ f $\frac{18l^2}{12lp} =$ ______

g $\frac{15s^2}{3rs} =$ ______ h $\frac{14g^2}{7hg} =$ ______

i $\frac{40e^2}{-5ej} =$ ______ j $\frac{-32y^2}{4yx} =$ ______

**5** Simplify the following:

a $\frac{10nh}{25n^2} = \frac{\square}{\square}$ b $\frac{16ab}{24a^2} = \frac{\square}{\square}$

c $\frac{20cd}{24c^2} =$ ______ d $\frac{18xy}{21x^2} =$ ______

e $\frac{30yz}{35z^2} =$ ______ f $\frac{24ab}{36a^2} =$ ______

g $\frac{36cd}{12c^2} =$ ______ h $\frac{24ap}{2a^2} =$ ______

i $\frac{45sb}{-9s^2} =$ ______ j $\frac{-36tk}{4k^2} =$ ______

**6** Simplify the following:

a $\frac{10x^2yz}{15xy^2} = \frac{\square}{\square}$ b $\frac{9a^2bc}{12ab^2} = \frac{\square}{\square}$

c $\frac{8m^2np}{12mn^2} =$ ______ d $\frac{12c^2de}{20cd^2} =$ ______

e $\frac{3b^2cd}{15bc^2} =$ ______ f $\frac{21q^2pr}{7qp^2} =$ ______

UNIT 10

**ALGEBRA**

# Writing algebraic expressions

**1** Write the following statements as algebraic expressions:

- **a** 5 more than $x$ $x +$ ______
- **b** 4 more than $y$ $y +$ ______
- **c** 3 more than $b$ ______
- **d** 7 more than $h$ ______
- **e** $y$ more than $x$ ______
- **f** Increase $x$ by 3 ______
- **g** Increase $c$ by 5 ______
- **h** Increase $w$ by 2 ______
- **i** Increase $2m$ by 7 ______
- **j** Increase $3p$ by $q$ ______
- **k** The sum of $3h$ and 9 ______
- **l** The sum of $7m$ and 6 ______
- **m** The sum of $4g$ and 3 ______
- **n** The sum of $6n$ and $2v$ ______
- **o** The sum of $2a$ and $5r$ ______
- **p** Three more than $x$ ______
- **q** Increase $4y$ by two ______
- **r** The sum of $7g$ and six ______
- **s** Add $9k$ and five ______

**2** Write the following statements as algebraic expressions:

- **a** 5 less than $x$ $x -$ ______
- **b** 4 less than y $y -$ ______
- **c** 3 less than $b$ ______
- **d** 7 less than $h$ ______
- **e** $y$ less than $x$ ______
- **f** Decrease 3 by $x$ $3 -$ ______
- **g** Decrease 5 by $c$ $5 -$ ______
- **h** Decrease 12 by $w$ ______
- **i** Decrease $7b$ by $m$ ______
- **j** Decrease $5p$ by $q$ ______
- **k** The difference of $3h$ and 9 ______
- **l** The difference of $4m$ and 7 ______
- **m** The difference of $6g$ and 4 ______
- **n** The difference of $2n$ and $3m$ ______
- **o** The difference of $8a$ and $4r$ ______
- **p** Three less than $x$ ______
- **q** Decrease $y$ by two ______
- **r** The difference of $2g$ and six ______
- **s** Subtract seven from $10x$ ______

**3** Write the following statements as algebraic expressions:

- **a** The product of $h$ and 9 9______
- **b** The product of $m$ and 7 7______
- **c** The product of $s$ and 3 ______
- **d** The product of $d$ and 4 ______
- **e** The product of $a$ and –6 ______
- **f** The product of $t$ and –2 ______

UNIT 11

ALGEBRA

# Writing algebraic expressions

**1** Write the following statements as algebraic expressions:

- **a** The product of $r$ and $-5$ ________
- **b** The product of $e$ and $-1$ ________
- **c** The product of $x$ and $y$ ________
- **d** The product of $m$ and $-n$ ________
- **e** Twice a number $x$ ________
- **f** Twice a number $n$ ________
- **g** Double the number $p$ ________
- **h** Triple the number $w$ ________
- **i** The product of $a$ itself $a \times a =$ ________
- **j** The product of $x$ itself ________
- **k** The product of $n$ itself ________
- **l** Square of $r$ ________
- **m** Square of $m$ ________
- **n** Half of $k$ $\frac{k}{\square}$
- **o** Half of $a$ ________
- **p** Half of $x$ ________
- **q** Quarter of $y$ ________
- **r** Third of $w$ ________

**2** Write the following statements as algebraic expressions:

- **a** $h$ divided by 3 $\frac{\square}{\square}$
- **b** $a$ divided by 5 ________
- **c** $w$ divided by $4y$ ________
- **d** $t$ divided by $6e$ ________
- **e** The quotient of $h$ and 9 $\frac{\square}{9}$
- **f** The quotient of $m$ and 7 ________
- **g** The quotient of $c$ and $5d$ ________
- **h** The quotient of $n$ and $2p$ ________
- **i** The quotient of $-5$ and $k$ ________
- **j** The quotient of $-p$ and $q$ ________

**3** Write the following statements as algebraic expressions:

- **a** Double the number $x$ plus $3y$ ________
- **b** Double the number $m$ plus $5n$ ________
- **c** Twice the number $p$ plus $4r$ ________
- **d** Triple the number $y$ minus $6z$ ________
- **e** Triple the number $c$ minus $7e$ ________
- **f** The quotient of $x$ and 3 decreased by 1 ________
- **g** The quotient of $y$ and 5 decreased by 3 ________
- **h** Half the number $a$ plus $3b$ ________
- **i** Half the number $t$ plus $6s$ ________
- **j** A third of the number $v$ plus $9w$ ________
- **k** The product of $x$ and $y$ increased by 3 ________
- **l** The product of $a$ and $b$ increased by 4 ________
- **m** The product of 2 and $p$ decreased by $5q$ ________

## ALGEBRA

# Writing algebraic expressions

1 Write an expression for the following:

a The total number of students if there are $x$ number of girls and y number of boys

b The total number of stationery items if there are $m$ number of pencils and $n$ number of pens

c The total number of animals if there are $r$ number of rabbits and $c$ number of cats

d The amount of change I receive if I pay \$20 for a pencil case that costs \$$y$

e The amount of change I receive if I pay \$10 for a sandwich that costs \$$x$

f The amount of change I receive if I pay \$40 for a bag that costs \$$w$

g The number of water bottles I can buy with \$50 if they cost \$$w$ each

h The number of tickets I can buy with \$25 if they cost \$$c$ each

i The number of books I can buy with \$30 if they cost \$$b$ each

j The total cost of 10 envelopes if the cost is \$$p$ per envelope

k The total cost if there are 15 children and it costs \$$x$ for each child to enter the swimming pool

l The total cost of a dozen eggs if each egg costs \$$e$

m The total cost of buying $x$ drinks if each drink costs \$9.50

n The total cost of buying $y$ snacks if each snack costs \$2.99

o The total cost of buying x drinks and $y$ snacks, when each drink costs \$5.40 and each snack costs \$3.10

p The total cost of pool entry for $m$ adults and $n$ children, when adult tickets are \$9.30 and child tickets are \$6

q The amount of change I receive if I pay \$60 for 5 tickets that cost \$$c$ each

r The amount of change I receive if I pay \$100 for 4 jackets that cost \$$j$ each

# One-step equations

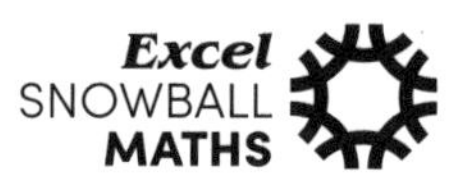

1 Solve the following equations:

**a** $x + 3 = 7$
$x = 7 -$ ____
$=$ ____

**b** $x + 4 = 6$
$x = 6 -$ ____
$=$ ____

**c** $x + 5 = 9$
$x = 9$ ____
$=$ ____

**d** $x + 6 = 11$
$x = 11$ ____
$=$ ____

**e** $x + 7 = 13$
$x =$ ____
$=$ ____

**f** $x + 9 = 14$
$x =$ ____
$=$ ____

**g** $x + 8 = 16$
____ $=$ ____
$=$ ____

**h** $x + 6 = 15$
____ $=$ ____
$=$ ____

**i** $x + 9 = 17$
____ $=$ ____
$=$ ____

**j** $x + 11 = 23$
____ $=$ ____
$=$ ____

**k** $x + 5 = 2$
$x = 2$ ____
$=$ ____

**l** $x + 4 = 3$
$x = 3$ ____
$=$ ____

**m** $x + 9 = 4$
$x =$ ____
$=$ ____

**n** $x + 7 = 3$
$x =$ ____
$=$ ____

**o** $x + 5 = 4$
____ $=$ ____
$=$ ____

**p** $x + 6 = 3$
____ $=$ ____
$=$ ____

**q** $x + 8 = 2$
____
____

**r** $x + 7 = 1$
____
____

**s** $x + 10 = 5$
____
____

**t** $x + 11 = 9$
____
____

2 Solve the following equations:

**a** $x + 4 = -3$
$x = -3$ ____
$=$ ____

**b** $x + 6 = -2$
$x = -2$ ____
$=$ ____

**c** $x + 7 = -1$
____ $=$ ____
$=$ ____

**d** $x + 10 = -5$
____ $=$ ____
$=$ ____

**e** $x + 6 = -6$
____ $=$ ____
$=$ ____

**f** $x + 3 = -9$
____ $=$ ____
$=$ ____

**g** $x + 4 = -5$
____
____

**h** $x + 5 = -7$
____
____

**i** $x + 9 = -6$
____
____

**j** $x + 8 = -4$
____
____

**k** $x + 3 = -12$
____
____

**l** $x + 8 = -15$
____
____

**m** $x + 10 = -15$
____
____

**n** $x + 12 = -14$
____
____

UNIT 14

ALGEBRA

# One-step equations

**1** Solve the following equations:

**a** $x - 3 = 4$

$x = 4 + $ ____

$= $ ____

**b** $x - 4 = 2$

$x = 2 + $ ____

$= $ ____

**c** $x - 5 = 3$

$x = 3$ ____

$= $ ____

**d** $x - 6 = 3$

$x = 3$ ____

$= $ ____

**e** $x - 7 = 6$

$x = $ ____

$= $ ____

**f** $x - 9 = 5$

$x = $ ____

$= $ ____

**g** $x - 6 = 7$

____ $= $ ____

$= $ ____

**h** $x - 6 = 9$

____ $= $ ____

$= $ ____

**i** $x - 8 = 5$

**j** $x - 12 = 13$

**k** $x - 5 = -2$

**l** $x - 4 = -3$

**m** $x - 6 = -4$

**n** $x - 7 = -2$

**o** $x - 5 = -4$

**p** $x - 9 = -6$

**2** Solve the following equations:

**a** $2x = 4$

$x = \frac{4}{\square}$

$= $ ____

**b** $4x = 8$

$x = \frac{8}{\square}$

$= $ ____

**c** $3x = 12$

$x = \frac{\square}{\square}$

$= $ ____

**d** $5x = 20$

$x = \frac{\square}{\square}$

$= $ ____

**e** $6x = 24$

**f** $3x = 21$

**g** $5x = 15$

**h** $9x = 18$

**i** $6x = -3$

$x = \frac{-3}{\square}$

**j** $10x = -5$

$x = \frac{-5}{\square}$

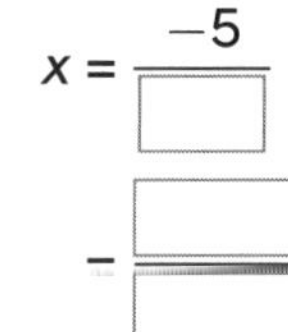

**k** $9x = -3$

**l** $25x = -5$

**m** $30x = -6$

**n** $27x = -3$

## ALGEBRA

# One-step equations

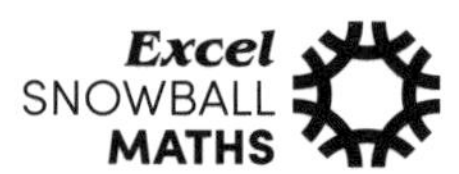

**1** Solve the following equations:

**a** $\frac{x}{4} = 3$

$x = 3 \times$ ____

= ____

**b** $\frac{x}{2} = 5$

$x = 5 \times$ ____

= ____

**c** $\frac{x}{7} = 2$

$x =$ ________

= ____

**d** $\frac{x}{3} = 6$

$x =$ ________

= ____

**e** $\frac{x}{4} = 5$

____ = ________

= ____

**f** $\frac{x}{3} = 12$

____ = ________

= ____

**g** $\frac{x}{2} = -4$

$x = -4 \times$ ____

= ____

**h** $\frac{x}{5} = -3$

$x = -3 \times$ ____

= ____

**i** $\frac{x}{2} = -6$

**j** $\frac{x}{4} = -4$

**k** $\frac{x}{4} = -6$

**l** $\frac{x}{7} = -5$

**m** $\frac{x}{9} = -4$

**n** $\frac{x}{8} = -6$

**o** $\frac{x}{10} = -5$

**p** $\frac{x}{12} = -3$

**2** Solve the following equations:

**a** $4 + x = 7$

$x = 7 -$ ____

= ____

**b** $3 + x = 6$

$x = 6 -$ ____

= ____

**c** $5 + x = 9$

**d** $2 + x = 8$

**e** $6 + x = 10$

**f** $4 + x = 12$

**g** $6 = x + 5$

**h** $12 = x + 7$

**i** $14 = x + 6$

**j** $13 = x + 5$

**k** $3 = x + 8$

**l** $4 = x + 6$

**m** $7 = x + 11$

**n** $4 = x + 10$

**o** $-3 = x + 9$

**p** $-6 = x + 7$

UNIT 16

**ALGEBRA**

# Two-step equations

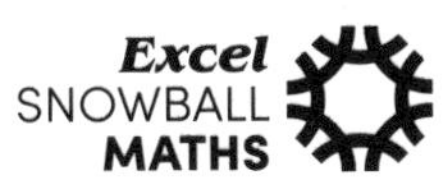


**1** Solve the following equations:

**a** $2x - 3 = 5$

$2x = 5 +$ ____

$=$ ____

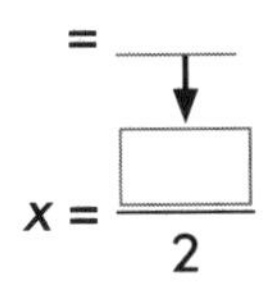

$x = \frac{\square}{2}$

$=$ ____

**b** $2x - 4 = 6$

$2x = 6 +$ ____

$=$ ____

$x = \frac{\square}{2}$

$=$ ____

**c** $2x - 5 = 3$

$2x = 3$ ____

$=$ ____

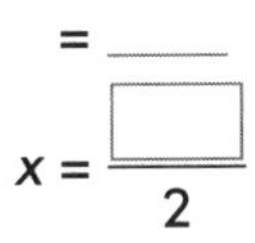

$x = \frac{\square}{2}$

$=$ ____

**d** $2x - 4 = 8$

$2x = 8$ ____

$=$ ____

$x = \frac{\square}{2}$

$=$ ____

**e** $2x - 1 = 5$

$2x =$ ____

$=$ ____

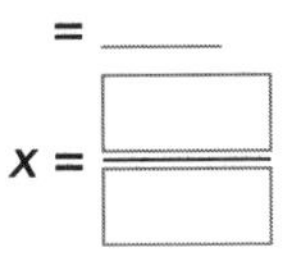

$x = \frac{\square}{\square}$

$=$ ____

**f** $2x - 4 = 10$

$2x =$ ____

$=$ ____

$x = \frac{\square}{\square}$

$=$ ____

**g** $2x - 7 = 9$

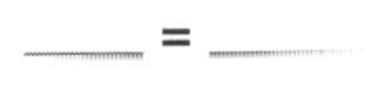

____ $=$ ____

$=$ ____

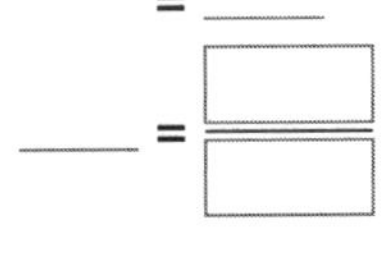

____ $= \frac{\square}{\square}$

$=$ ____

**h** $2x - 5 = 9$

____ $=$ ____

$=$ ____

____ $= \frac{\square}{\square}$

$=$ ____

**i** $2x - 11 = 9$

____

____

____

____

**j** $2x - 4 = 14$

____

____

____

____

**2** Solve the following equations:

**a** $3x + 5 = 11$

$3x = 11 -$ ____

$=$ ____

$x = \frac{\square}{3}$

$=$ ____

**b** $3x + 4 = 13$

$3x = 13 -$ ____

$=$ ____

$x = \frac{\square}{3}$

$=$ ____

**c** $3x + 3 = 15$

$3x = 15$ ____

$=$ ____

$x = \frac{\square}{3}$

$=$ ____

**d** $3x + 6 = 9$

$3x = 9$ ____

$=$ ____

$x = \frac{\square}{3}$

$=$ ____

**e** $3x + 1 = 13$

$3x =$ ____

$=$ ____

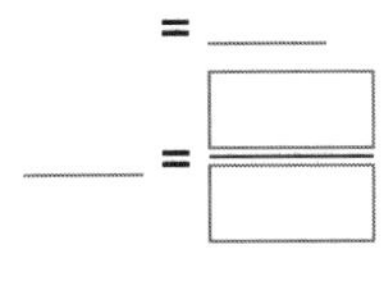

____ $= \frac{\square}{\square}$

$=$ ____

**f** $3x + 2 = 17$

$3x =$ ____

$=$ ____

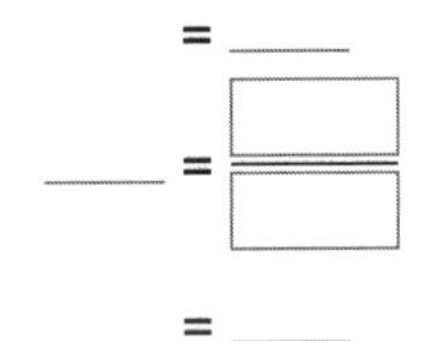

____ $= \frac{\square}{\square}$

$=$ ____

**g** $3x + 6 = 15$

____ $=$ ____

$=$ ____

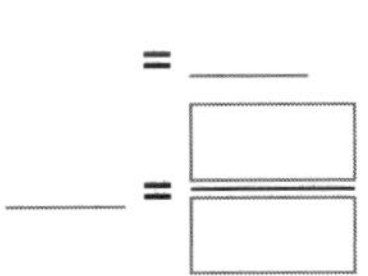

____ $= \frac{\square}{\square}$

$=$ ____

**h** $3x + 4 = 22$

____ $=$ ____

$=$ ____

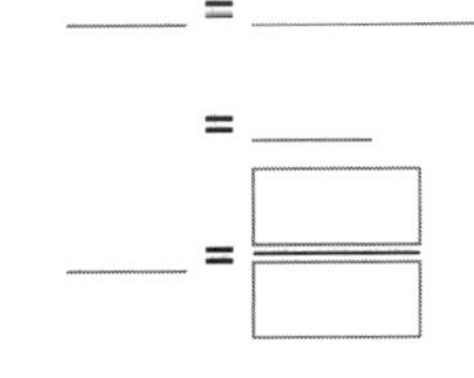

____ $= \frac{\square}{\square}$

$=$ ____

**i** $3x + 9 = 15$

____

____

____

____

**j** $3x + 6 = 27$

____

____

____

____

UNIT 17

**ALGEBRA**

# Two-step equations

**1** Solve the following equations:

**a** $4x - 2 = -10$

$4x = -10 + \_\_\_$

$= \_\_\_$

$x = \frac{\square}{4}$

$= \_\_\_$

**b** $4x - 5 = -17$

$4x = -17 + \_\_\_$

$= \_\_\_$

$x = \frac{\square}{4}$

$= \_\_\_$

**c** $4x - 3 = -7$

$4x = \_\_\_$

$= \_\_\_$

$x = \frac{\square}{4}$

$= \_\_\_$

**d** $4x - 2 = -18$

$4x = \_\_\_$

$= \_\_\_$

$x = \frac{\square}{4}$

$= \_\_\_$

**e** $4x - 1 = -9$

**f** $4x - 7 = -27$

**g** $4x - 8 = -20$

**h** $4x - 11 = -15$

**i** $5x - 4 = -19$

**j** $5x - 7 = -27$

**2** Solve the following equations:

**a** $5x + 3 = -12$

**b** $5x + 4 = -6$

**c** $5x + 3 = -2$

**d** $5x + 6 = -14$

**e** $5x + 3 = -7$

**f** $5x + 3 = -22$

**g** $6x + 4 = -8$

**h** $6x + 7 = -11$

**i** $7x + 6 = -15$

**j** $7x + 8 = -20$

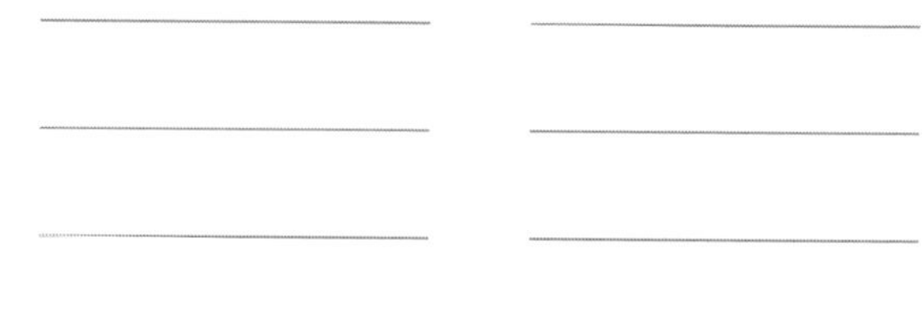

UNIT 18

**ALGEBRA**

# Two-step equations

**1** Solve for $x$:

**a** $\frac{x}{2} + 3 = 7$

$\frac{x}{2} = 7 -$ ____

$=$ ____

$x =$ ____ $\times 2$

$=$ ____

**b** $\frac{x}{2} + 5 = 8$

$\frac{x}{2} = 8 -$ ____

$=$ ____

$x =$ ____ $\times 2$

$=$ ____

**c** $\frac{x}{2} + 4 = 10$

$\frac{x}{2} = 10$ ____

$=$ ____

$x =$ ____ $\times 2$

$=$ ____

**d** $\frac{x}{2} + 7 = 8$

$\frac{x}{2} = 8$ ____

$=$ ____

$x =$ ____ $\times 2$

$=$ ____

**e** $\frac{x}{2} + 6 = 9$

$\frac{x}{2} =$ ____

$=$ ____

$x =$ ____ $\times$ ____

$=$ ____

**f** $\frac{x}{2} + 5 = 12$

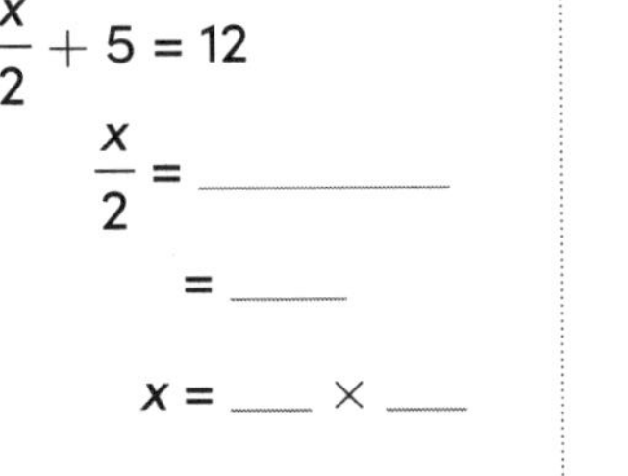

$\frac{x}{2} =$ ____

$=$ ____

$x =$ ____ $\times$ ____

$=$ ____

**g** $\frac{x}{2} + 6 = 12$

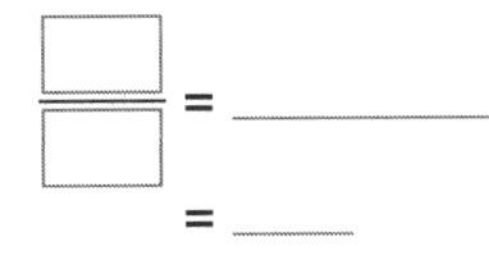

$=$ ____

$=$ ____

____ $=$ ____

$=$ ____

**h** $\frac{x}{2} + 7 = 11$

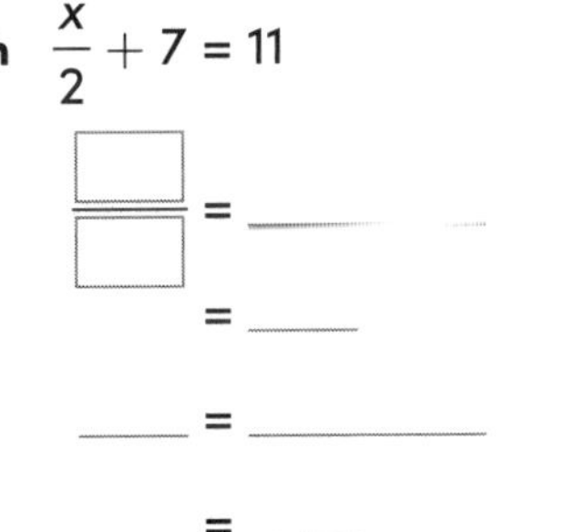

$=$ ____

$=$ ____

____ $=$ ____

$=$ ____

**i** $\frac{x}{2} + 3 = 14$

**j** $\frac{x}{2} + 5 = 13$

**2** Solve for $x$:

**a** $\frac{x}{3} - 4 = 1$

$\frac{x}{3} = 1 +$ ____

$=$ ____

$x =$ ____ $\times 3$

$=$ ____

**b** $\frac{x}{3} - 6 = 2$

$\frac{x}{3} = 2 +$ ____

$=$ ____

$x =$ ____ $\times 3$

$=$ ____

**c** $\frac{x}{3} - 3 = 4$

$\frac{x}{3} = 4$ ____

$=$ ____

$x =$ ____ $\times 3$

$=$ ____

**d** $\frac{x}{3} - 6 = 3$

$\frac{x}{3} = 3$ ____

$=$ ____

$x =$ ____ $\times 3$

$=$ ____

**e** $\frac{x}{3} - 5 = 5$

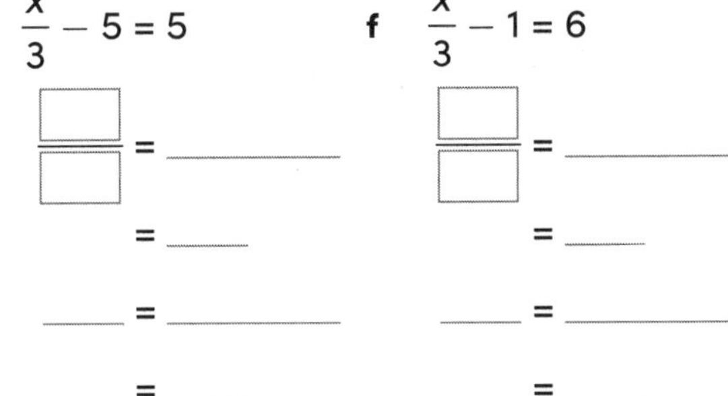

$=$ ____

$=$ ____

____ $=$ ____

$=$ ____

**f** $\frac{x}{3} - 1 = 6$

$=$ ____

$=$ ____

____ $=$ ____

$=$ ____

**g** $\frac{x}{3} - 9 = 2$

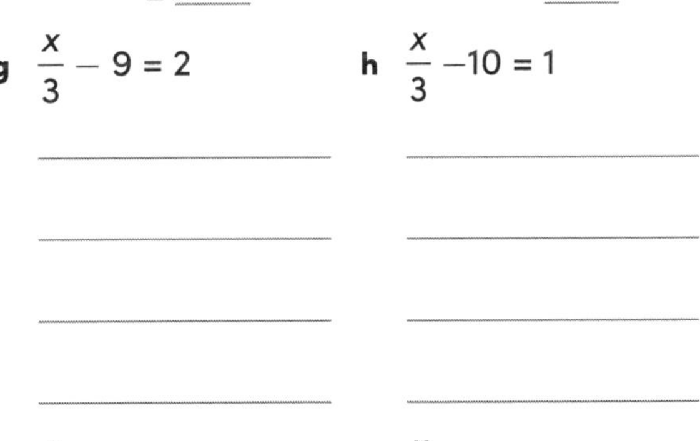

**h** $\frac{x}{3} -10 = 1$

**i** $\frac{x}{4} - 3 = 8$

**j** $\frac{x}{5} - 6 = 2$

ALGEBRA

# Two-step equations

**1** Solve for $x$:

**a** $\frac{3x}{2} = 5$

$3x = 5 \times$ ____

$=$ ____

$x = \frac{\square}{\square}$

**b** $\frac{3x}{4} = 2$

$3x = 2 \times$ ____

$=$ ____

$x = \frac{\square}{\square}$

**c** $\frac{3x}{7} = 4$

$3x = 4$ ____

$=$ ____

$x = \frac{\square}{\square}$

**d** $\frac{3x}{8} = 5$

$3x = 5$ ____

$=$ ____

$x = \frac{\square}{\square}$

**e** $\frac{3x}{9} = 1$

$3x =$ ____

$=$ ____

$x = \frac{\square}{\square}$

$=$ ____

**f** $\frac{3x}{5} = 6$

$3x =$ ____

$=$ ____

$x = \frac{\square}{\square}$

$=$ ____

**g** $\frac{3x}{4} = 6$

____ $=$ ____

$=$ ____

____ $= \frac{\square}{\square}$

$=$ ____

**h** $\frac{3x}{8} = 9$

____ $=$ ____

$=$ ____

____ $= \frac{\square}{\square}$

$=$ ____

**i** $\frac{3x}{7} = 9$

**j** $\frac{3x}{10} = 2$

**2** Solve for $x$:

**a** $\frac{2x}{5} = 3$

**b** $\frac{2x}{7} = 5$

**c** $\frac{2x}{3} = -4$

**d** $\frac{2x}{7} = -6$

**e** $\frac{3x}{7} = -5$

**f** $\frac{3x}{8} = -6$

**g** $\frac{4x}{5} = -8$

**h** $\frac{4x}{9} = -3$

**i** $\frac{5x}{6} = -4$

**j** $\frac{6x}{7} = -12$

UNIT 20

# ALGEBRA
## Two-step equations

**1** Solve:

**a** $\frac{y-1}{2} = 5$

$y - 1 = 5 \times$ ____

= ____

$y =$ ____ + ____

= ____

**b** $\frac{y-2}{5} = 3$

$y - 2 = 3 \times$ ____

= ____

$y =$ ____ + ____

= ____

**c** $\frac{y-6}{3} = 4$

$y - 6 = 4$ ____

= ____

$y =$ ____ + ____

= ____

**d** $\frac{y-5}{3} = 2$

$y - 5 = 2$ ____

= ____

$y =$ ____ + ____

= ____

**e** $\frac{y-4}{7} = 3$

$y - 4 =$ ____

= ____

$y =$ ____

= ____

**f** $\frac{y-7}{4} = 3$

$y - 7 =$ ____

= ____

$y =$ ____

= ____

**g** $\frac{y-5}{9} = 2$

**h** $\frac{y-6}{5} = 4$

**i** $\frac{y-8}{6} = 2$

**j** $\frac{y-9}{5} = 7$

**k** $\frac{y-6}{2} = -7$

**l** $\frac{y-5}{3} = -9$

**m** $\frac{y-5}{5} = -3$

**n** $\frac{y-9}{10} = -2$

**o** $\frac{a+7}{6} = 8$

**p** $\frac{a+5}{8} = 3$

**q** $\frac{4+n}{7} = 9$

**r** $\frac{11+n}{5} = 7$

**s** $\frac{2+m}{5} = -4$

**t** $\frac{3+m}{7} = -6$

UNIT 21

CHAPTER 5 REVIEW

# Algebra

1 Simplify the following expressions:

a $3a + 4a =$ ______ b $4t + 2t =$ ______

c $9x^2 - 4x^2 =$ ______ d $10a^2 - a^2 =$ ______

e $6xy + 2xy =$ ______ f $5ab + 2ba =$ ______

2 Simplify the following expressions:

a $8x + 2x + 3 =$ ______

b $2x + x + 5 =$ ______

c $5a + 3d + 2a =$ ______

d $7a + 4d + a =$ ______

e $4b + 3y - 5b =$ ______

f $2b + 5y - 8b =$ ______

g $6x - 4y - 2x + 3y =$ ______

h $8x - 5y - 3x + 2y =$ ______

i $-4ab - 3a + 5a - 2ab =$ ______

j $-7ab - a + 4a - 3ab =$ ______

3 From the term $-8h$ state:

a the pronumeral ______ b the coefficient ______

4 Simplify the following:

a $5t \times 2r =$ ______ b $10w \times 3v =$ ______

c $4h \times (-3x) =$ ______ d $-5p \times 2q =$ ______

e $7a \times 5a =$ ______ f $8y \times (-4y) =$ ______

5 Simplify the following:

a $10m \div 2 =$ ______ b $12n \div (-3) =$ ______

c $\frac{20xy}{4x} =$ ______ d $-\frac{16cd}{2c} =$ ______

e $\frac{30x^2}{5x} =$ ______ f $-\frac{15a^2}{3a} =$ ______

6 Write the following statements as an algebraic expression:

a 2 more than $x$ ______

b 4 less than $y$ ______

c The product of $r$ and 6 ______

d Double the number $k$ ______

e The quotient of $d$ and 3 ______

f Half of $m$ ______

g The difference of $5x$ and 10 ______

h The sum of $7s$ and 6 ______

7 Solve the following equations:

a $a + 2 = 7$ ______ b $t + 3 = 10$ ______ c $x - 2 = 6$ ______

d $w - 1 = 4$ ______ e $3b = 15$ ______ f $4x = 32$ ______

g $\frac{c}{4} = 7$ ______ h $\frac{v}{3} = -8$ ______ i $2a - 6 = 4$ ______

j $3b - 7 = 11$ ______ k $2y + 3 = 11$ ______ l $3e + 5 = 26$ ______

m $4x - 2 = 10$ ______ n $5m - 4 = 21$ ______ o $\frac{a}{2} + 3 = 9$ ______

p $\frac{b}{2} + 7 = 10$ ______ q $\frac{c}{3} - 4 = 5$ ______ r $\frac{p}{3} - 2 = 8$ ______

s $\frac{3w}{2} = 6$ ______ t $\frac{4z}{5} = -8$ ______ u $\frac{h-3}{6} = 2$ ______

v $\frac{n-5}{7} = 3$ ______ w $\frac{m+3}{4} = -2$ ______ x $\frac{a+5}{4} = -3$ ______

UNIT 1

## MEASUREMENT
# Length and perimeter

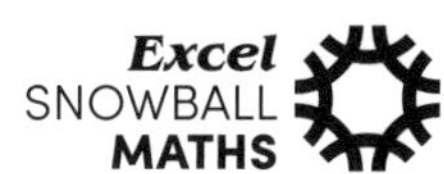

1 Measure the following lines in centimetres (cm):

a ________ cm

b ________ cm

c ________ cm

d ________ cm

2 Measure the following lines in centimetres, correct to one decimal place:

a ________ cm

b ________ cm

c ________ cm

d ________ cm

e ________ cm

f ________ cm

3 Convert all the lengths above to millimetres (mm):

(1) a 12 cm × 10 = ________ mm

b 4 cm × 10 = ________ mm

c ________ × ________ = ________

d ________ × ________ = ________

(2) a ________ × ________ = ________

b ________ × ________ = ________

c ________ × ________ = ________

d ________ × ________ = ________

e ________ × ________ = ________

f ________ × ________ = ________

4 Draw a horizontal line of each length:

a 3 cm

b 5.5 cm

c 1.9 cm

d 63 mm

e 8 mm

UNIT 2

**MEASUREMENT**

# Length and perimeter

**1** Measure the sides of each shape, in centimetres, and find its perimeter:

**Squares**

**a** Perimeter = ______

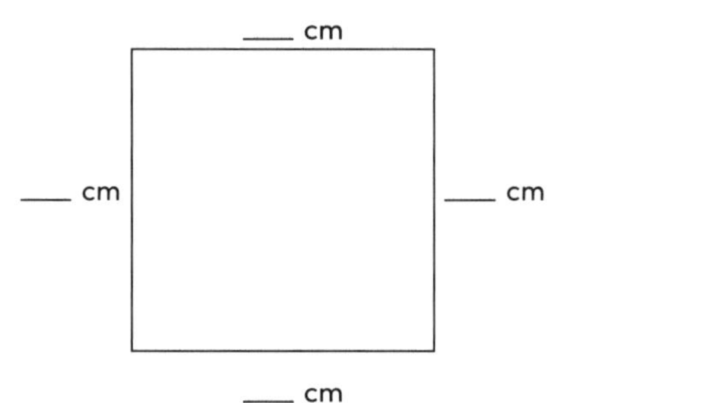


**b** Perimeter = ______

**Rectangles**

**c**

**d**

Perimeter = ______ Perimeter = ______

**Parallelograms**

**e**

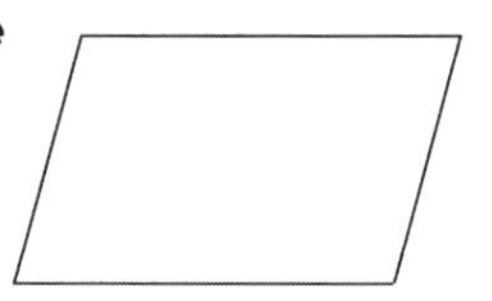

**f**

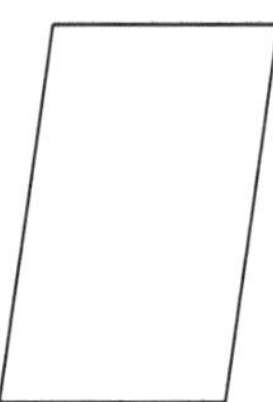

Perimeter = ______ Perimeter = ______

**Kites**

**g** **h**

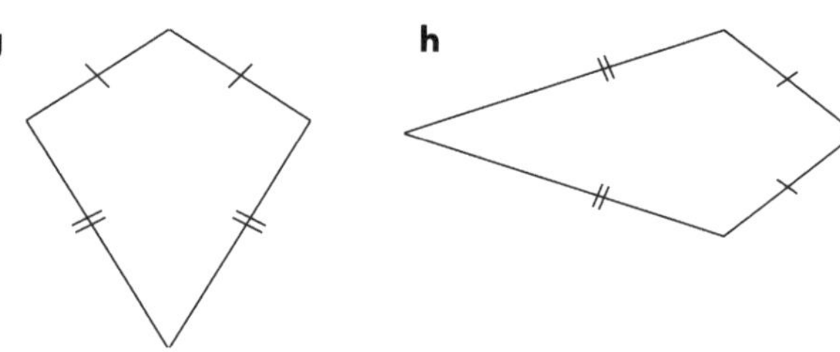

Perimeter = ______ Perimeter = ______

**2** Find the perimeter of each composite figure:

**a**

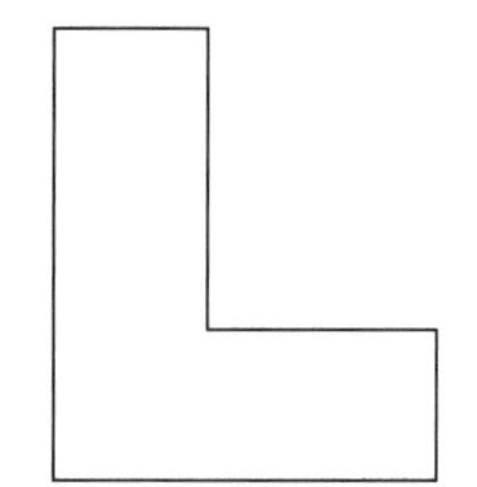

**b**

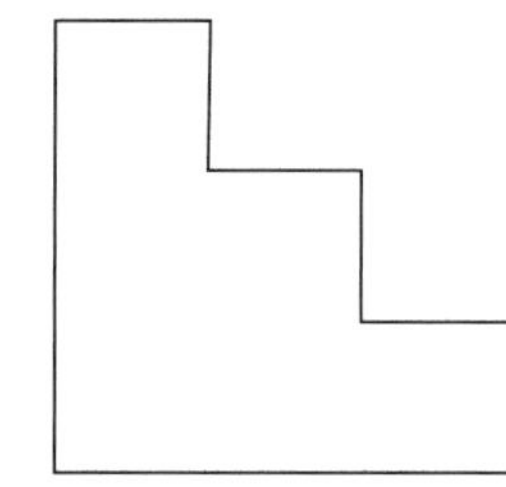

Perimeter = ______ Perimeter = ______

**c**

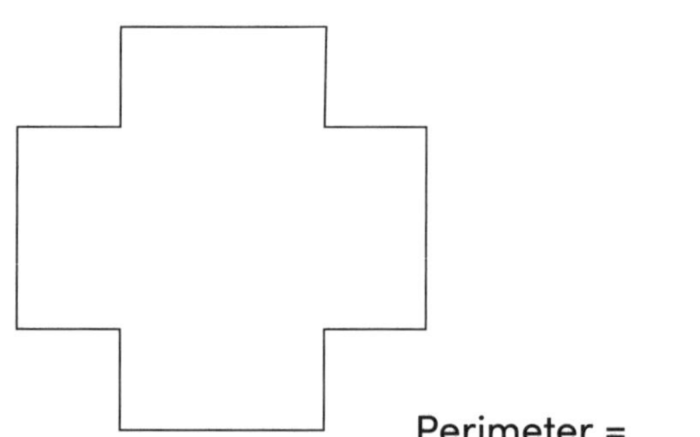

Perimeter = ______

**3** **a** Draw below a square with a side length of 2.5 cm:

**b** What is the perimeter of the square?

______

UNIT 3

## MEASUREMENT

# Length and perimeter

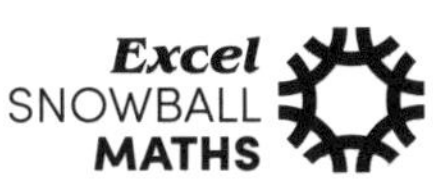


**1** A square has a perimeter of 8 cm.

**a** What is the side length?

8 ÷ 4 = ______ cm

**b** Draw this square:

**2** A square has a perimeter of 20 cm.

**a** What is the side length?

20 ÷ 4 = ______ cm

**b** Draw this square:

**3** A square has a perimeter of 16 cm.

**a** What is the side length? ______

**b** Draw this square:

**4** A square has a perimeter of 10 cm.

**a** What is the side length? ______

**b** Draw this square below:

**5** **a** Draw a rectangle with a length of 4 cm and a width of 3 cm:

**b** What is the perimeter? ______

**6** Draw a rectangle with a perimeter of 10 cm:

**MEASUREMENT**

# Perimeter

**1** Find the perimeter of each of the following shapes, in centimetres:

**a**

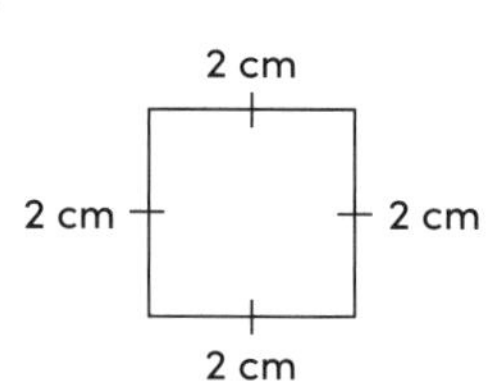


**b**

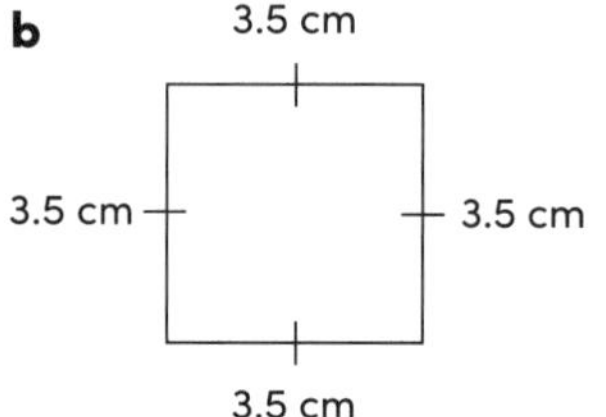


**c**

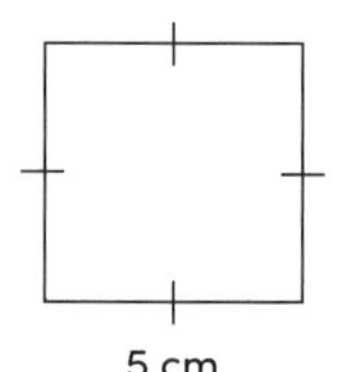


**d**

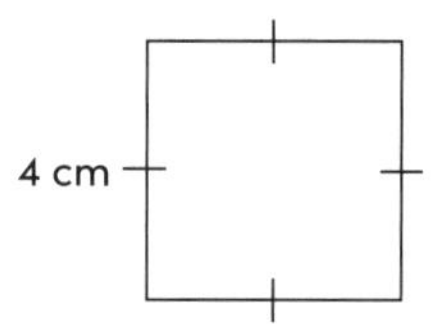


**e**

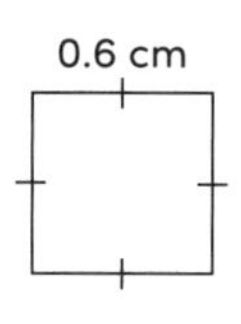


**f**

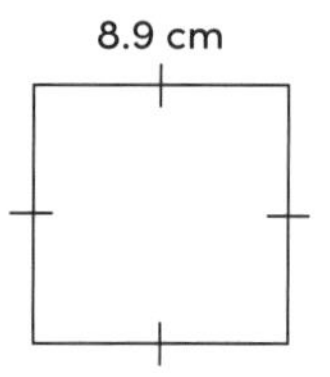


**g** **h**

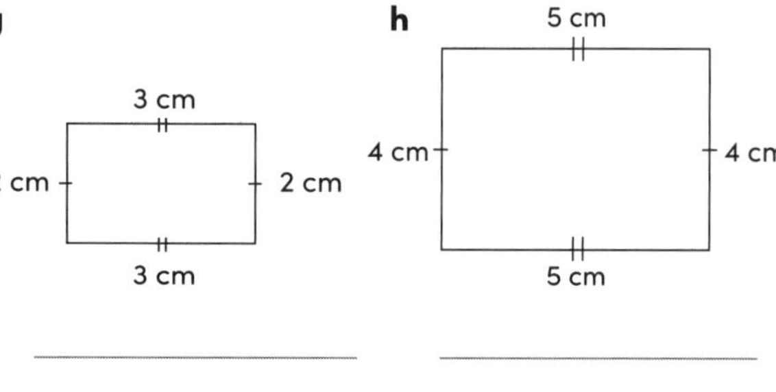


**i**

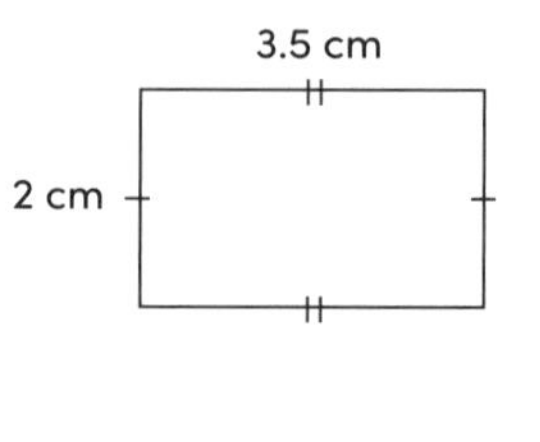


**j**

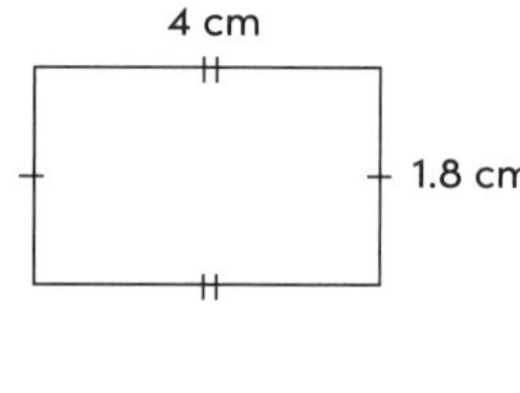


**k**

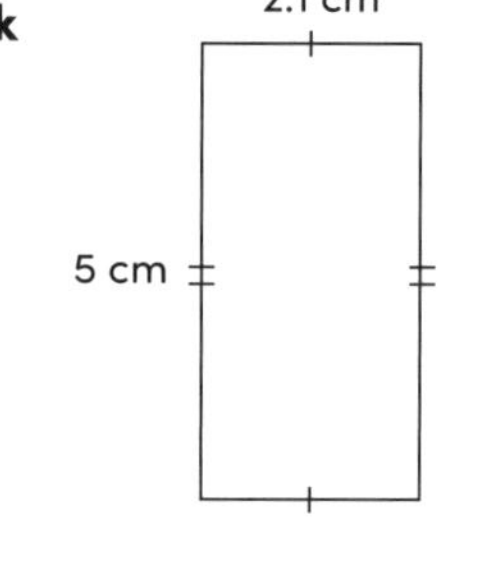


**l**

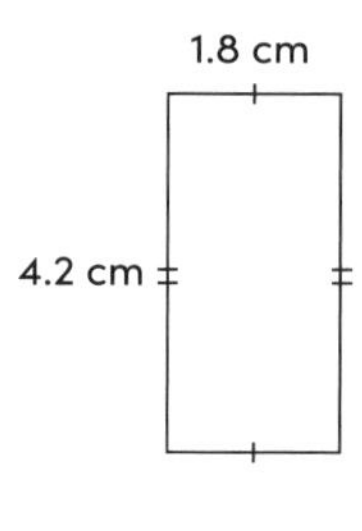


**m**

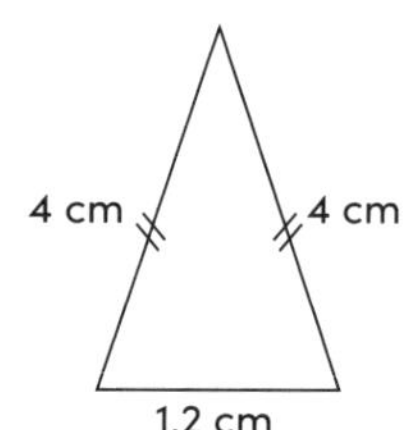


**n**

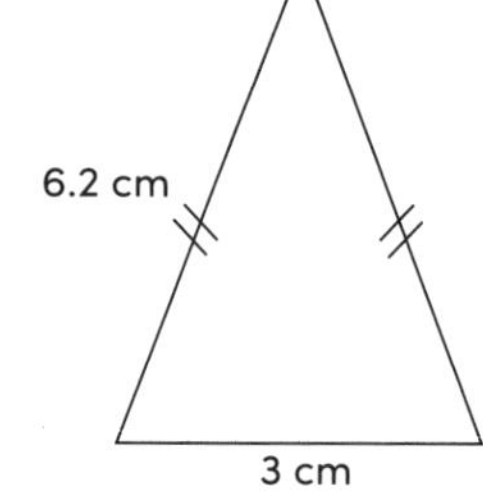


**o**

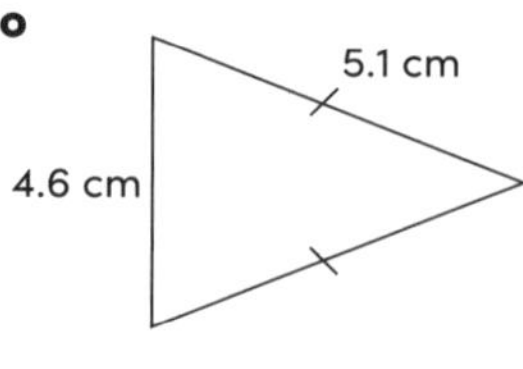


**p**

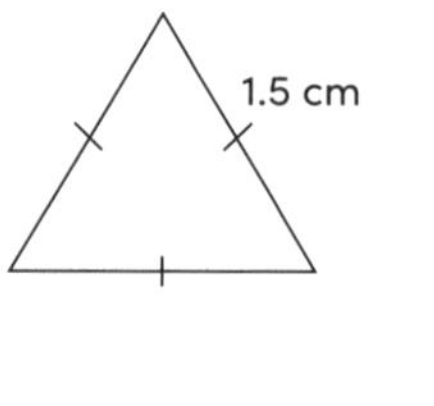

UNIT 5

**MEASUREMENT**

# Perimeter

**1** Find the perimeter of each of the following shapes, in centimetres:

**a**

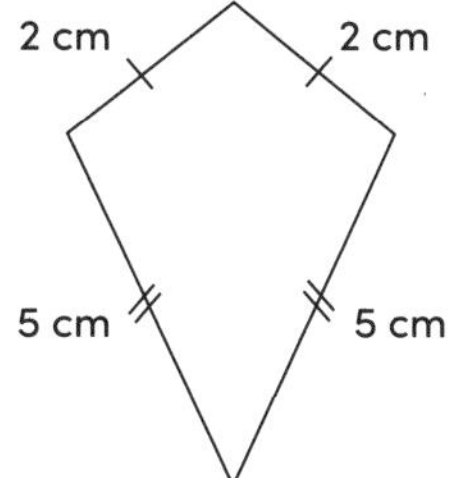


**b**

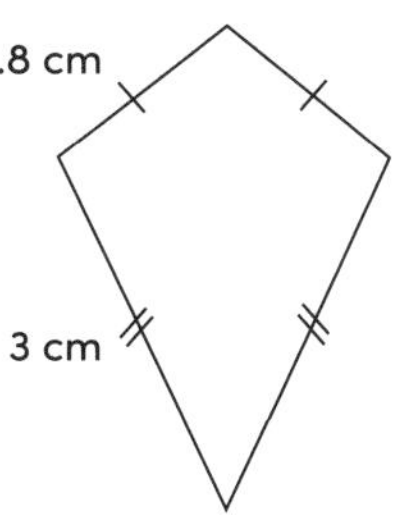


**c**

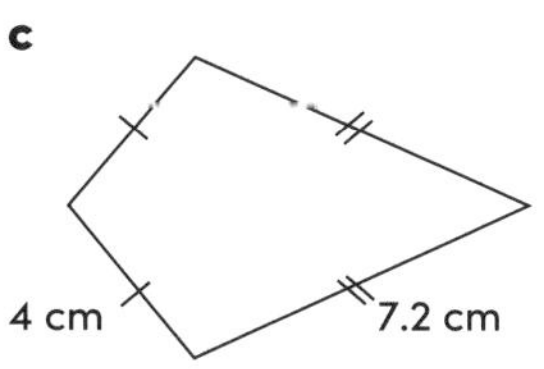


**d**

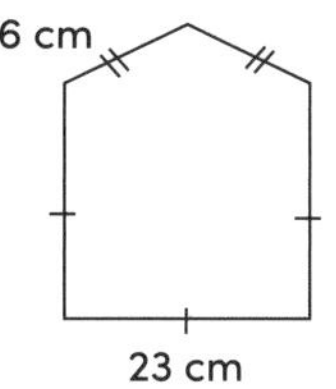


**e**

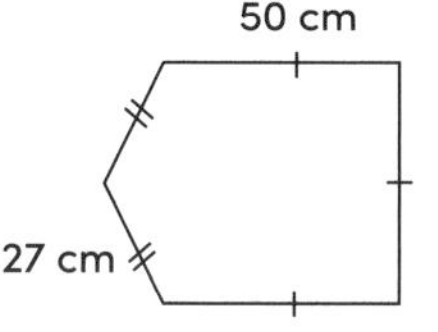


**f**

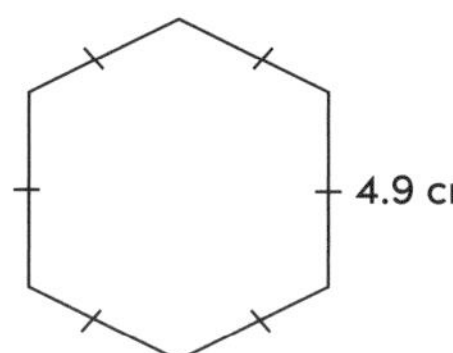


**g**

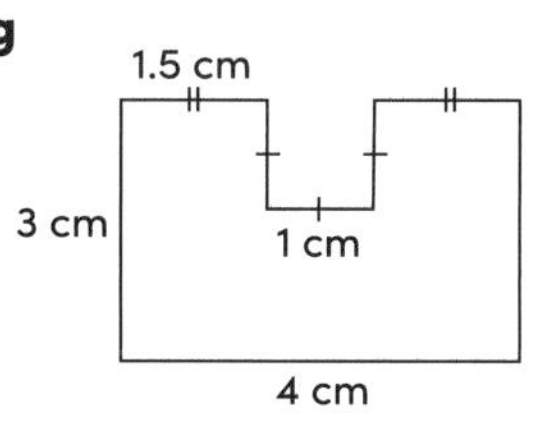


**h**

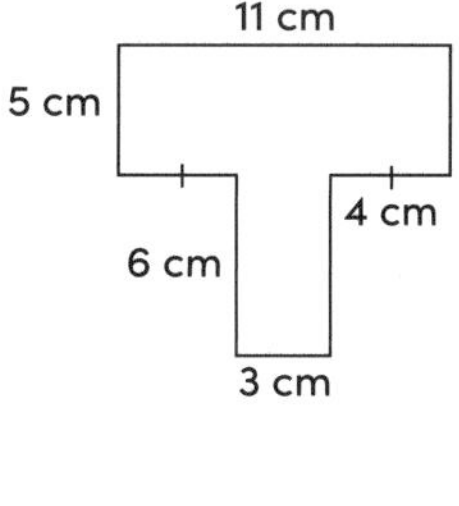


**2** Find the unknown length for each of the figures:

**a**

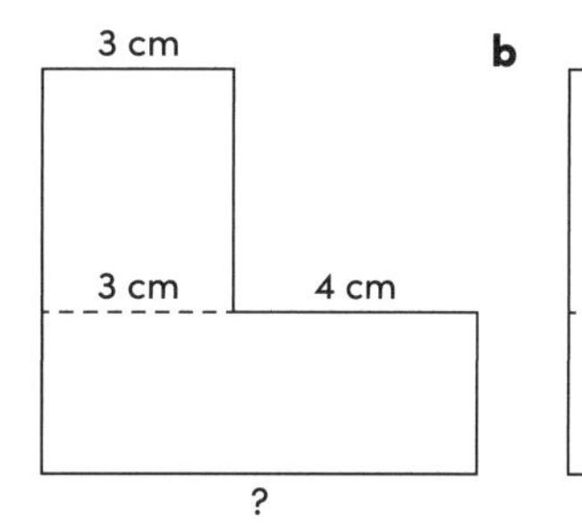


? = ____ + ____

= ____ cm

**b**

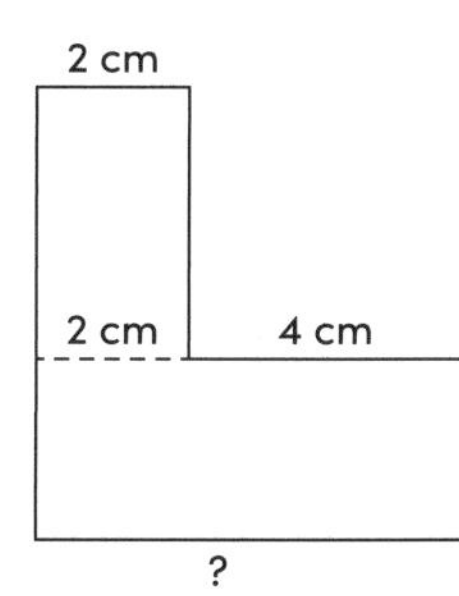


? = ____ + ____

= ____ cm

**c**

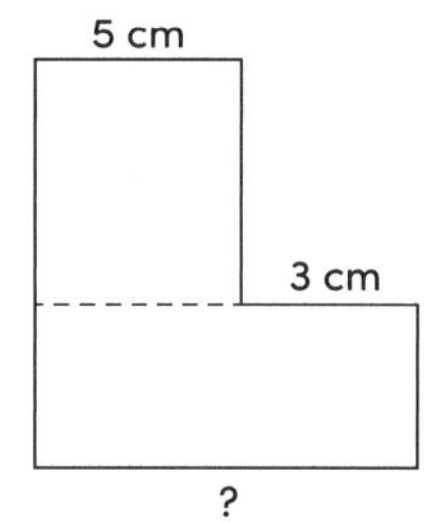


**d**

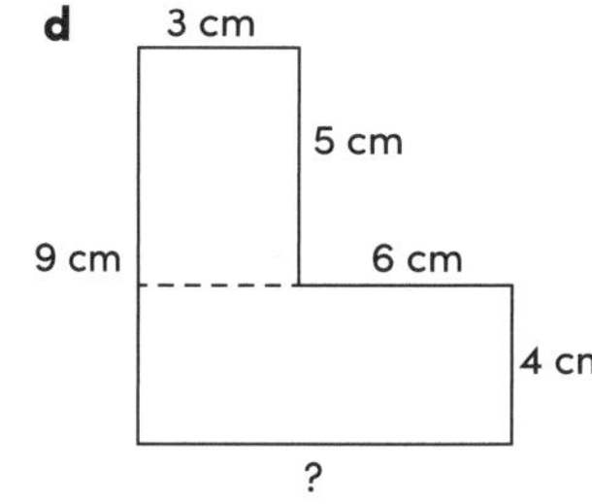


**e**

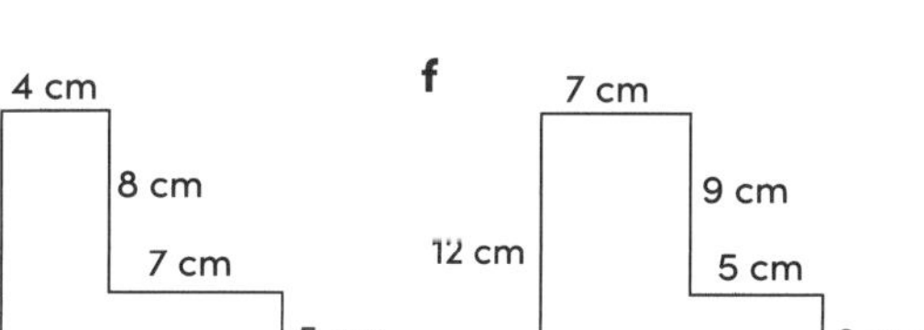


**f**

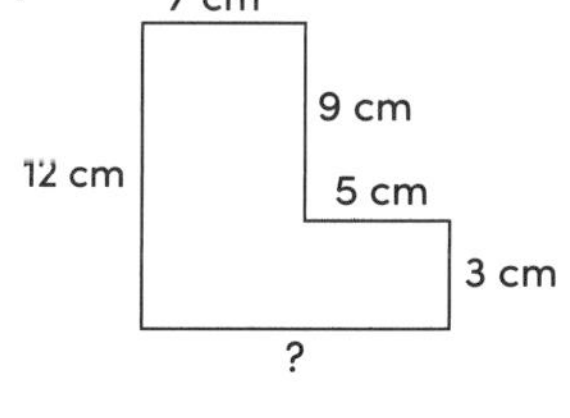

# Perimeter

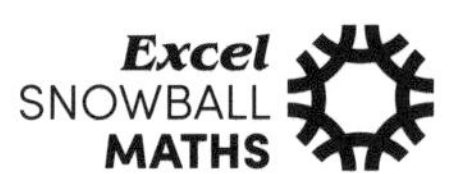


1 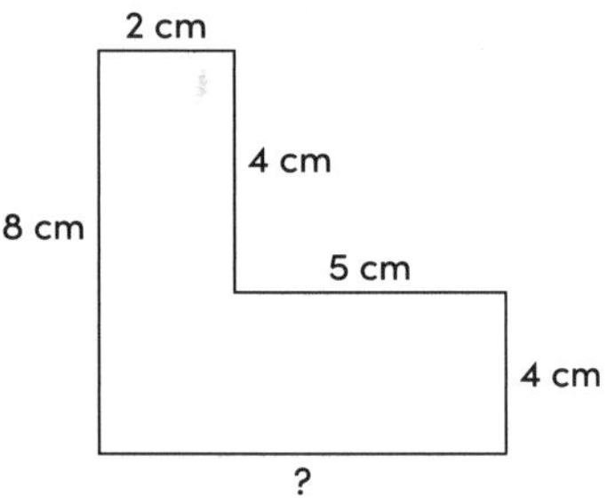


a What is the unknown length? ________

b Calculate the perimeter:

________

________

________

2 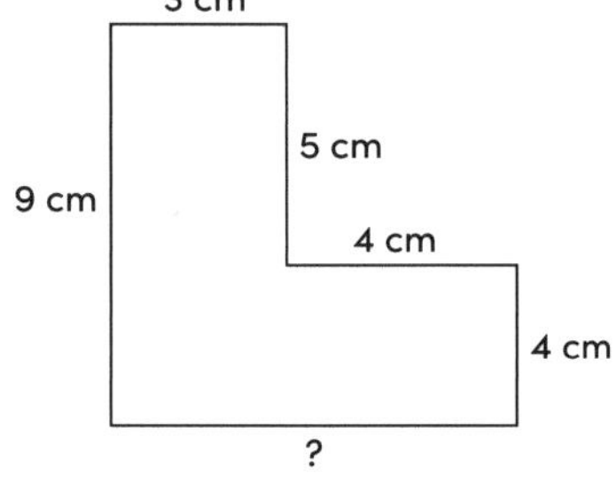


a What is the unknown length? ________

b Calculate the perimeter:

________

________

________

3 Find the perimeter of each of the figures below:

a 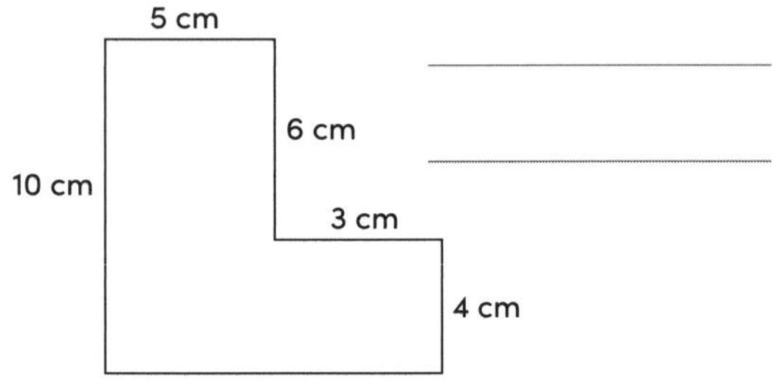


________

________

b 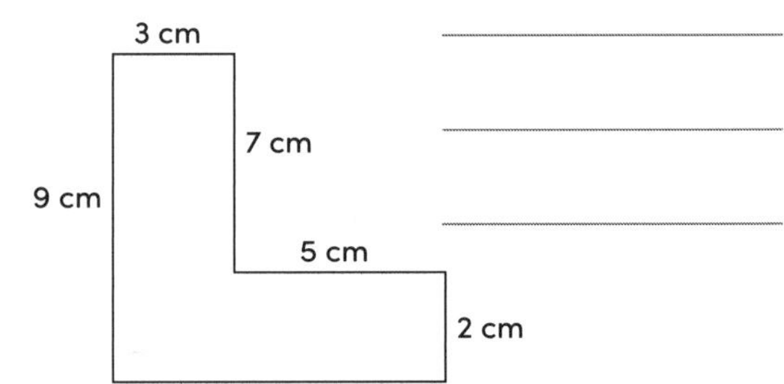


________

________

________

c 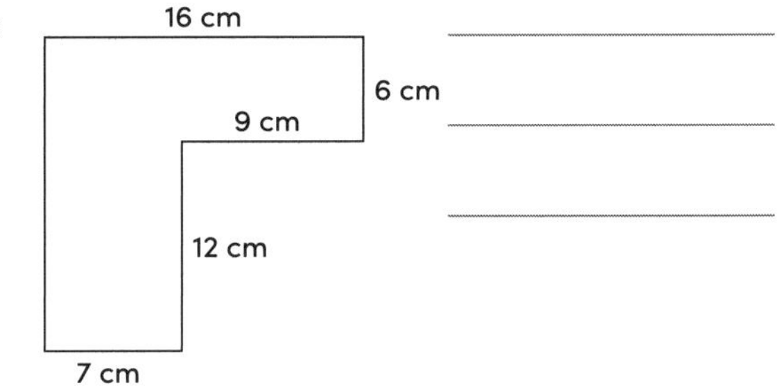


________

________

________

d 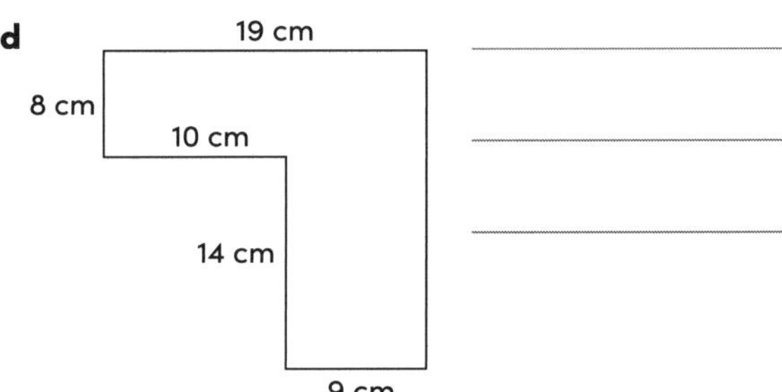


________

________

________

e 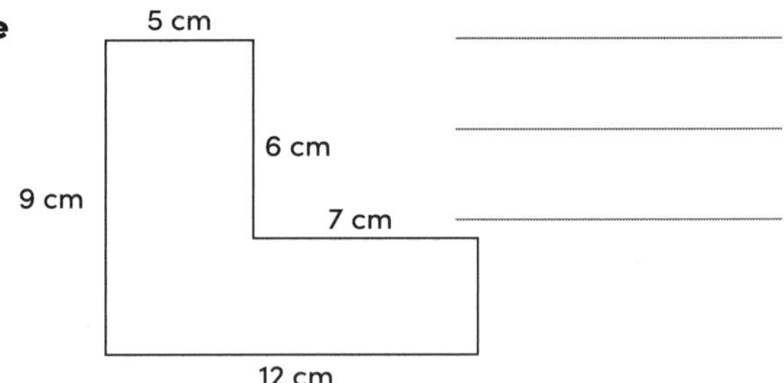


________

________

________

f 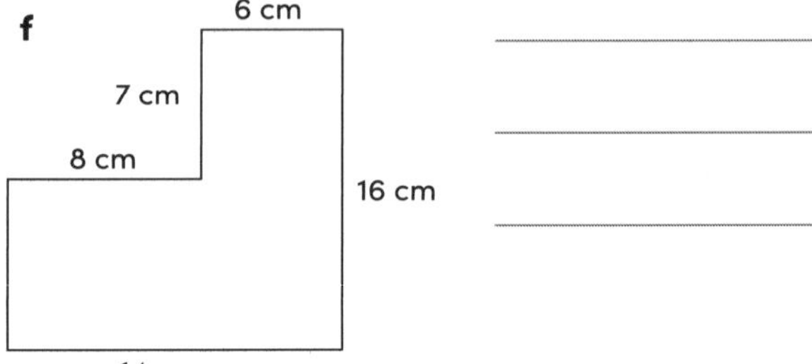


________

________

________

UNIT 7

## MEASUREMENT

# Units of length

mm  cm  m  km

mm → cm → m → km: ÷ 10, ÷ 100, ÷ 1000

km → m → cm → mm: × 1000, × 100, × 10

**1** What unit of length would be appropriate to use to measure:

**a** a small pin? ____________

**b** a screw? ____________

**c** a pencil? ____________

**d** a book? ____________

**e** a door? ____________

**f** a table? ____________

**g** the distance between two suburbs? ____________

**h** the distance of a road trip? ____________

**i** a small insect? ____________

**j** a small shirt button? ____________

**k** a pencil case? ____________

**l** a bathtub? ____________

**m** a bed mattress? ____________

**n** the distance between two countries? ____________

**o** a television screen? ____________

**p** a football field? ____________

**2** Convert each length to centimetres:

**a** 2 m
= 2 × ______
= ______cm

**b** 4.5 m
= 4.5 × ______
= ______cm

**c** 0.1 m
= ______ × ______
= ____________

**d** 0.27 m
= ______ × ______
= ____________

**e** 8.26 m
= ____________
= ____________

**f** 7.01 m
= ____________
= ____________

**g** 0.125 m
= ____________
= ____________

**h** 0.678 m
= ____________
= ____________

**3** Convert each length to millimetres:

**a** 5 cm
= 5 × ______
= ______mm

**b** 6 cm
= 6 × ______
= ______mm

**c** 0.8 cm
= ____________
= ____________

**d** 0.2 cm
= ____________
= ____________

**e** 7.6 cm
= ____________
= ____________

**f** 10.1 cm
= ____________
= ____________

UNIT 8

**MEASUREMENT**

# Units of length

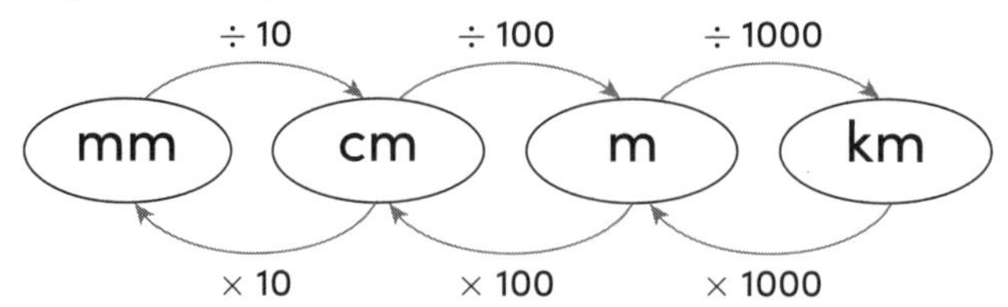


**1** Convert each length to metres:

a 2 km

= 2 × ______

= ______ m

b 5 km

= 5 × ______

= ______ m

c 3.8 km

= ______ × ______

= ______

d 4.75 km

= ______ × ______

= ______

e 6.01 km

= ______

= ______

f 0.9 km

= ______

= ______

g 0.84 km

= ______

= ______

h 0.15 km

= ______

= ______

**2** Convert each length to centimetres:

a 30 mm

= 30 ÷ ______

= ______ cm

b 50 mm

= 50 ÷ ______

= ______ cm

c 65 mm

= ______ ÷ ______

= ______

d 97 mm

= ______ ÷ ______

= ______

e 105 mm

= ______

= ______

f 125 mm

= ______

= ______

**3** Convert:

a 400 cm to m

= ______

= ______

b 520 cm to m

= ______

= ______

c 90 cm to m

= ______

= ______

d 85 cm to m

= ______

= ______

e 3400 m to km

= ______

= ______

f 2500 m to km

= ______

= ______

g 720 mm to cm

= ______

= ______

h 201 mm to cm

= ______

= ______

**4** Convert:

a 0.5 m to mm

= 0.5 × ______ × ______ = ______ mm

b 0.4 m to mm

= 0.4 × ______ × ______ = ______ mm

c 0.35 m to mm

= ______ = ______

d 2 m to mm

= ______

= ______

e 6.8 m to mm

= ______

= ______

UNIT 9

**MEASUREMENT**

# Units of length

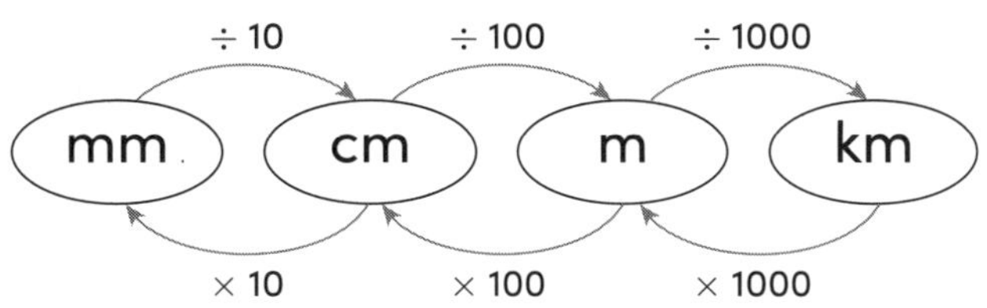


**1** Convert:

**a** 7000 mm to m

= 7000 ÷ ______ ÷ ______ = ______ m

**b** 5200 mm to m

= 5200 ÷ ______ ÷ ______ = ______ m

**c** 600 mm to m

= ______________ = ______

**d** 410 mm to m

= ______

= ______

**e** 505 mm to m

= ______

= ______

**2** Convert each length to metres:

**a** 520 cm

= ______

= ______

**b** 460 cm

= ______

= ______

**c** 3 km

= ______

= ______

**d** 7.6 km

= ______

= ______

**e** 3000 mm

= ______

= ______

**f** 4200 mm

= ______

= ______

**g** 1300 cm

= ______

= ______

**h** 2560 cm

= ______

= ______

**3** A $1 Australian coin is 2.8 mm in thickness. What is this in centimetres?

______________

**4** A $100 note is 158 mm in length. What is this in centimetres?

______________

**5** A bookshelf is 91.4 cm long.

**a** How long is the bookshelf in millimetres?

______________

**b** How many 20-mm-thick books can fit on this bookshelf?

______________

**6** A couch is 2.13 m long.

**a** What is this length in centimetres?

______________

**b** If each person occupies about 60 cm, how many people can sit on this couch?

______________

**7** A book is 18 mm thick.

**a** How thick is the book in centimetres?

______________

**b** If 100 of these books are stacked on top of each other, how high will they reach, in centimetres?

______________

UNIT 10

**MEASUREMENT**

# Area of squares and rectangles

**1** Find the area of each square, correct to one decimal place:

**a**

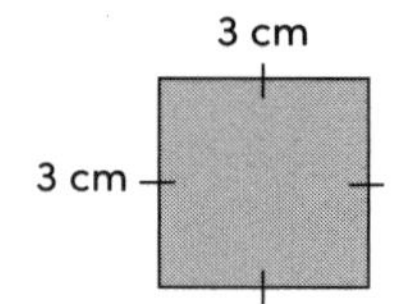


$A = s \times s$

$= \_\_\_\_ \times \_\_\_\_$

$= \_\_\_\_\_\_ \text{cm}^2$

**b**

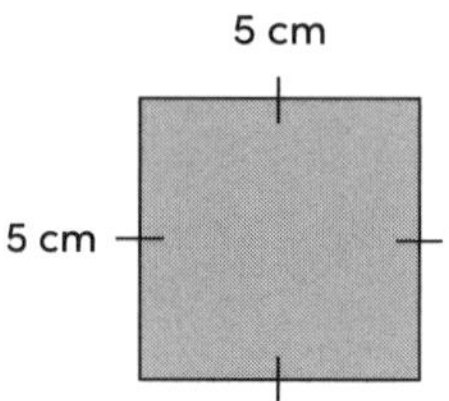


$A = s \times s$

$= \_\_\_\_ \times \_\_\_\_$

$= \_\_\_\_\_\_ \text{cm}^2$

**c**

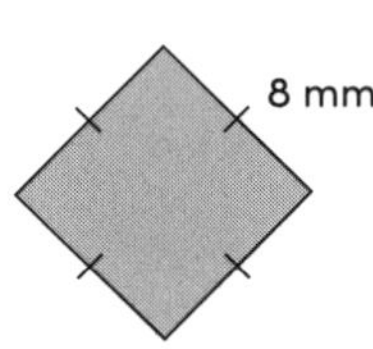


$A =$ ______

$=$ ______

$=$ ______

**d**

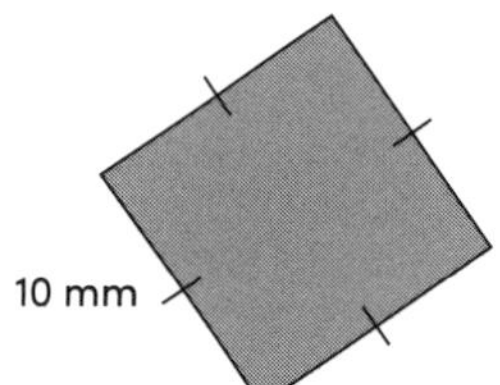


$A =$ ______

$=$ ______

$=$ ______

**e**

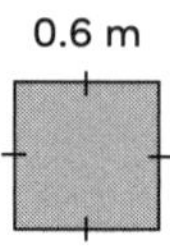


______

______

______

**f**

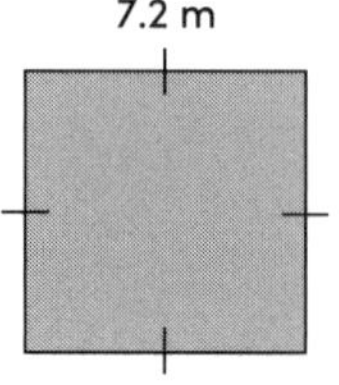


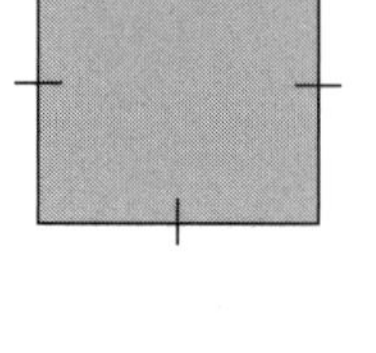

______

______

______

**2** Find the area of each rectangle, correct to one decimal place:

**a** **b**

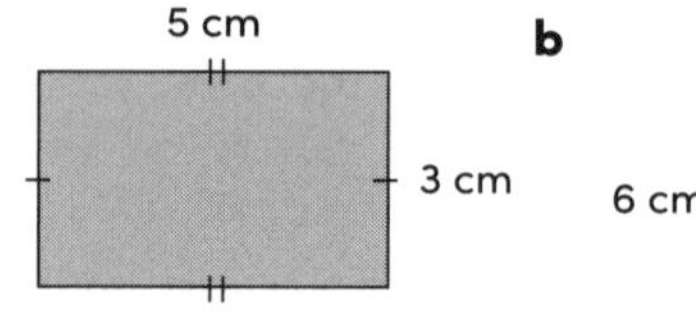


**a**

$A = lw$

$= \_\_\_\_ \times \_\_\_\_$

$= \_\_\_\_\_\_ \text{cm}^2$

**b**

$A = lw$

$= \_\_\_\_ \times \_\_\_\_$

$= \_\_\_\_\_\_ \text{cm}^2$

**c**

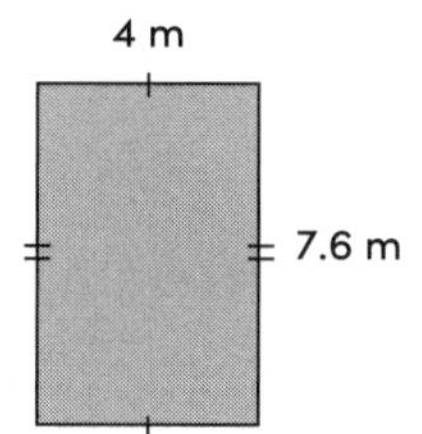


$A =$ ______

$=$ ______

$=$ ______

**d**

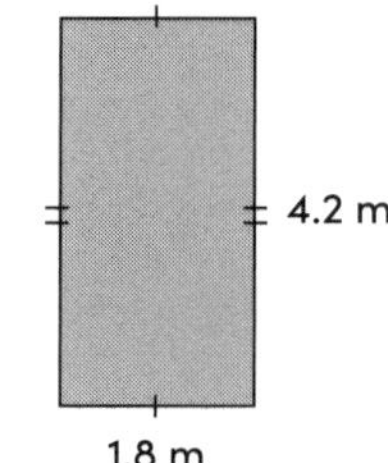


$A =$ ______

$=$ ______

$=$ ______

**e**

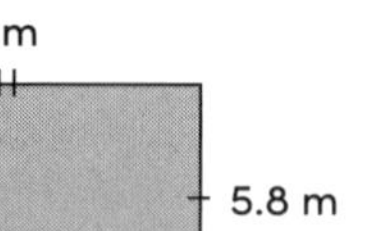


______

______

______

**f**

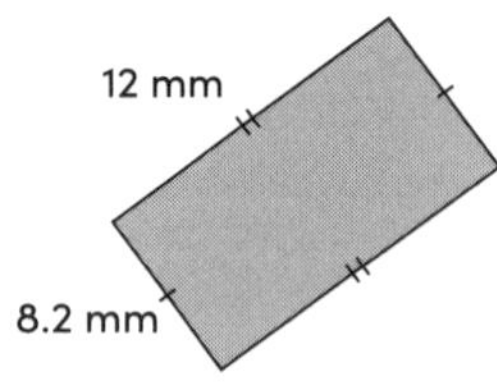


______

______

______

UNIT 11

**MEASUREMENT**

# Area of squares and rectangles

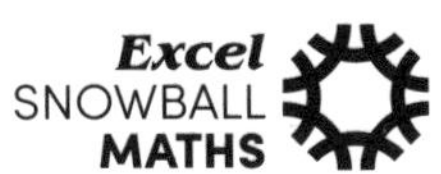


**1** A rectangular table measures 1.83 m in length and 0.76 m in width. What is the area of the table correct to one decimal place?

______

______

______

**2** A swimming pool is 50 m long and 25 m wide. What is the area of the pool?

______

______

______

**3** A square garden has a side length of 1.2 m. What is the area of the garden correct to one decimal place?

______

______

______

**4** A square dining table has a side length of 0.91 m. What is the area of the table correct to one decimal place?

______

______

______

**5** Find the unknown length and the perimeter of each rectangle:

**a**

7 cm

Area = 21 cm$^2$ ?

? = 21 ÷ ____

= ____ cm

$P = 7 + 7 +$ __ + __

= ______

**b**

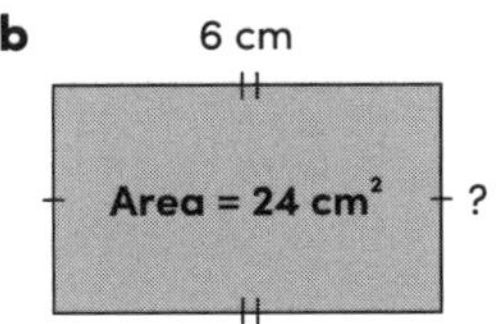


? = 24 ÷ ____

= ____ cm

$P = 6 + 6 +$ __ + __

= ______

**c**

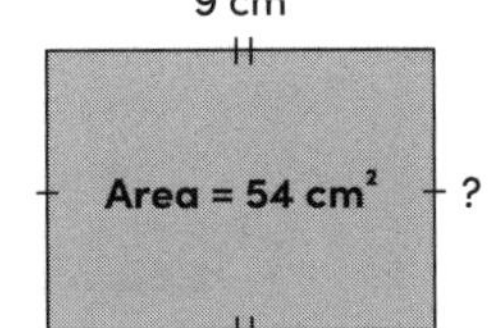


? = ______

= ______

$P =$ ______

= ______

**d**

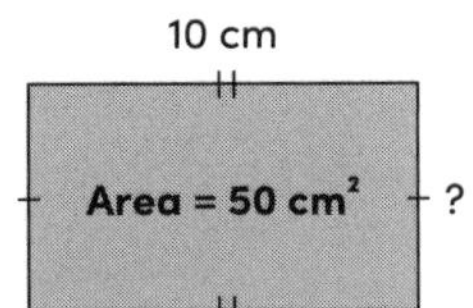


? = ______

= ______

$P =$ ______

= ______

**6** A rectangle has a length of 8 m and an area of 40 m$^2$.

**a** What is the width?

______

______

**b** Calculate the perimeter.

______

______

UNIT 12

**MEASUREMENT**

# Area of squares

**1** Find the unknown side length and the area of each square:

**a**

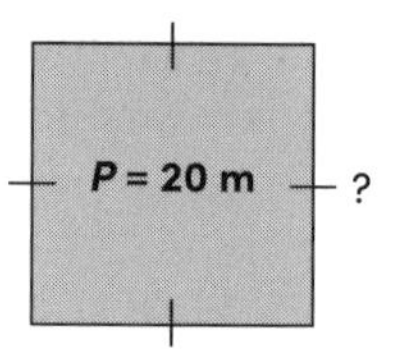


? = 20 ÷ ____

= ____ m

$A = s \times s$

= ____ × ____

= ____ $m^2$

**b**

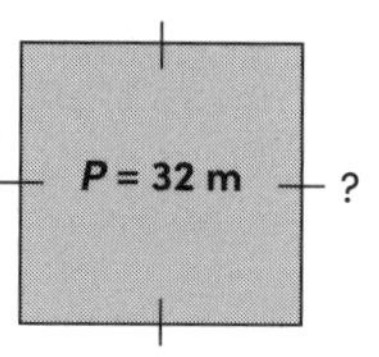


? = 32 ÷ ____

= ____ m

$A = s \times s$

= ____ × ____

= ____ $m^2$

**c**

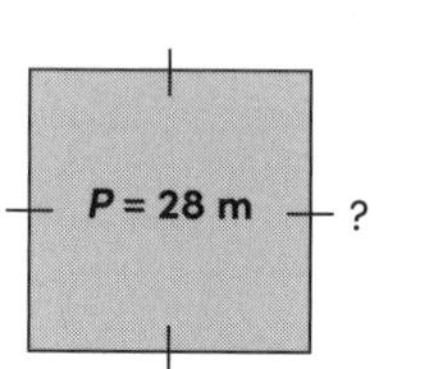


? = ________

= ________

$A$ = ________

= ________

= ________

**d**

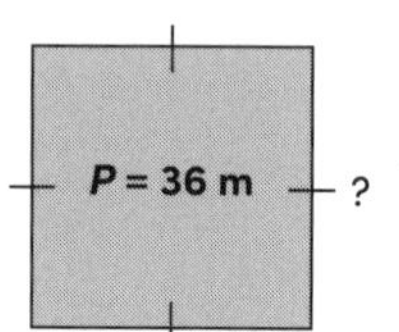


? = ________

= ________

$A$ = ________

= ________

= ________

**e**

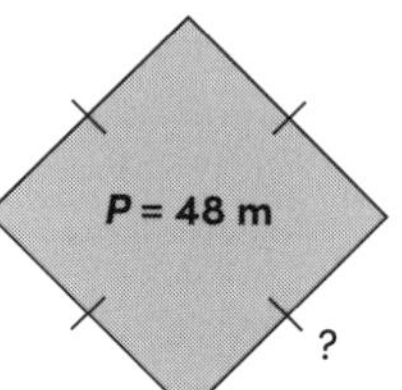


? = ________

= ________

$A$ = ________

= ________

= ________

**f**

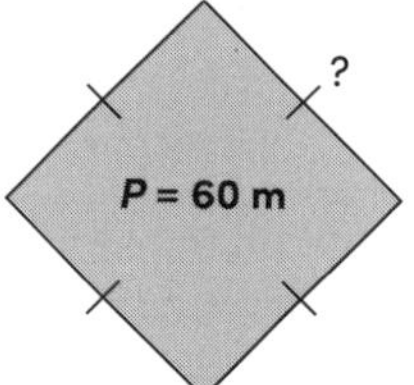


? = ________

= ________

$A$ = ________

= ________

= ________

**2** A square has a perimeter of 36 m.

**a** What is the side length?

________________

**b** Calculate the area.

________________

________________

________________

**3** A square has a perimeter of 40 m.

**a** What is the side length?

________________

**b** Calculate the area.

________________

________________

________________

**4** A square has an **area** of 25 $m^2$.

**a** What is the side length?

________________

**b** Calculate the perimeter.

________________

**5** A square has an **area** of 49 $m^2$.

**a** What is the side length?

________________

**b** Calculate the perimeter.

________________

UNIT 13

# MEASUREMENT

## Area of triangles

Note: Answers should be correct to one decimal place.

**1** Find the area of the following triangles:

**a**

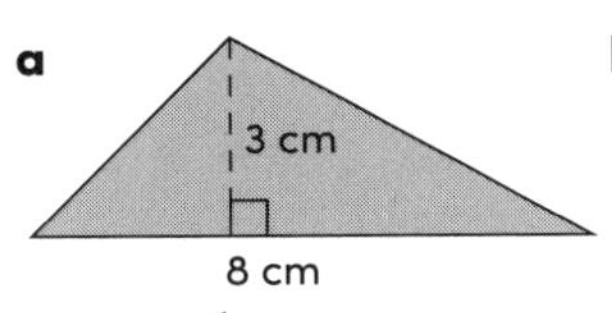


$A = \frac{1}{2}bh$

$= \frac{1}{2} \times$ ___ $\times$ ___

$=$ ________ $\text{cm}^2$

**b**

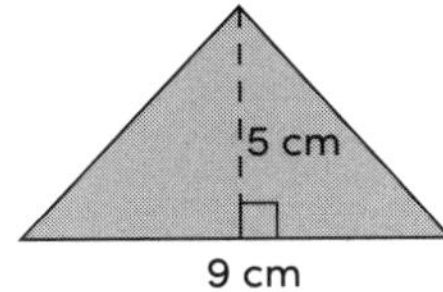


$A = \frac{1}{2}bh$

$= \frac{1}{2} \times$ ___ $\times$ ___

$=$ ________ $\text{cm}^2$

**c**

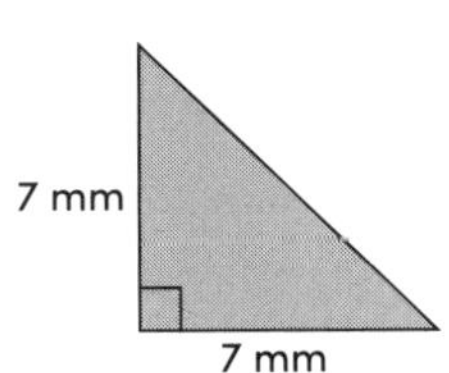


$A =$ ________

$=$ ________

$=$ ________

**d**

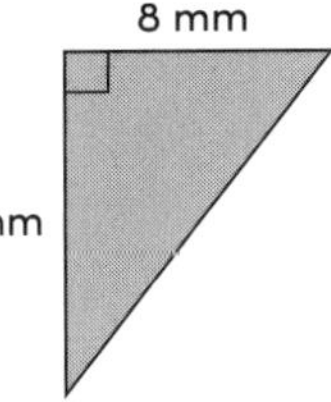


$A =$ ________

$=$ ________

$=$ ________

**e**

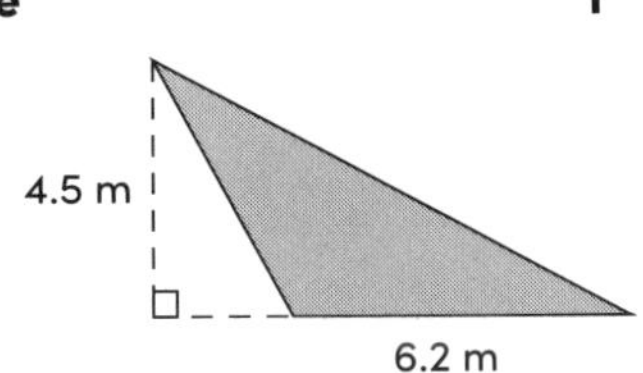


**f**

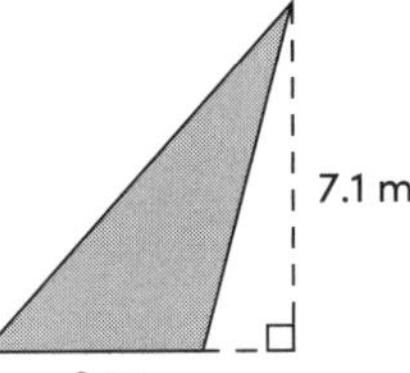


**g**

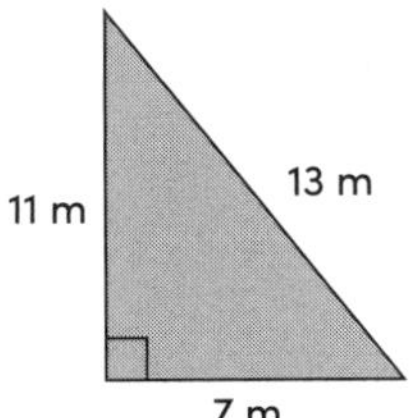


**h**

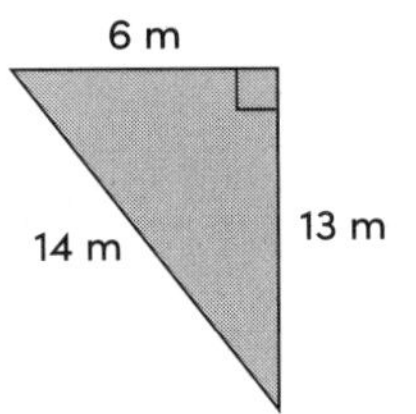


**i**

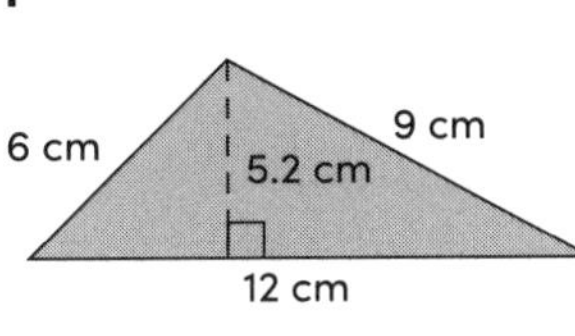


**j**

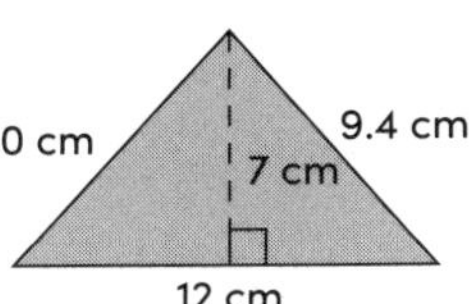


**k**

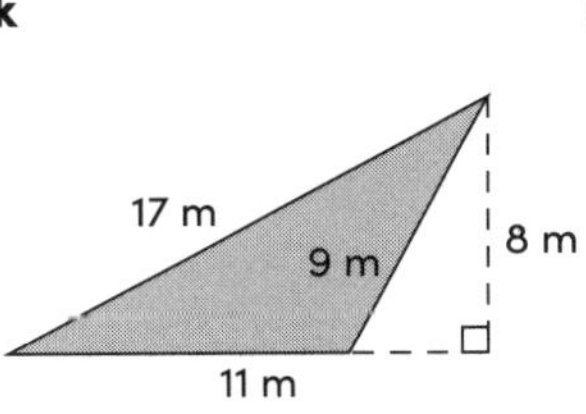


**l**

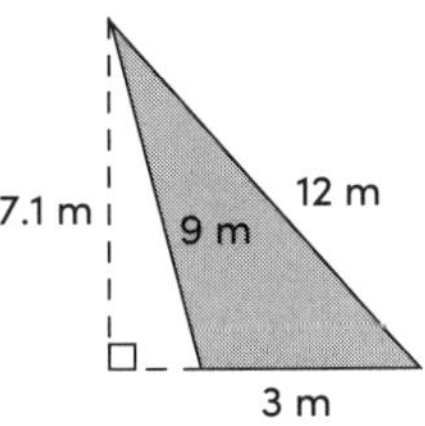


**2** A triangular sandwich has a base length of 15 cm and a height of 10 cm. What is the area of the sandwich?

**3** A triangular traffic sign has a base length of 0.6 m and a height of 0.76 m. What is the area of the traffic sign?

## MEASUREMENT

# Area of triangles

**1** The Great Pyramid of Giza has a base length of 230 m and a slant of height 187 m. What is the area of the triangular face?

______________________________

______________________________

______________________________

**2** A house roof is triangular and has a height of 1.5 m and a base of 6 m. Calculate the area of the triangular face of the roof.

______________________________

______________________________

______________________________

**3** A yacht has a triangular sail with 8.8 m in height and 4 m in base length.

**a** Calculate the area of the sail.

______________________________

______________________________

______________________________

**b** If sail cloth costs $122 per $m^2$, how much will it cost to replace the whole sail?

______________________________

______________________________

**4** Find the unknown perpendicular height for each of the triangles:

**a**

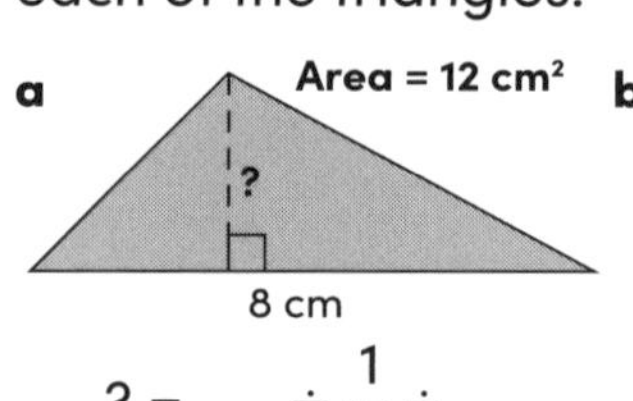


$? = \_\_ \div \frac{1}{2} \div \_\_$

= ________cm

**b**

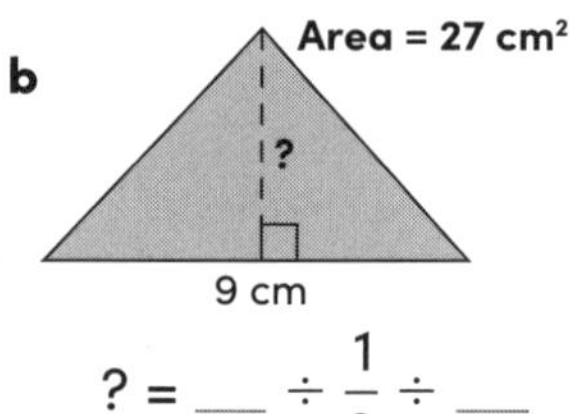


$? = \_\_ \div \frac{1}{2} \div \_\_$

= ________cm

**c**

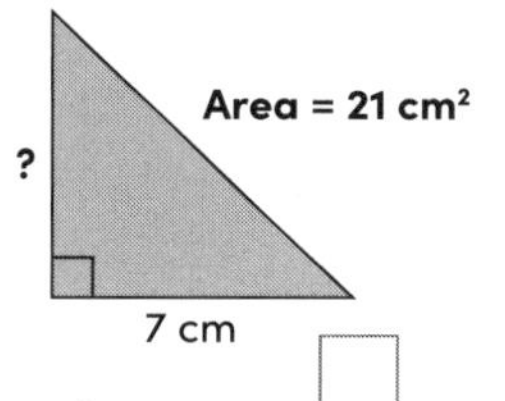


$? = \_\_ \div \frac{\square}{\square} \div \_\_$

= ________

**d**

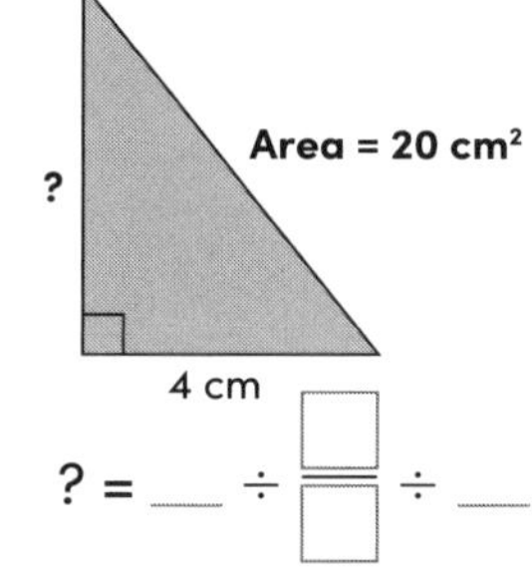


$? = \_\_ \div \frac{\square}{\square} \div \_\_$

= ________

**e** **f**

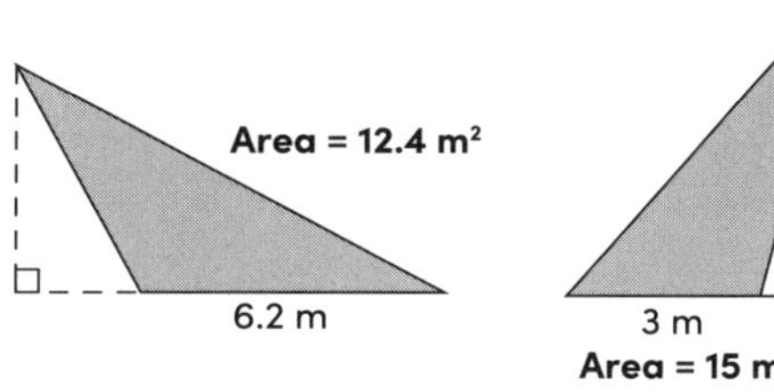


______________________________

______________________________

**5** A triangle has an area of 27 $cm^2$ with a base length of 9 cm. What is the perpendicular height of the triangle?

______________________________

______________________________

______________________________

UNIT 15

**MEASUREMENT**

# Area of parallelograms

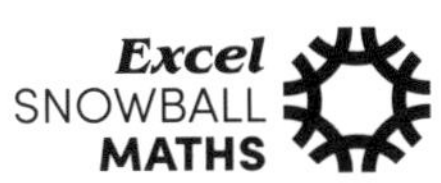


**1** Find the area, correct to one decimal place, of the following parallelograms:

**a**

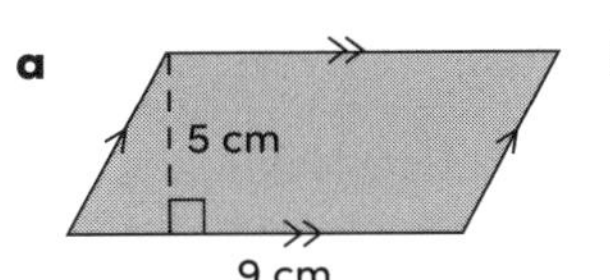


$A = bh$

= ______ × ______

= ________ $\text{cm}^2$

**b**

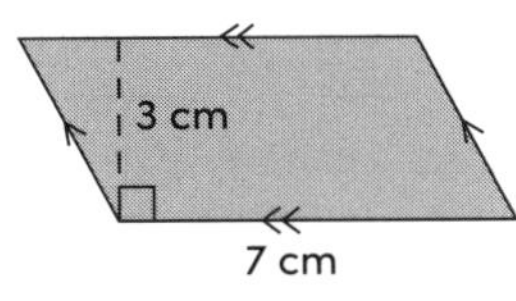


$A = bh$

= ______ × ______

= ________ $\text{cm}^2$

**c**

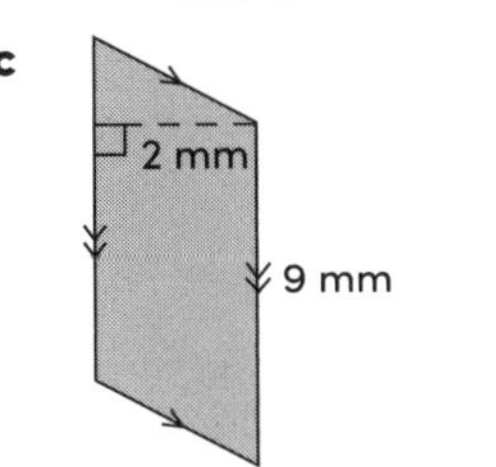


______

______

______

**d**

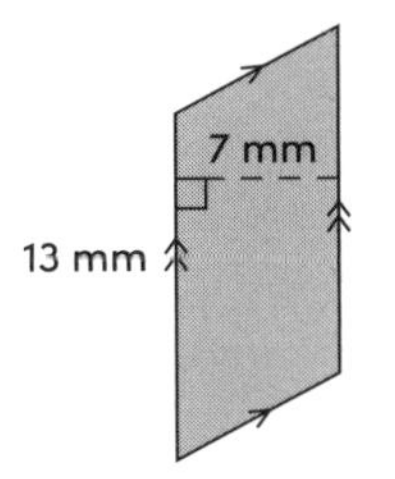


______

______

______

**e** **f**

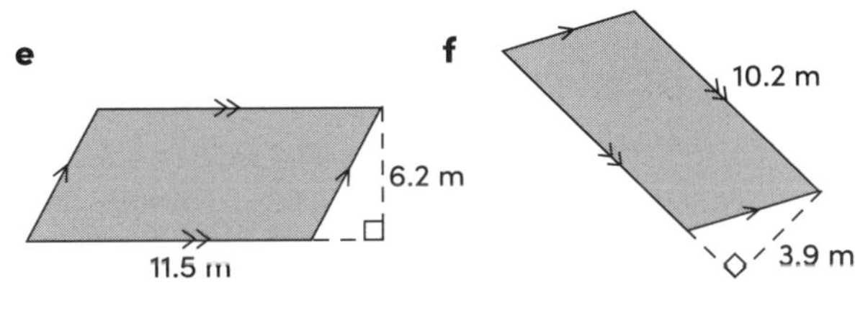


______

______

______

**g** **h**

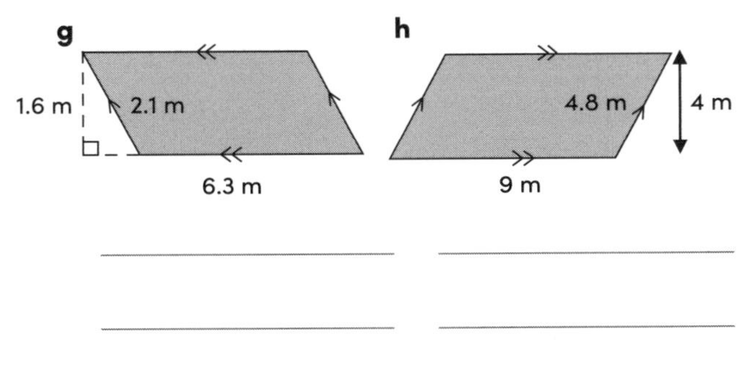


______

______

______

**2** Find the area of a parallelogram with a perpendicular height of 4.6 m and a base length of 10.3 m, correct to one decimal place.

______

______

**3** A house has a window in the shape of a parallelogram that needs to be fitted with glass. If the window has a perpendicular height of 0.91 metres and a base length of 0.6 metres, find:

**a** the area of the window, correct to two decimal places

______

______

______

**b** the cost of the glass if it costs $145 per square metre

______

______

**4** A building has a wall in the shape of a parallelogram that has a height of 5 metres and a base length of 26.3 metres.

**a** What is the area of the wall?

______

______

______

**b** If a one-litre bucket of paint covers 12 square metres, how many buckets of paint are needed to paint the wall?

______

______

# MEASUREMENT
## Area of parallelograms

**1** Find the unknown length for each parallelogram:

**a**

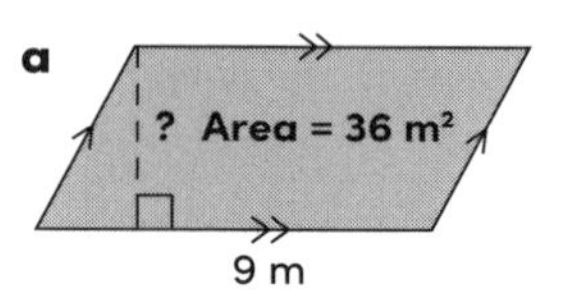


? = ______ ÷ ______

= ________ m

**b**

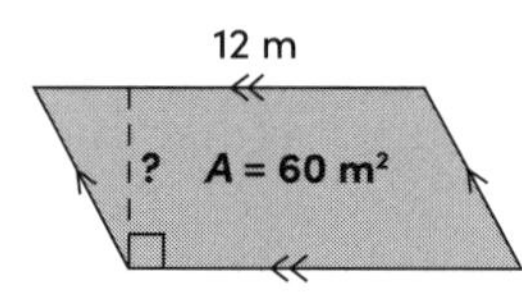


? = ______ ÷ ______

= ________ m

**c**

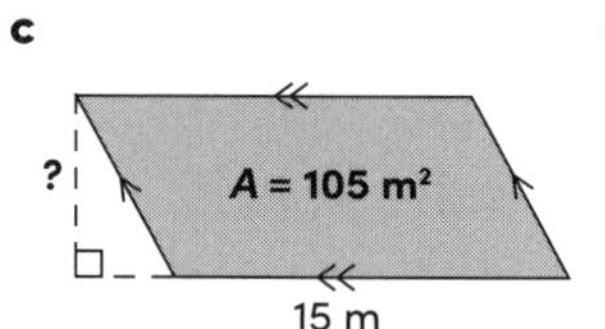


**d**

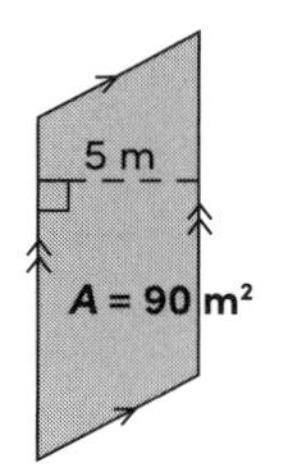


**e**

**f**

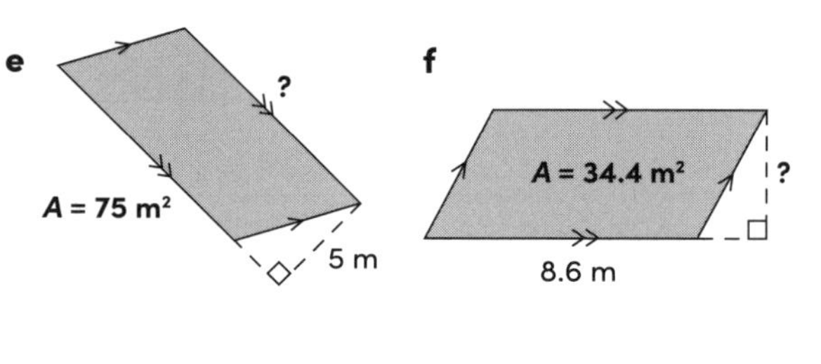


**2** What is the height of a parallelogram with an area of 36 $m^2$ and a base length of 9 m?

**3** What is the base length of a parallelogram with an area of 44.5 $m^2$ and a height of 5 m?

**4** Find the perpendicular height of a parallelogram with an area of 106.4 $m^2$ and a base length of 7.6 m.

**5** The front of the bench leg is a parallelogram which has an area of 160 $cm^2$ with a perpendicular height of 20.8 cm as shown. Find the length of the base, correct to one decimal place.

20.8 cm

**6** A wall tile is made up of black and white parallelograms as shown. If each parallelogram has an area of 480 $cm^2$ with a base of 16 cm, find the perpendicular height.

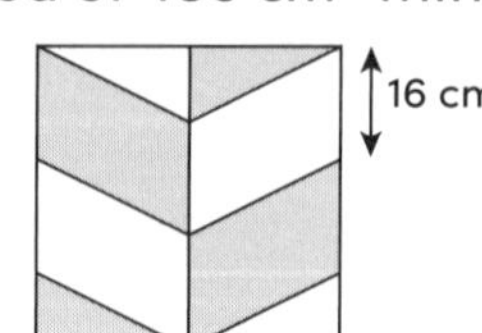

MEASUREMENT

# Area of composite shapes

**1** Find the area of the following composite shapes:

**a**

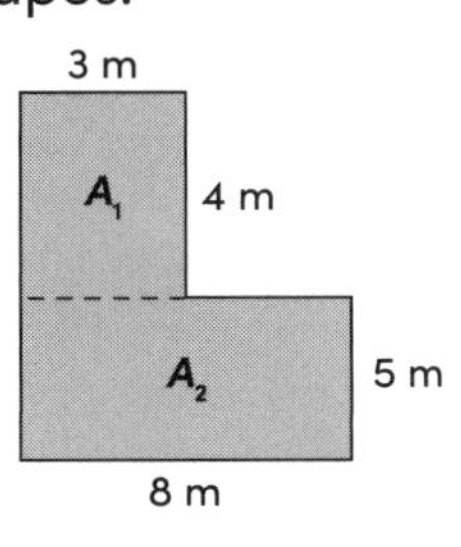


$A_1$ = ____ × ____

= ________ m²

$A_2$ = ____ × ____

= ________ m²

Total

$A$ = ____ + ____

= ________ m²

**b**

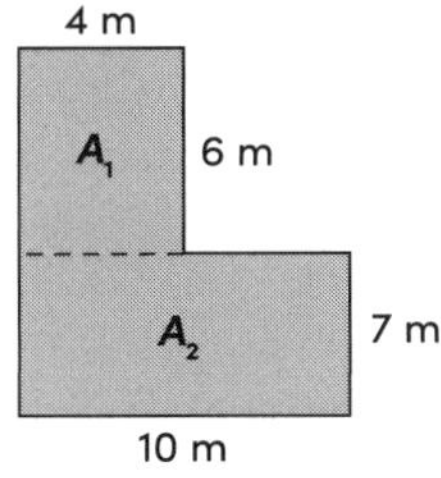


$A_1$ = ____ × ____

= ________ m²

$A_2$ = ____ × ____

= ________ m²

Total

$A$ = ____ + ____

= ________ m²

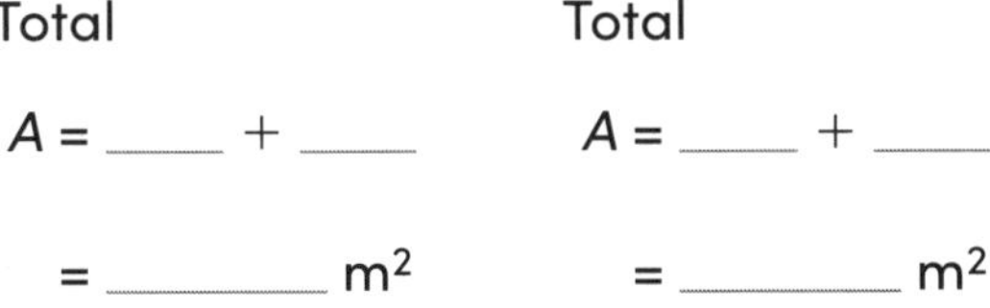

**c**

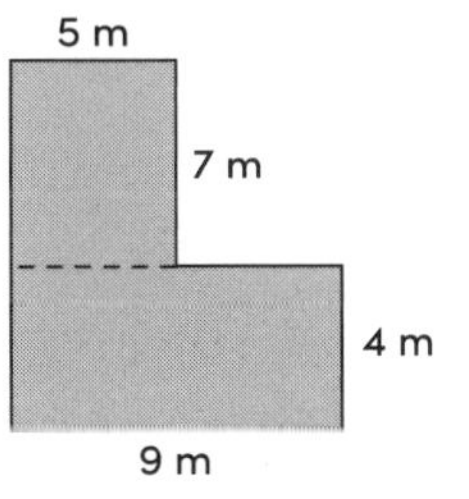


________________

________________

________________

________________

________________

________________

**d**

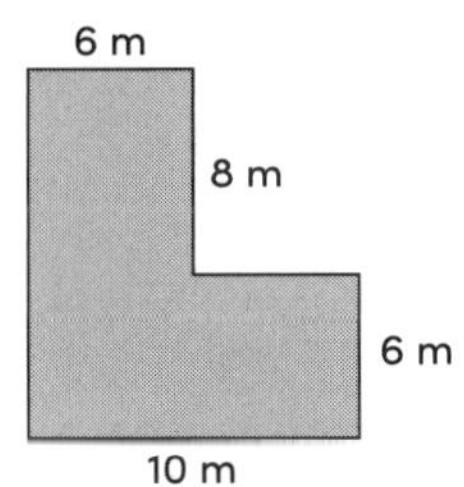


________________

________________

________________

________________

________________

________________

**2** Find the area of the following composite shapes:

**a**

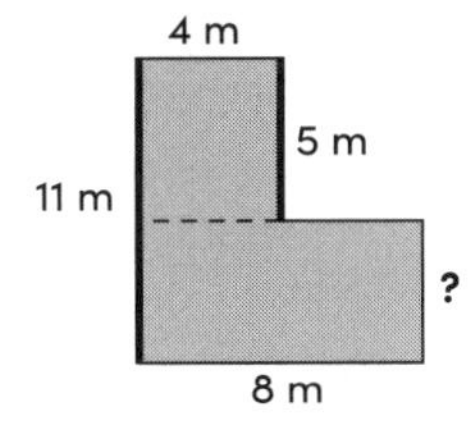


$A_1$ = ____ × ____

= ________ m²

? = 11 − ____

= ________

↓

$A_2$ = ____ × ____

= ________ m²

Total

$A$ = ____________

= ________ m²

**b**

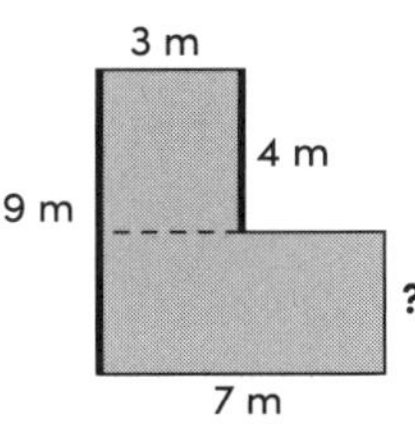


$A_1$ = ____ × ____

= ________ m²

? = 9 − ____

= ________ m²

↓

$A_2$ = ____ × ____

= ________ m²

Total

$A$ = ____________

= ________ m²

**c**

5 m

6 m

13 m

?

9 m

$A_1$ = ____________

= ____________

? = ____________

= ____________

$A_2$ = ____________

= ____________

Total

$A$ = ____________

= ________

**d**

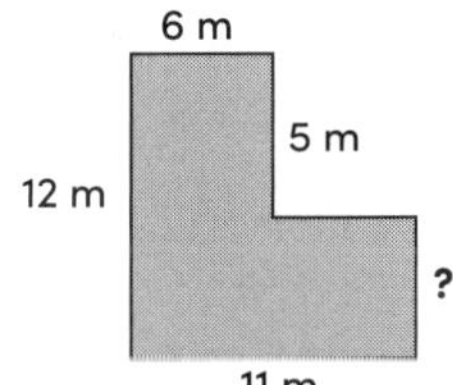


$A_1$ = ____________

= ____________

? = ____________

= ____________

$A_2$ = ____________

= ____________

Total

$A$ = ____________

= ________

# Area of composite shapes

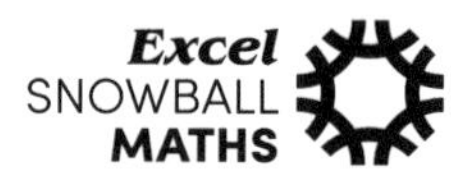


**1** Find the area of the following composite shapes:

**a**

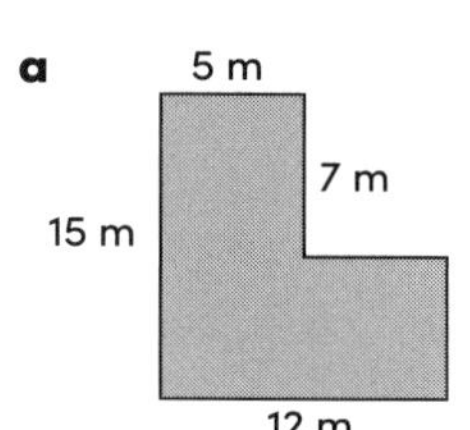


**b**

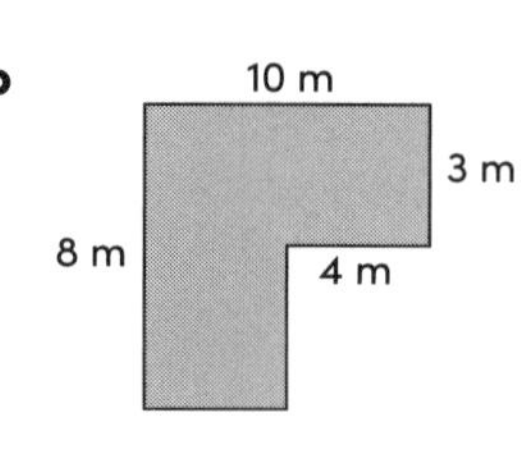


**2** Find the area of the following composite shapes:

**a**

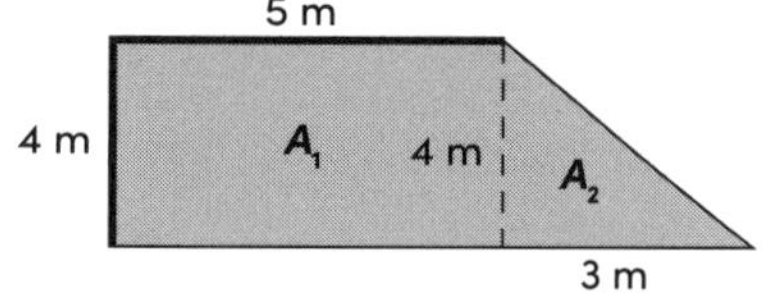


$A_1$ = ______ × ______ = ________

$A_2 = \frac{1}{2}$ × ______ × ______ = ________

Total $A$ = ______ + ______ = ________

**b**

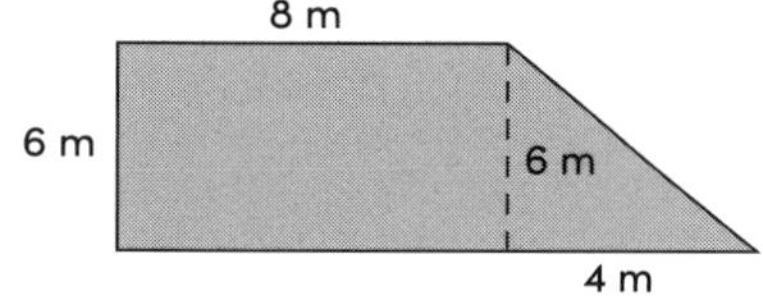


$A_1$ = ______ × ______ = ________

$A_2 = \frac{1}{2}$ × ______ × ______ = ________

Total $A$ = ______ + ______ = ________

**c**

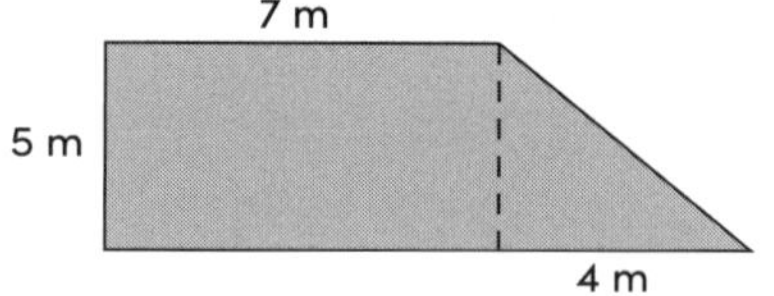


$A_1$ = ________________

$A_2$ = ________________

Total $A$ = ________________

**d**

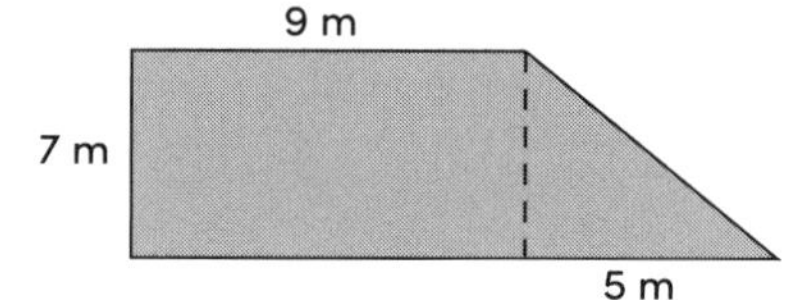


$A_1$ = ________________

$A_2$ = ________________

Total $A$ = ________________

**e**

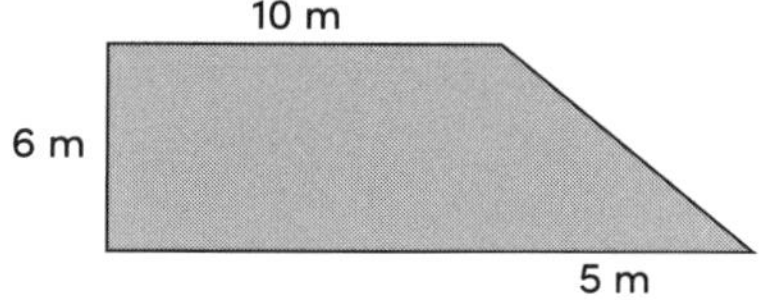


**f**

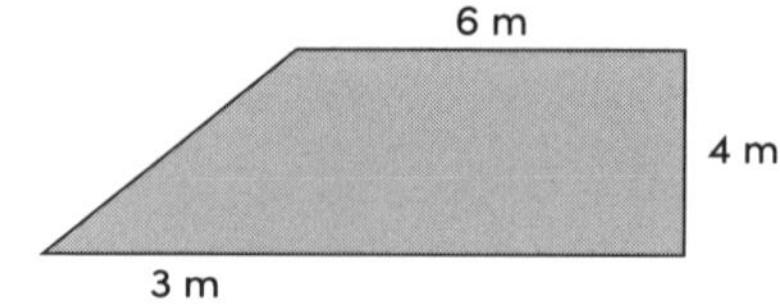

# MEASUREMENT
## Area of composite shapes

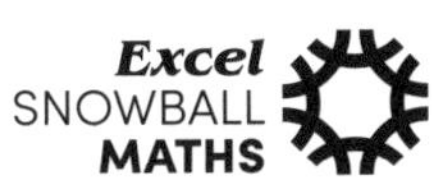


**1** Find the shaded area for the following composite shapes:

**a**

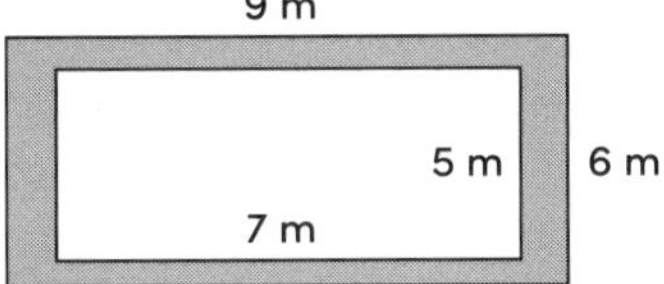


$A_1$ = ____ × ____ = ________ m²
(Outside numbers)

$A_2$ = ____ × ____ = ________ m²

Shaded $A$ = ________ − ________

= ________ m²

**b**

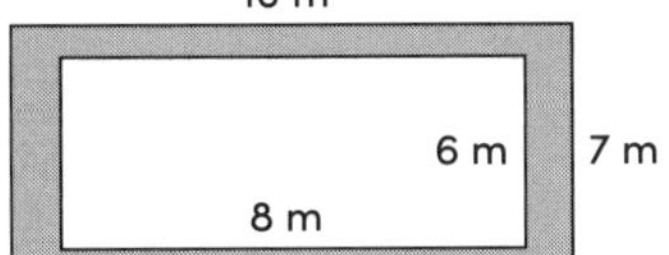


$A_1$ = ____ × ____ = ________ m²
(Outside numbers)

$A_2$ = ____ × ____ = ________ m²

Shaded $A$ = ________ − ________

= ________ m²

**c**

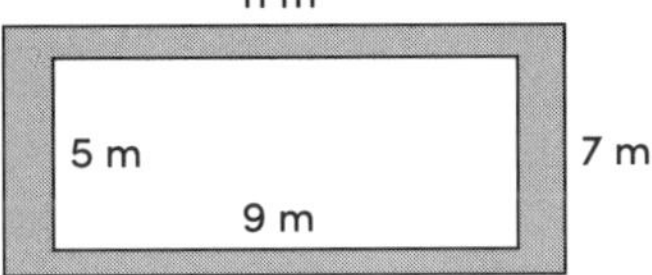


$A_1$ = ________________

$A_2$ = ________________

Shaded $A$ = ________________

**d**

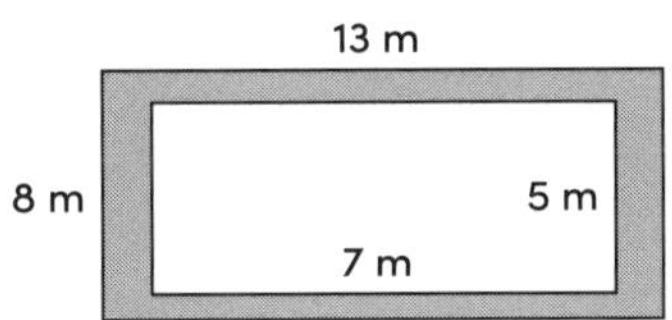


$A_1$ = ________________

$A_2$ = ________________

Shaded $A$ = ________________

**e**

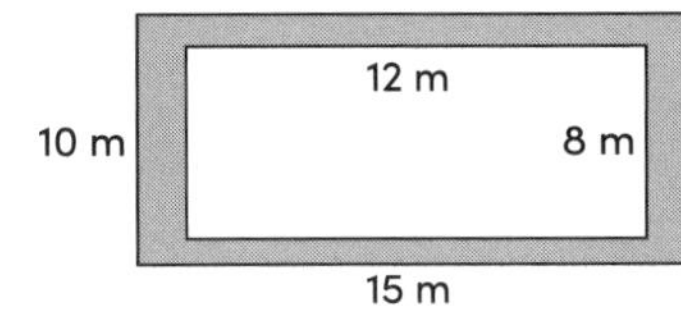


________________

________________

________________

**f**

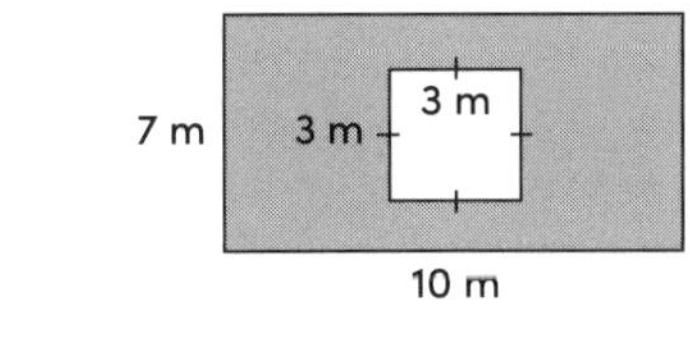


________________

________________

________________

**g**

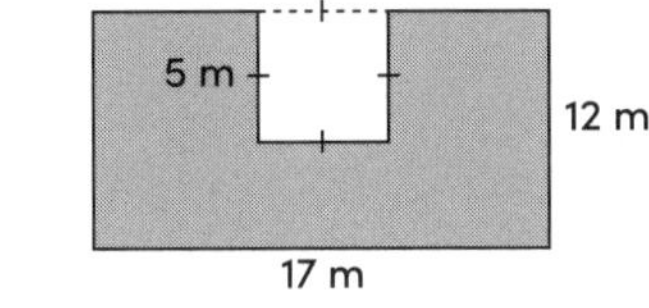


________________

________________

________________

**h**

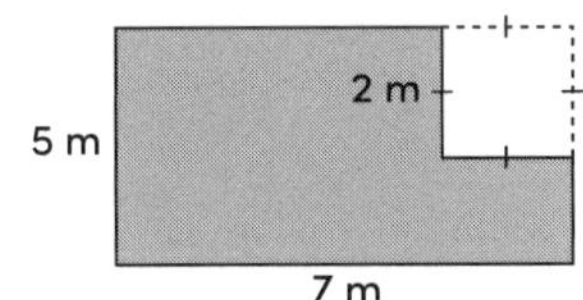


________________

________________

________________

# MEASUREMENT
## Area of composite shapes

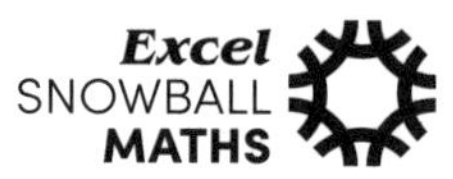


**1** Find the shaded area for the following composite shapes:

**a**

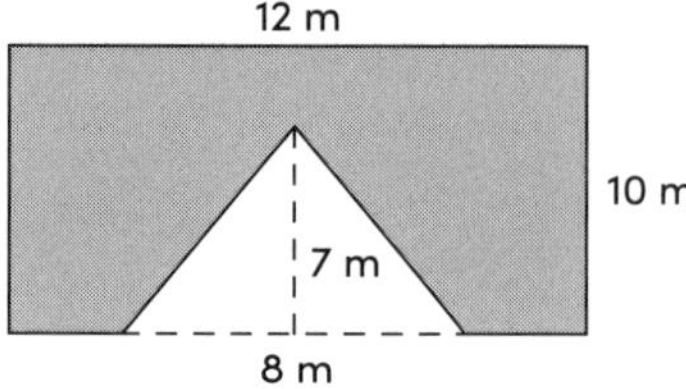


$A_1$ = ____ × ____ = ________ m²

$A_2$ = $\frac{\square}{\square}$ × ____ × ____ = ________ m²

Shaded $A$ = ________ − ________

= ________ m²

**b**

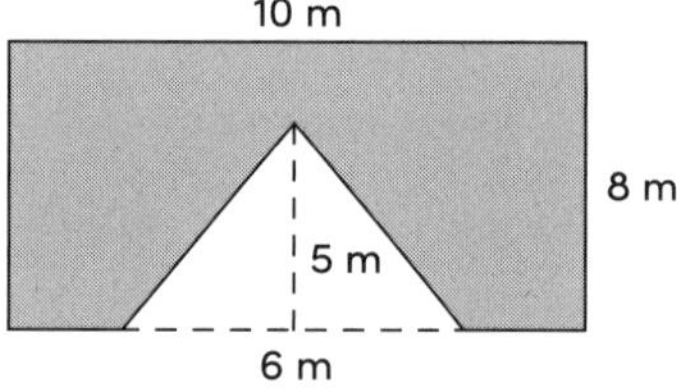


$A_1$ = ____ × ____ = ________ m²

$A_2$ = $\frac{\square}{\square}$ × ____ × ____ = ________ m²

Shaded $A$ = ________ − ________

= ________

**c**

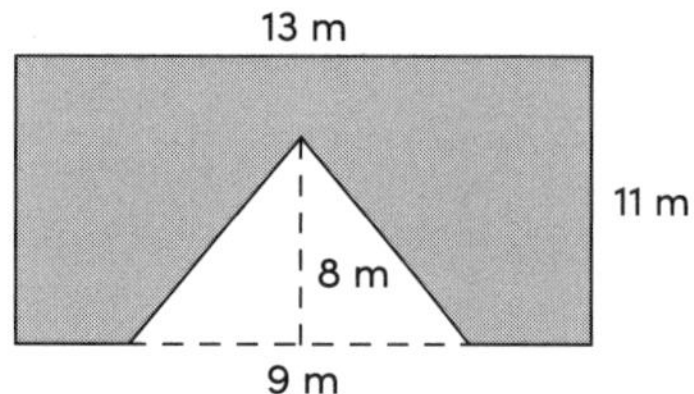


$A_1$ = ________________________________

$A_2$ = ________________________________

Shaded $A$ = ________________________________

**d**

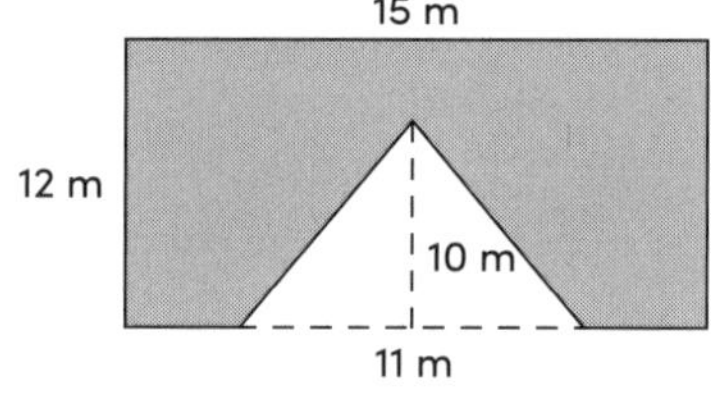


________________________________

________________________________

________________________________

**e**

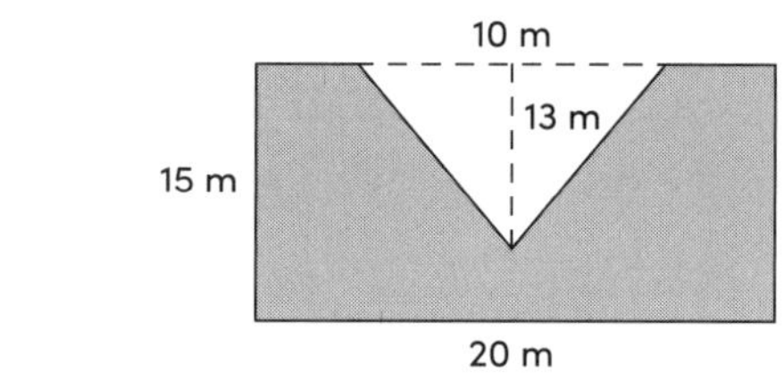


________________________________

________________________________

________________________________

**f**

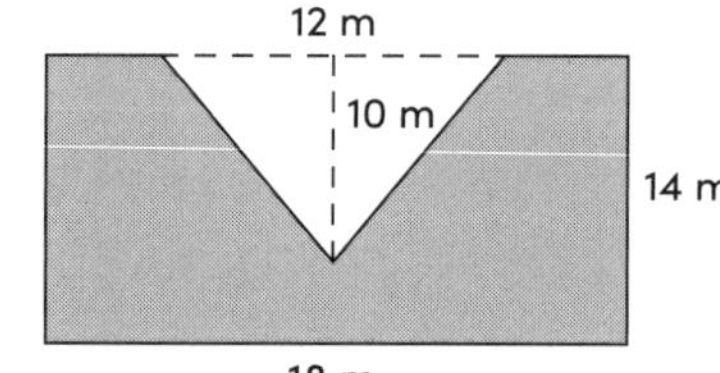


________________________________

________________________________

________________________________

**MEASUREMENT**

# Area of composite shapes

**1** Find the area of the following composite shapes:

**a**

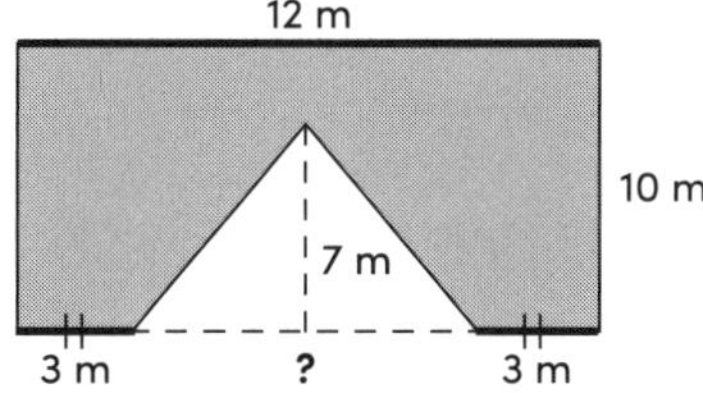


$A_1$ = ____ × ____ = ________

? = 12 − ____ − ____ = ________

$A_2$ = $\frac{\square}{\square}$ × ____ × ____ = ________

Shaded $A$ = ____ − ____ = ________ m²

**b**

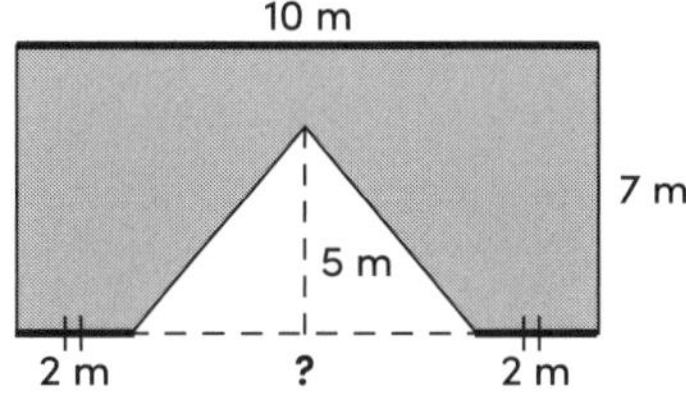


$A_1$ = ____ × ____ = ________

? = 10 − ____ − ____ = ________

$A_2$ = $\frac{\square}{\square}$ × ____ × ____ = ________

Shaded $A$ = ____ − ____ = ________ m²

**c**

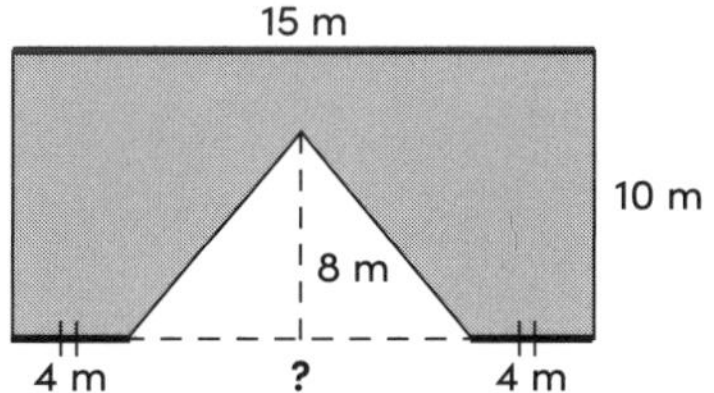


$A_1$ = ____ × ____ = ________

? = ____ − ____ − ____ = ________

$A_2$ = $\frac{\square}{\square}$ × ____ × ____ = ________

Shaded $A$ = ________________

**d**

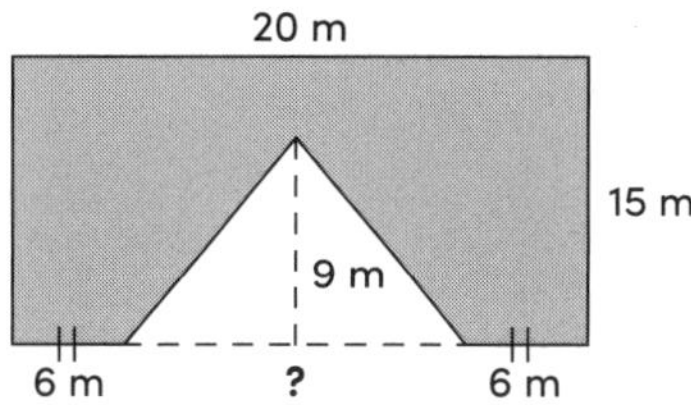


$A_1$ = ____ × ____ = ________

? = ____ − ____ − ____ = ________

$A_2$ = $\frac{\square}{\square}$ × ____ × ____ = ________

Shaded $A$ = ________________

**e**

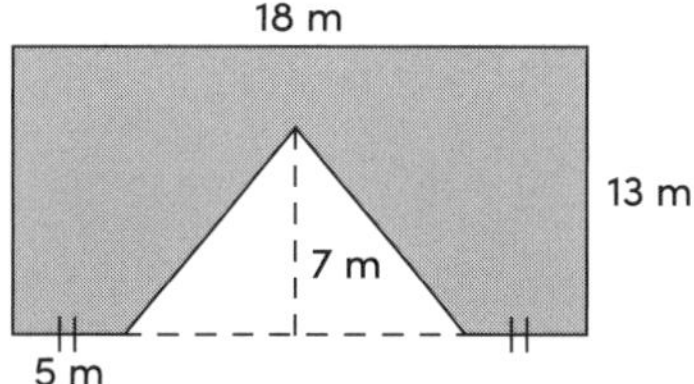


________________________________

________________________________

________________________________

________________________________

**f**

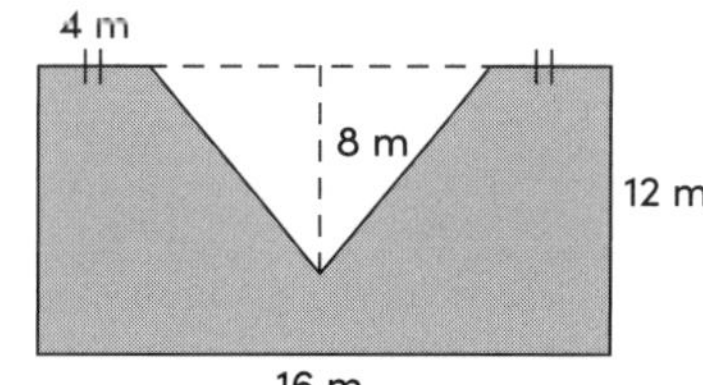


________________________________

________________________________

________________________________

________________________________

# Area of composite shapes

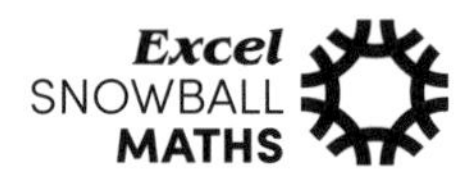


**1** Find the area of the following composite shapes:

**a**

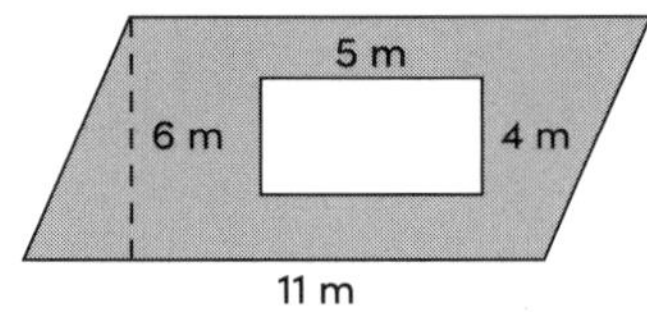


$A_1 = 11 \times$ _____ = ________ $m^2$

$A_2 = 5 \times$ _____ = ________ $m^2$

Shaded $A$ = ________ − ________

= ________ $m^2$

**b**

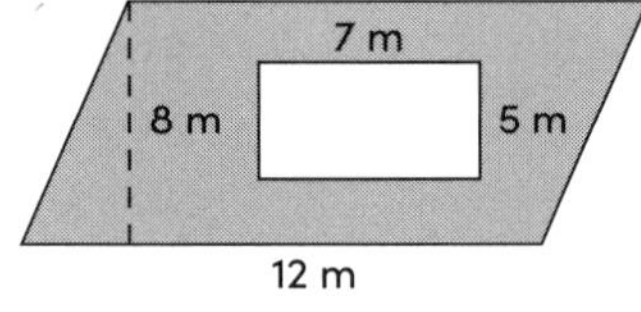


**c**

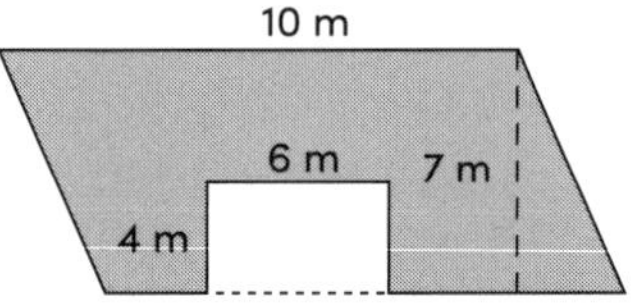


**d**

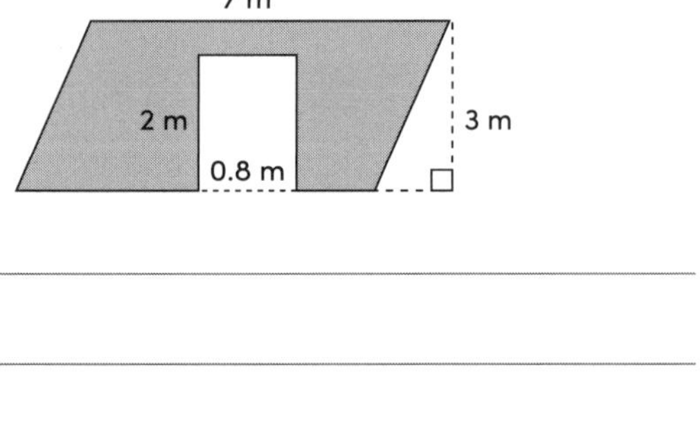


**2** Find the shaded area for the following shapes:

**a**

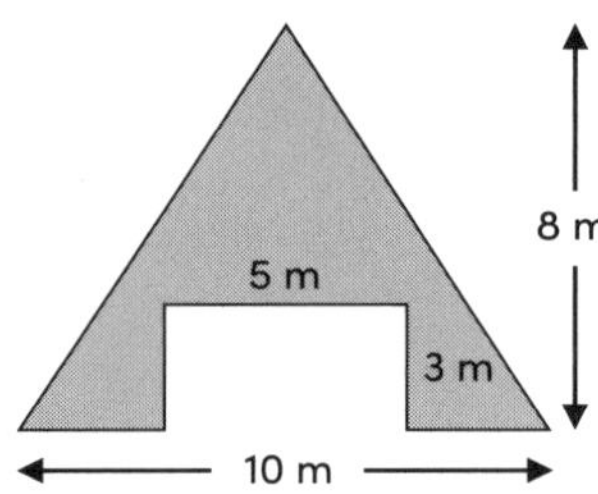


$A_1 = \frac{1}{2} \times 10 \times$ _____ = ________ $m^2$

$A_2 = 5 \times$ _____ = ________ $m^2$

Shaded $A$ = ________ − ________

= ________ $m^2$

**b**

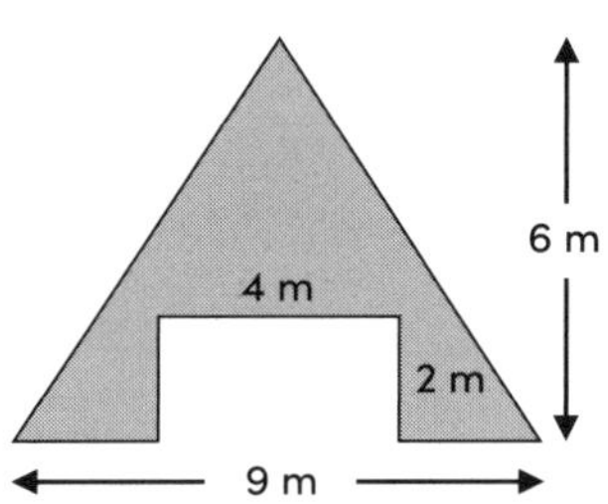


**c**

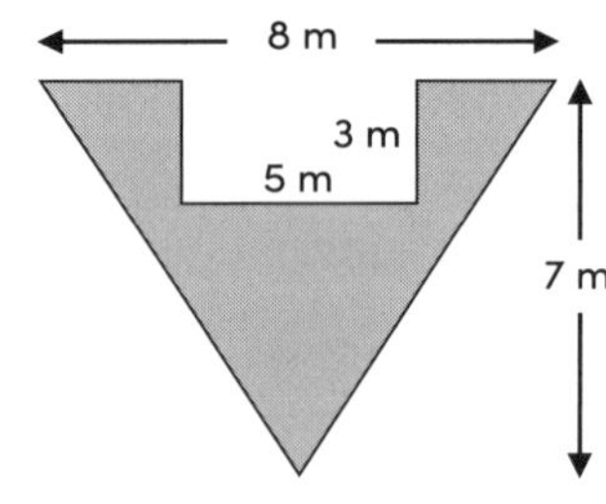

UNIT 23

MEASUREMENT

# Units of area

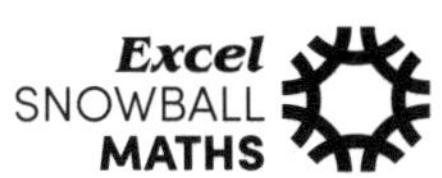


$mm^2$ $cm^2$ $m^2$ ha $km^2$

$mm^2$ → (÷ 100) → $cm^2$ → (÷ 10 000) → $m^2$ → (÷ 1 000 000) → $km^2$

$km^2$ → (× 1 000 000) → $m^2$ → (× 10 000) → $cm^2$ → (× 100) → $mm^2$

**1** What unit of area would be appropriate to use to measure:

**a** a pen tip? ________

**b** the head of a nail? ________

**c** a folder? ________

**d** a laptop? ________

**e** a house? ________

**f** a tennis court? ________

**g** a football field? ________

**h** a small farm? ________

**i** NSW? ________

**j** Sydney? ________

**k** an ice-skating rink? ________

**l** a calendar? ________

**m** a desk? ________

**n** a shopping centre? ________

**o** a city? ________

**p** the tip of a needle? ________

**q** a small painting ________

**r** a computer monitor? ________

**2** Convert the following:

**a** 530 $mm^2$ to $cm^2$
= 530 ÷ ____
= ____ $cm^2$

**b** 954 $mm^2$ to $cm^2$
= 954 ÷ ____
= ____ $cm^2$

**c** 384 $mm^2$ to $cm^2$
= ____
= ____

**d** 508 $mm^2$ to $cm^2$
= ____
= ____

**e** 1603 $mm^2$ to $cm^2$
= ____
= ____

**f** 72 $mm^2$ to $cm^2$
= ____
= ____

**g** 6.2 $cm^2$ to $mm^2$
= 6.2 × ____
= ____ $mm^2$

**h** 9.54 $cm^2$ to $mm^2$
= 9.54 × ____
= ____ $mm^2$

**i** 16.03 $cm^2$ to $mm^2$
= ____
= ____

**j** 0.72 $cm^2$ to $mm^2$
= ____
= ____

**k** 3.84 $cm^2$ to $mm^2$
= ____
= ____

**l** 24.1 $cm^2$ to $mm^2$
= ____
= ____

**m** 450 $mm^2$ to $cm^2$
= ____
= ____

**n** 5.8 $cm^2$ to $mm^2$
= ____
= ____

UNIT 24

MEASUREMENT

# Units of area

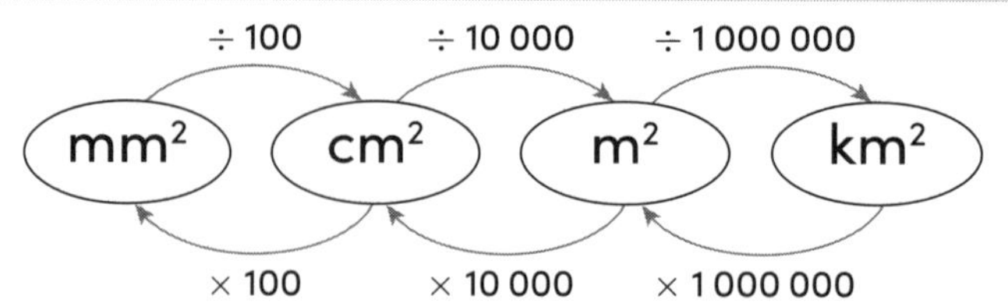


**1** Convert the following:

**a** 46 500 cm² to m²
= 46 500 ÷ ______
= ______

**b** 742 000 cm² to m²
= 742 000 ÷ ______
= ______

**c** 160 300 cm² to m²
= ______
= ______

**d** 95 000 cm² to m²
= ______
= ______

**e** 2100 cm² to m²
= ______
= ______

**f** 6500 cm² to m²
= ______
= ______

**g** 0.007 46 m² to cm²
= ______
= ______

**h** 0.055 m² to cm²
= ______
= ______

**i** 0.13 m² to cm²
= ______
= ______

**j** 0.21 m² to cm²
= ______
= ______

**2** Convert the following:

**a** 32 000 m² to ha
= ______ ÷ 10 000
= ______ ha

**b** 92 000 m² to ha
= ______ ÷ 10 000
= ______ ha

**c** 724 000 m² to ha
= ______
= ______

**d** 1600 m² to ha
= ______
= ______

**e** 0.092 ha to m²
= ______ × 10 000
= ______

**f** 0.064 ha to m²
= ______ × 10 000
= ______

**g** 0.0415 ha to m²
= ______
= ______

**h** 0.35 ha to m²
= ______
= ______

**i** 0.02 ha to m²
= ______
= ______

**j** 86 000 m² to ha
= ______
= ______

**3** Convert the following:

**a** 946 000 m² to km²
= ______
= ______

**b** 3 200 000 m² to km²
= ______
= ______

**c** 160 000 m² to km²
= ______
= ______

**d** 7 400 000 m² to km²
= ______
= ______

**e** 0.000 45 km² to m²
= ______
= ______

**f** 0.0032 km² to m²
= ______
= ______

**g** 0.015 km² to m²
= ______
= ______

**h** 186 000 m² to km²
= ______
= ______

**i** 0.000 048 km² to m²
= ______
= ______

MEASUREMENT

# Units of area

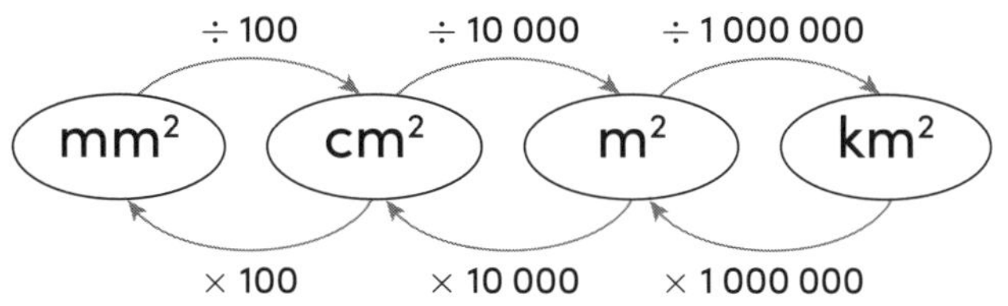


**1** A card has an area of 56 cm². What is this in square millimetres?

56 cm² to mm²

______________________

______________________

**2** A postage stamp has an area of 4 cm². What is this in square millimetres?

4 cm² to mm²

______________________

______________________

**3** A tablet device has an area of 400 cm². What is this in square millimetres?

______________________

______________________

**4** A parking space has an area of 15 m². What is this in square centimetres?

______________________

______________________

**5** A king-size bed has an area of 4.03 m². What is this in square centimetres?

______________________

______________________

**6** Sydney Opera House is 183 m long and 120 m wide.

**a** What is its area?

________ × ________

= ________

**b** What is this area in hectares, correct to two decimal places?

______________________

______________________

**7** A football field measures 109 m by 49 m.

**a** What is its area?

______________________

______________________

**b** What is this area in hectares, correct to two decimal places?

______________________

______________________

**8** A small rectangular farm has an area of 93 ha.

**a** What is this area in square metres?

______________________

______________________

**b** If the length is 1500 m, what is the width of the farm?

______________________

______________________

CHAPTER 6 REVIEW

# Measurement



**1** Find the perimeter of the following in cm (all measurements are in cm):

**a**

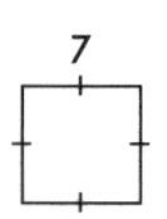


**b**

**c**

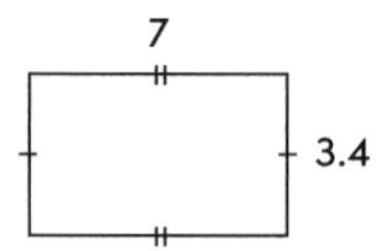


**d**

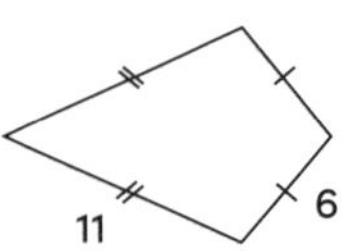


**e**

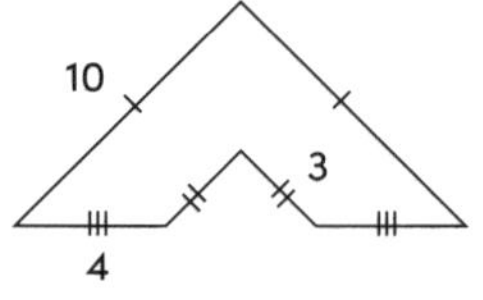


**f**

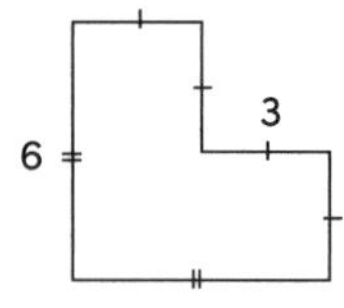


**g**

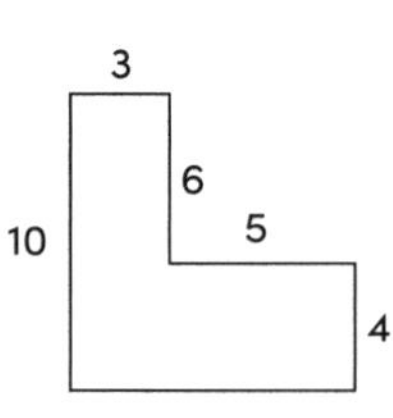


**h**

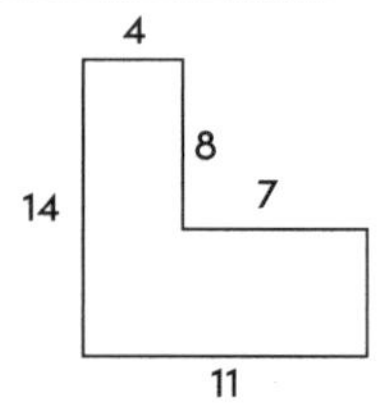


**2** Convert the following units of length:

**a** 5 cm to mm

**b** 12.6 cm to mm

**c** 150 mm to cm

**d** 75 mm to cm

**e** 4.5 km to m

**f** 0.95 km to m

**g** 180 cm to m

**h** 60 cm to m

**i** 1200 m to km

**j** 5000 m to km

**k** 0.04 m to mm

**l** 3500 mm to m

**3** Find the area of the following in square centimetres:

**a**

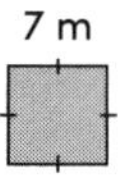


**b**

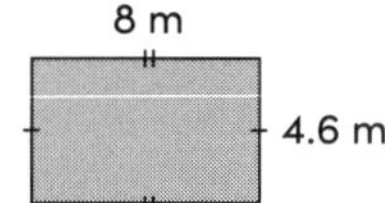


**c**

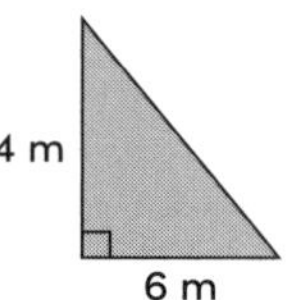


**d**

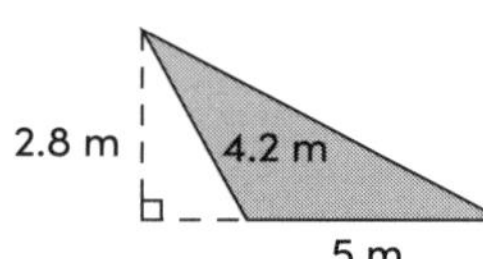


**e**

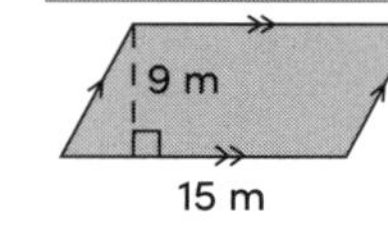


**f**

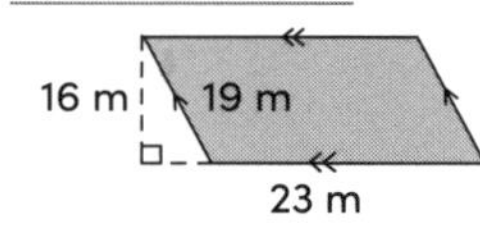


**4** A rectangular pool table measures 2.5 m in length and 1.3 m in width. What is the area of the table?

**5** Find the area of a triangle that has a perpendicular height of 0.8 m and base length of 3.1 m.

**6** Find the base length of a parallelogram with an area of 94.5 $m^2$ and perpendicular height of 6.3 m.

**7** Find the area of the following in square metres (all measurements are in metres):

**a**

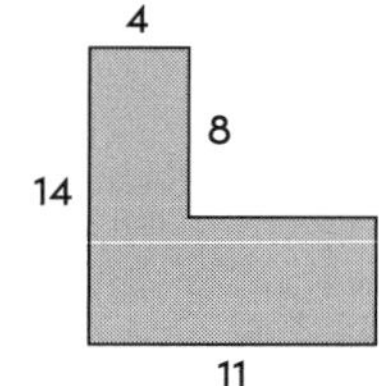


**b**

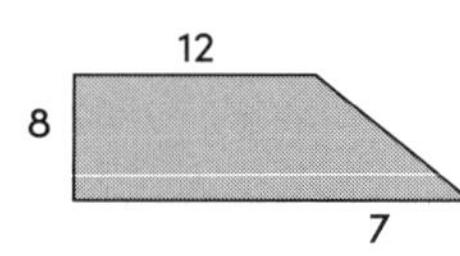


**c**

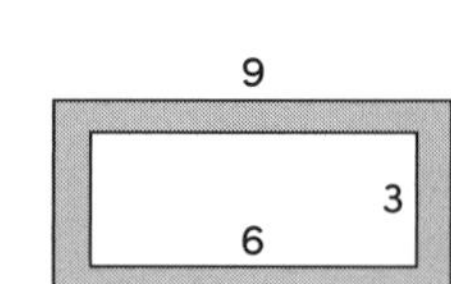


**d**

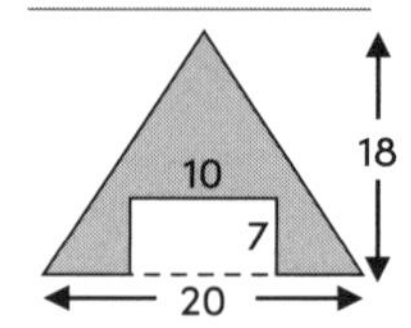


**8** Convert the following units of area:

**a** 240 $mm^2$ to $cm^2$

**b** 8.5 $cm^2$ to $mm^2$

**c** 56 000 $cm^2$ to $m^2$

**d** 1.7 ha to $m^2$

**GEOMETRY**

# Naming angles

**1** Name the following intervals:

**a**

______________

**b**

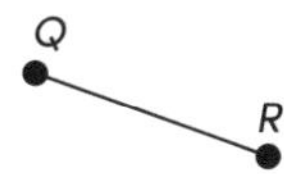


______________

**c**

______________

**d**

______________

**e**

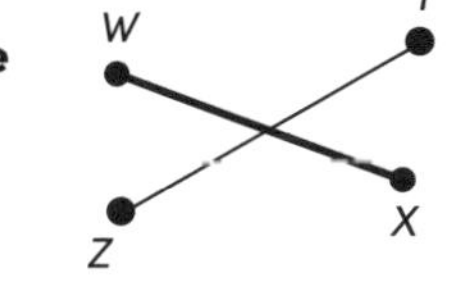


______________

______________

**f**

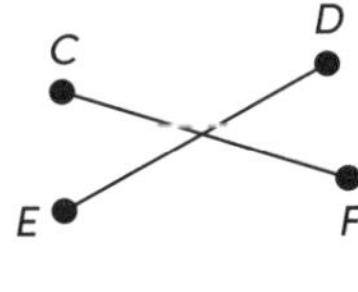


______________

______________

**2** Draw and label the following intervals:

**a** *AB*

**b** *PQ*

**3** Name the following marked angles:

**a**

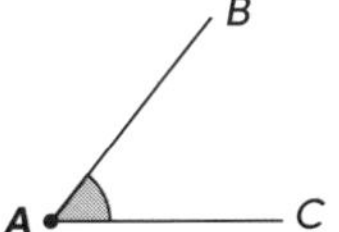


∠ ___ A ___

**b**

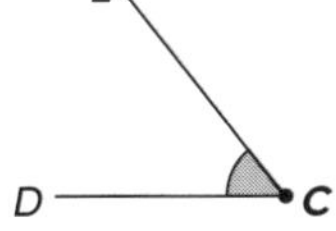

∠ ___ C ___

**c**

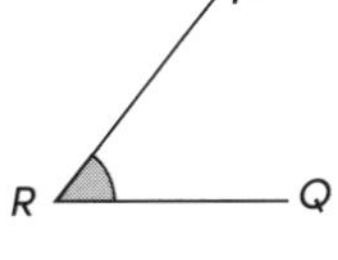


∠ ___ ___ ___

**d**

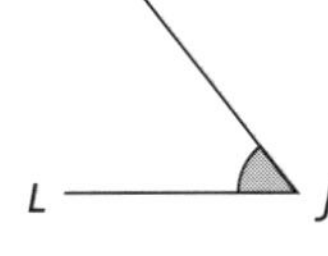

∠ ___ ___ ___

**e**

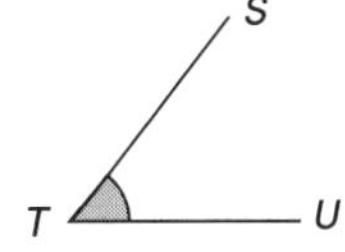


______________

**f**

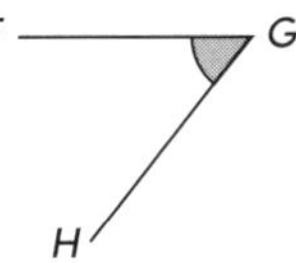


______________

**g**

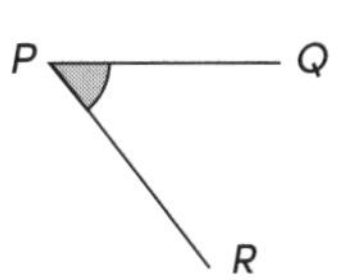


______________

**h**

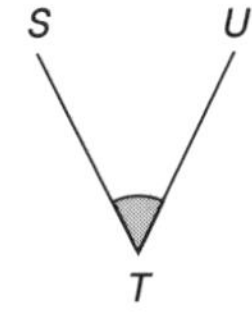


______________

**i**

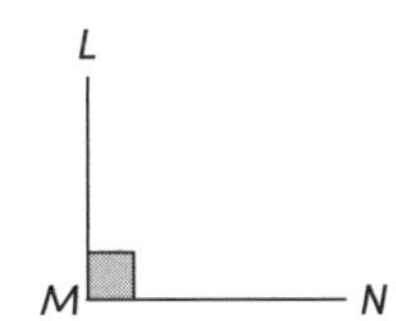


______________

**j**

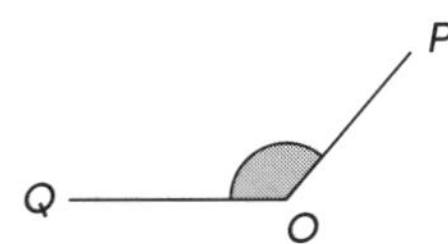


______________

**k**

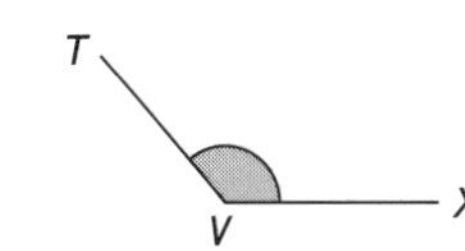


______________

**l**

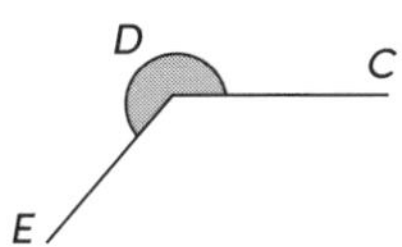


______________

**4** Draw and label the following angles:

**a** ∠*ABC*

**b** ∠*OPQ*

**c** ∠*XYZ*

**d** ∠*JKL*

UNIT 2

**GEOMETRY**

# Naming angles

**1** Name the following angles:

**a**

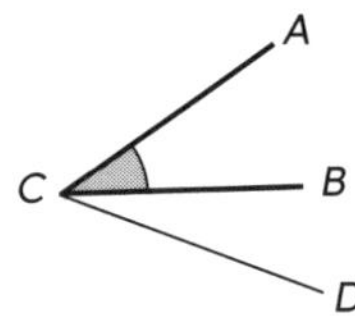


Marked angle:

∠ ______ C ______

Adjacent angle:

∠ ______ C ______

**b**

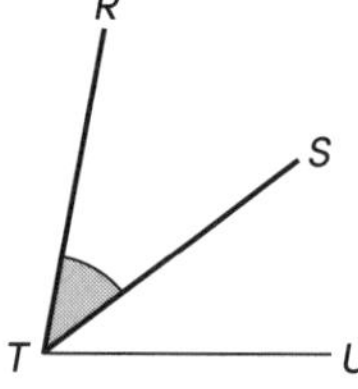


Marked angle:

∠ ______ T ______

Adjacent angle:

∠ ______ T ______

**c**

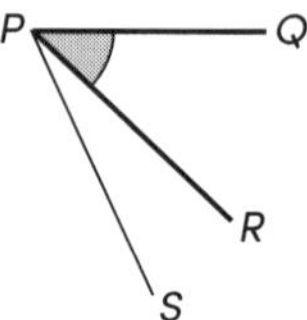


Marked angle:

∠ ____________

Adjacent angle:

∠ ____________

**d**

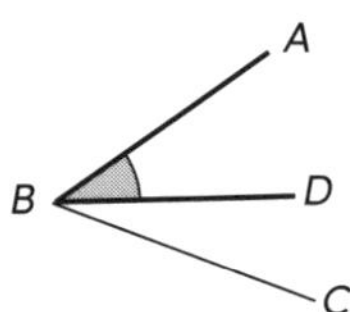


Marked angle:

∠ ____________

Adjacent angle:

∠ ____________

**e**

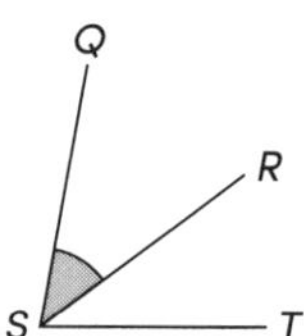


Marked angle:

∠ ____________

Adjacent angle:

∠ ____________

**f**

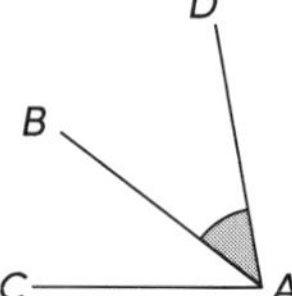


Marked angle:

∠ ____________

Adjacent angle:

∠ ____________

**g**

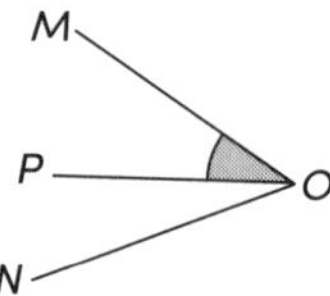


Marked angle:

____________

Adjacent angle:

____________

**h**

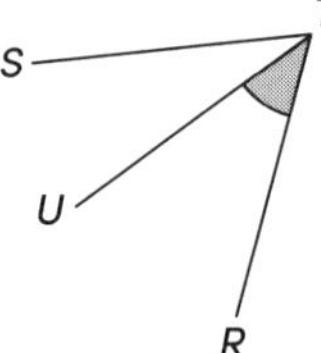


Marked angle:

____________

Adjacent angle:

____________

**i**

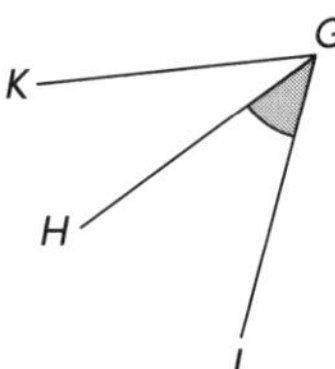


Marked angle:

____________

Adjacent angle:

____________

**j**

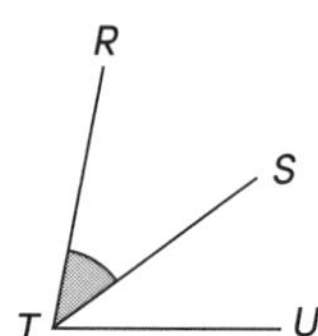


Marked angle:

____________

Adjacent angle:

____________

**k**

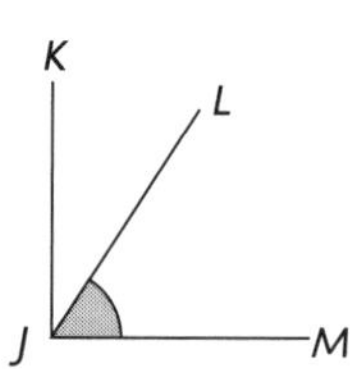


Marked angle:

____________

Adjacent angle:

____________

**l**

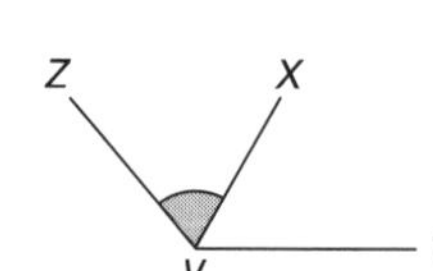


Marked angle:

____________

Adjacent angle:

____________

UNIT 3

GEOMETRY

# Naming angles

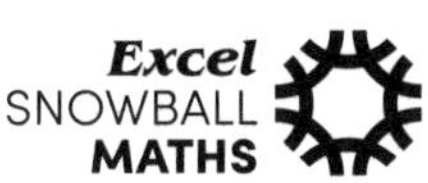


**1** Name the following marked angles:

**a** ∠ ______ A ______

∠ ______ B ______

∠ ______ C ______

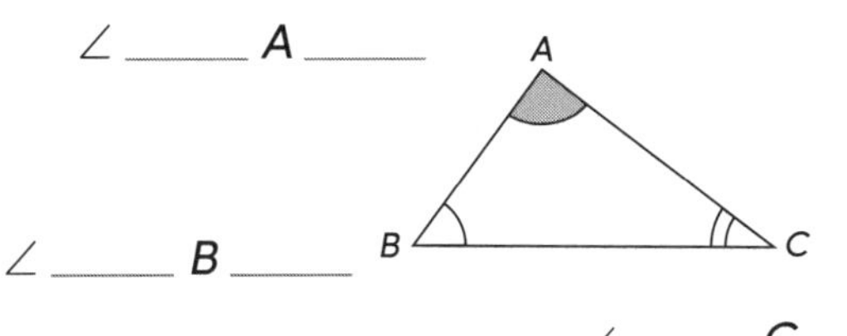


**b** ∠ ______ P ______

∠ ______ R ______

∠ ______ Q ______

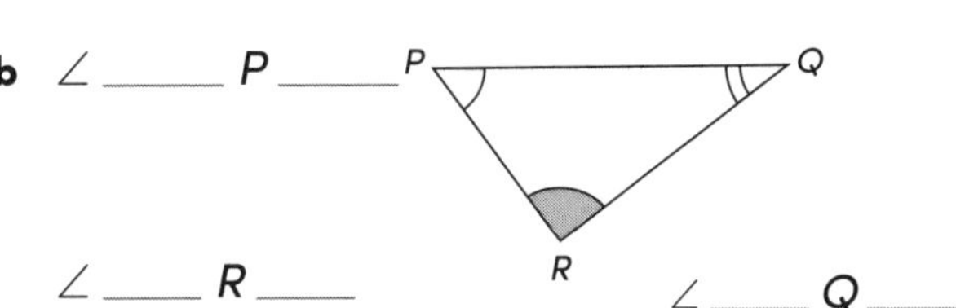


**c** ∠ ____________

∠ ____________

∠ ____________

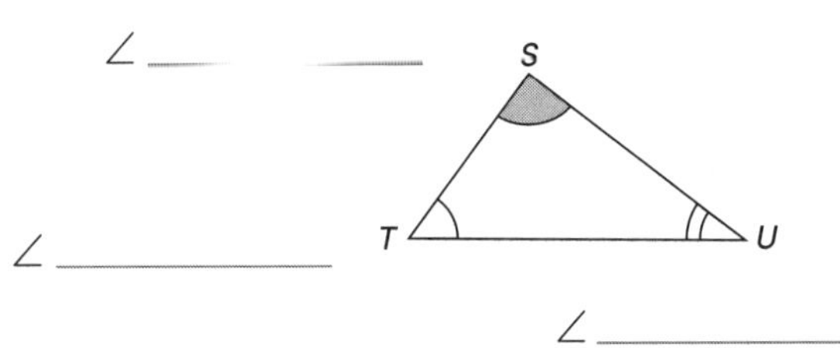


**d** ∠ ____________

∠ ____________

∠ ____________

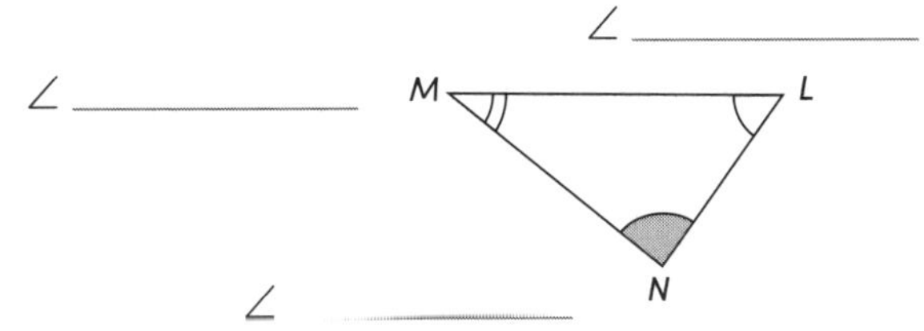


**e** ____________

____________

____________

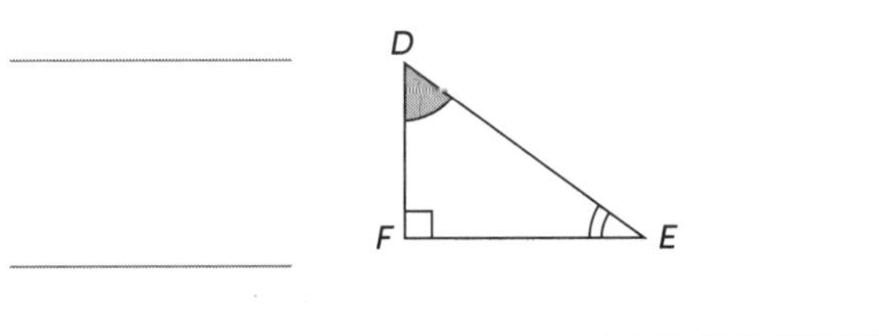


**f** ____________

____________

____________

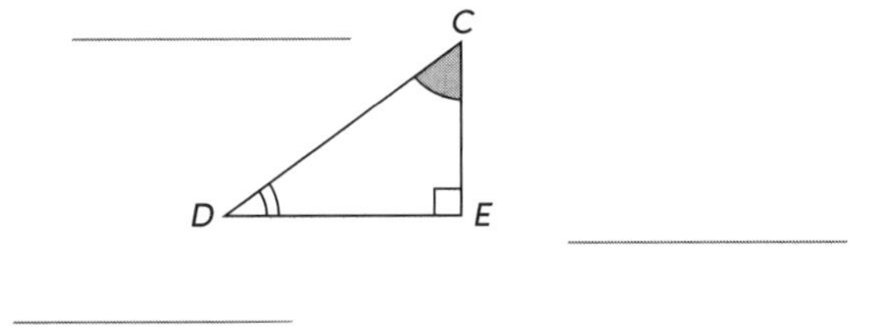


**2** Name the following marked angles:

**a** ∠ ______ E ______

∠ ______ E ______

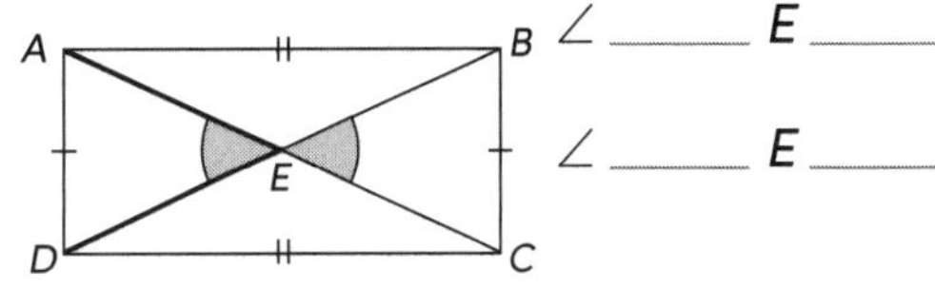


**b** ∠ ______ P ______

∠ ______ P ______

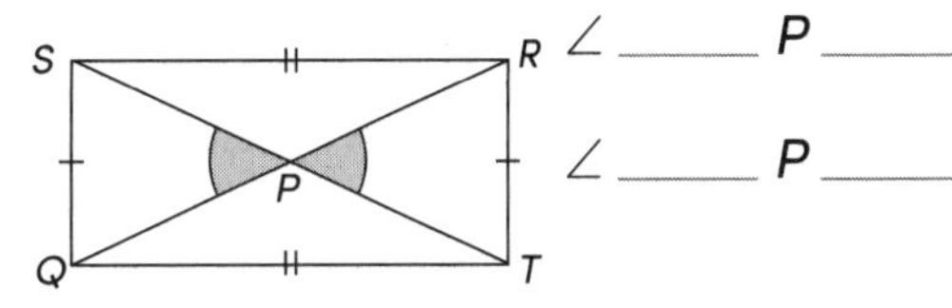


**c** ∠ ____________

∠ ____________

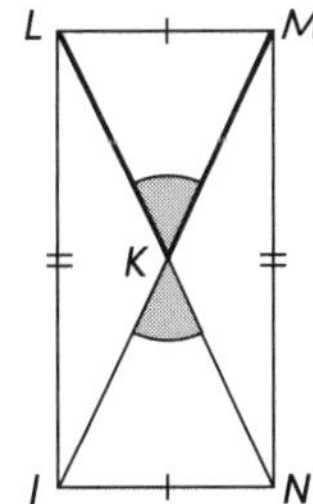


**d** ∠ ____________

∠ ____________

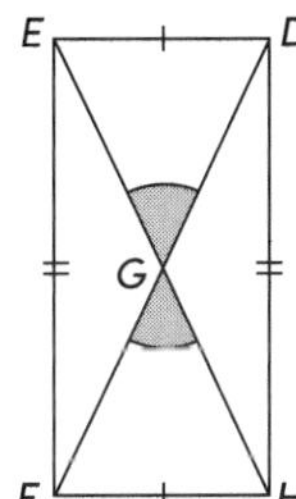


**e** ____________

____________

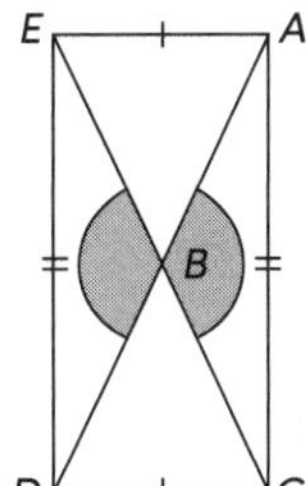


**f** ____________

____________

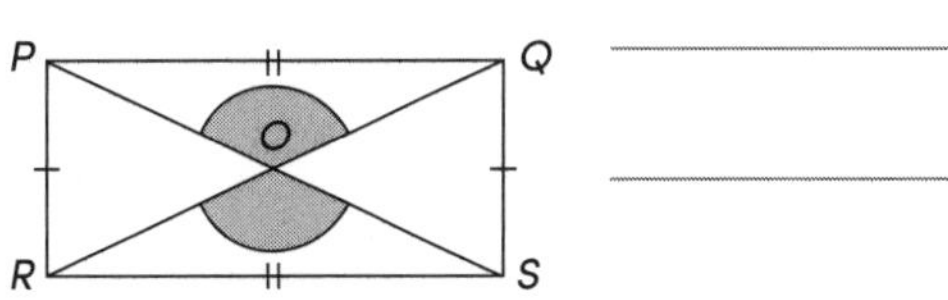

UNIT 4

**GEOMETRY**

# Naming angles

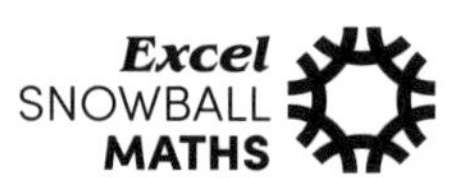


**1** Name the following marked angles:

**a**

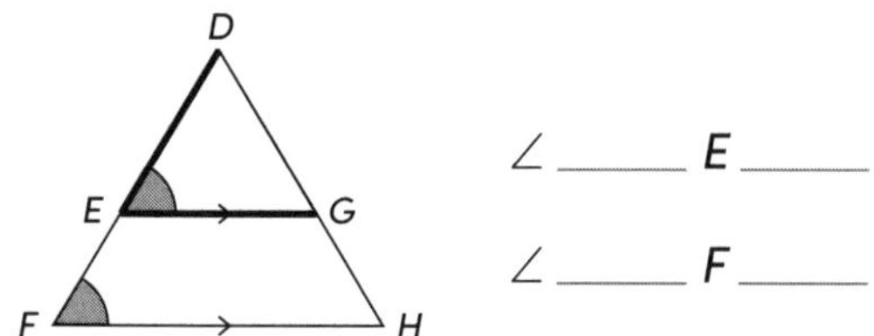


∠ ____ E ____

∠ ____ F ____

**b**

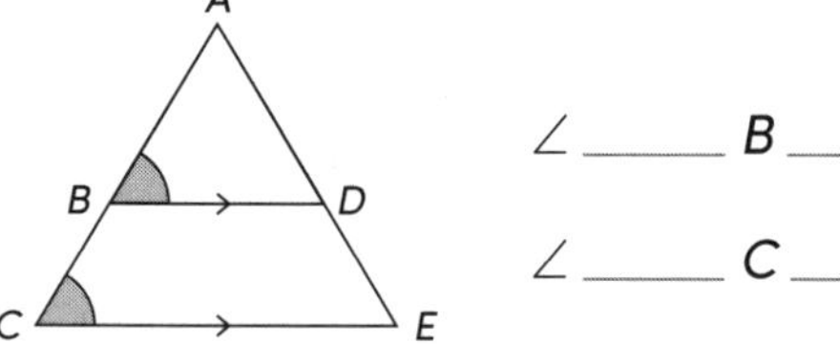


∠ ____ B ____

∠ ____ C ____

**c**

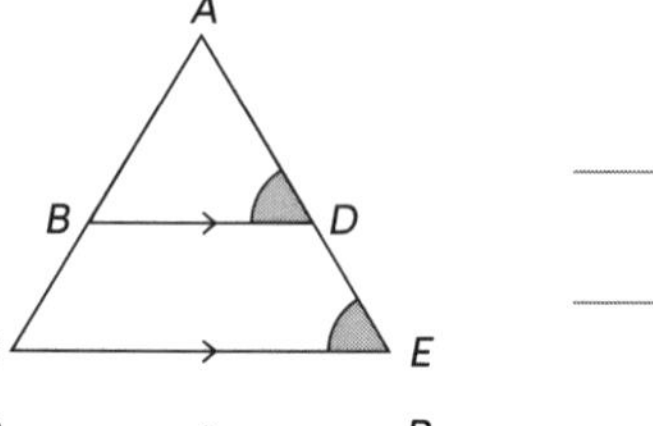


______

______

**d**

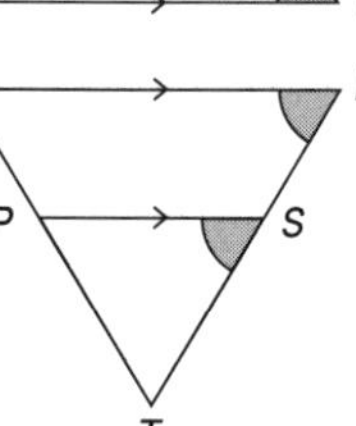


______

______

**2** Name the following marked angles:

**a**

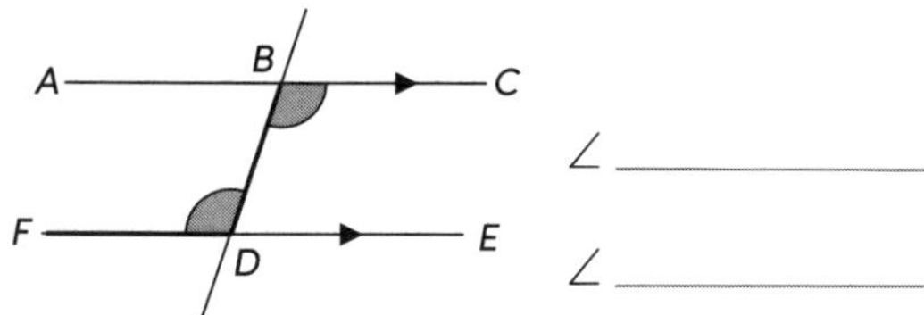


∠ ______

∠ ______

**b**

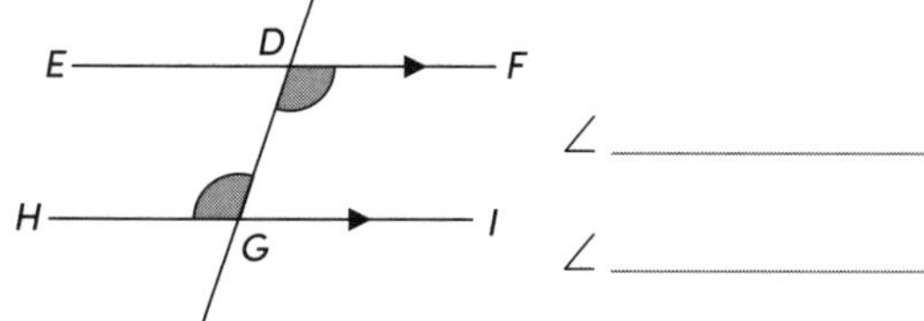


∠ ______

∠ ______

**c**

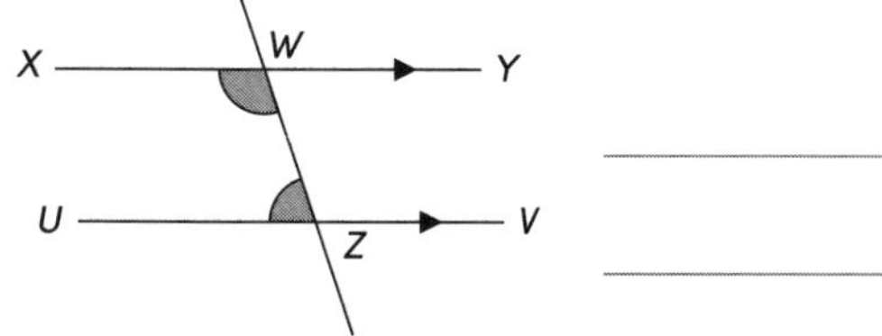


______

______

**d**

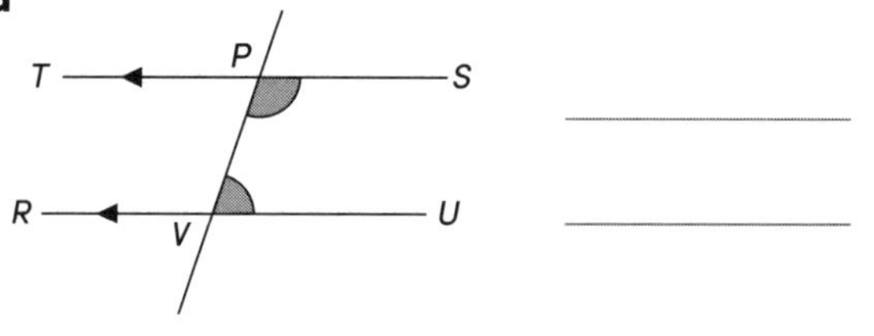


______

______

**3** Name the following marked angles:

**a**

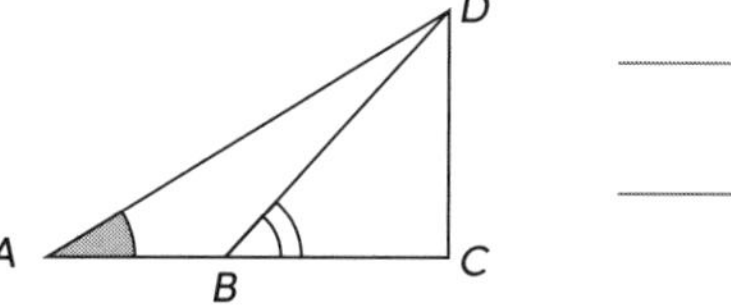


______

______

**b**

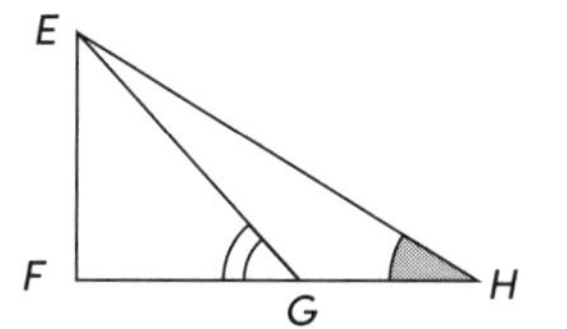


______

______

**c**

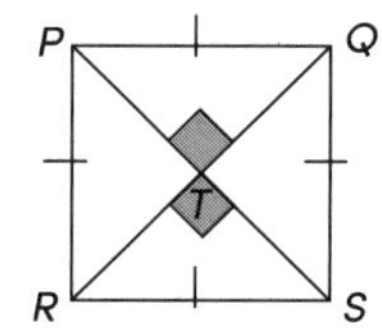


______

______

**d**

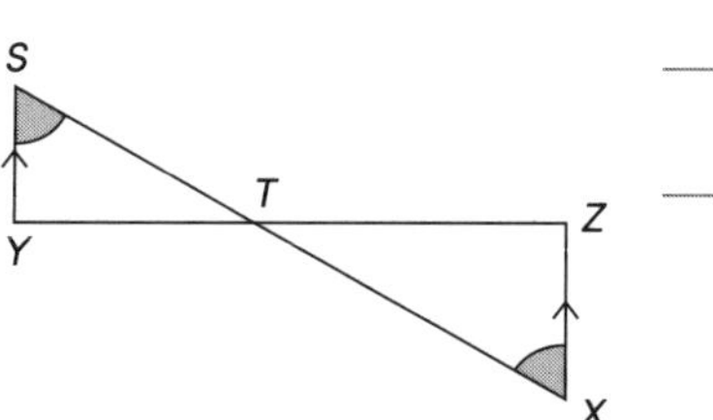


______

______

**e**

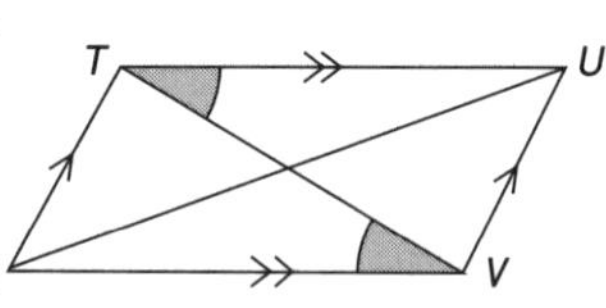


______

______

**f**

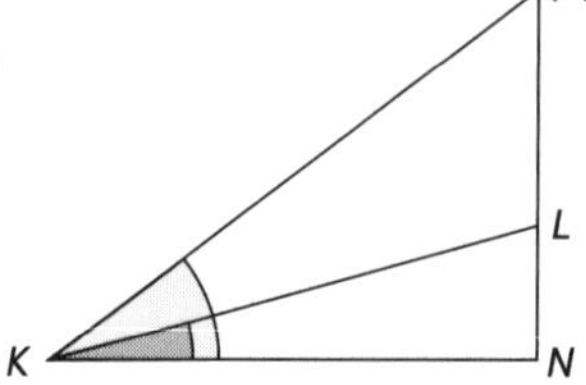


______

______

UNIT 5

**GEOMETRY**

# Types of angles

**1** Choose from the box to correctly name each angle:

| Acute | Straight | Obtuse |
|---|---|---|
| Right | Reflex | Revolution |

**a**

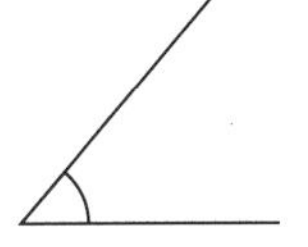

________________

**b**

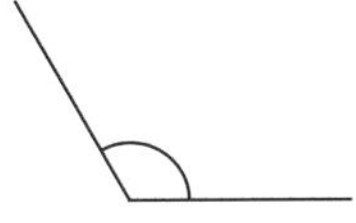

________________

**c**

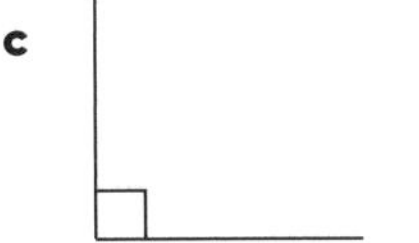

________________

**d**

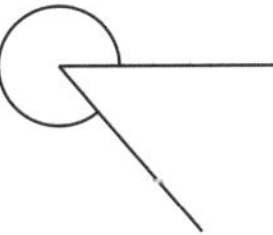

________________

**e**

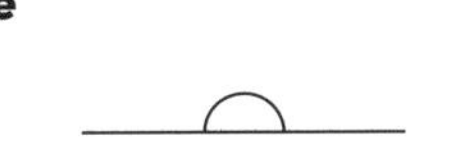

________________

**f**

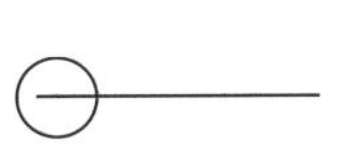

________________

**g**

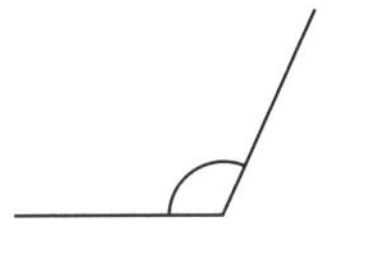

________________

**h**

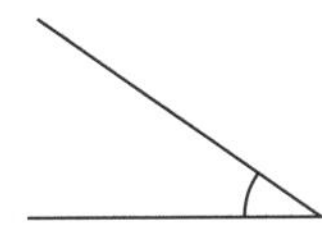

________________

**i**

________________

**j**

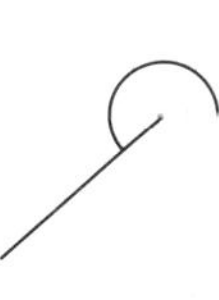

________________

**2** Name the type of angle in the diagrams below:

**a**

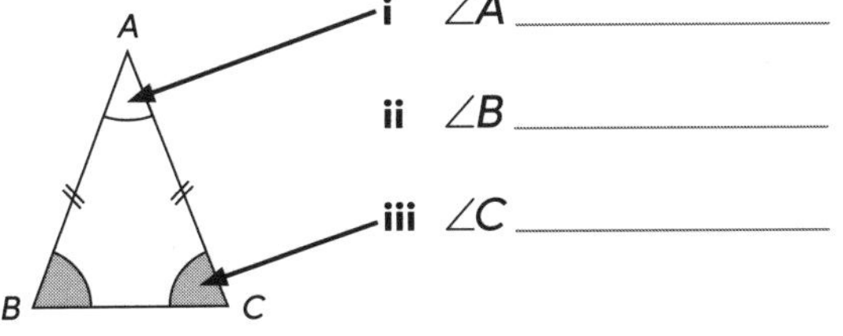


**i** $\angle A$ ________________

**ii** $\angle B$ ________________

**iii** $\angle C$ ________________

**b**

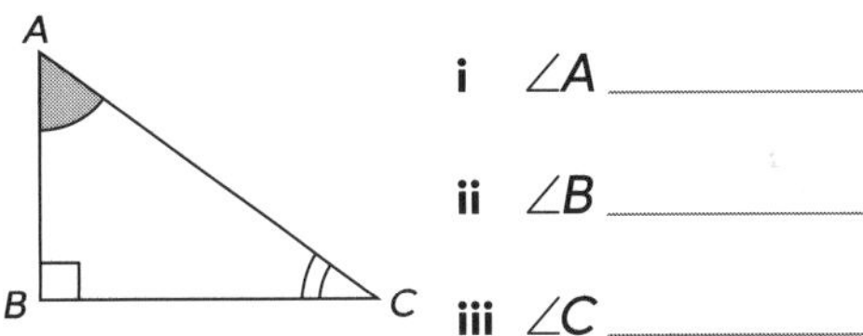


**i** $\angle A$ ________________

**ii** $\angle B$ ________________

**iii** $\angle C$ ________________

**c**

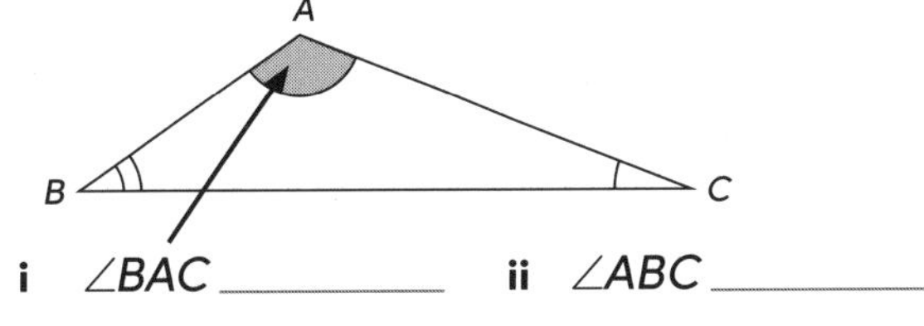


**i** $\angle BAC$ ____________ **ii** $\angle ABC$ ____________

**iii** $\angle BCA$ ____________

**d**

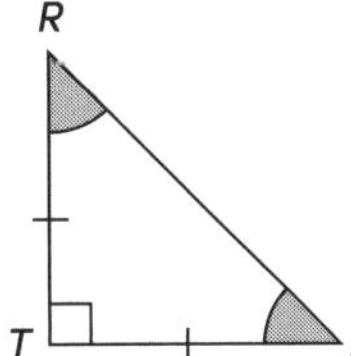


**i** $\angle SRT$ ____________

**ii** $\angle TSR$ ____________

**iii** $\angle RTS$ ____________

**e**

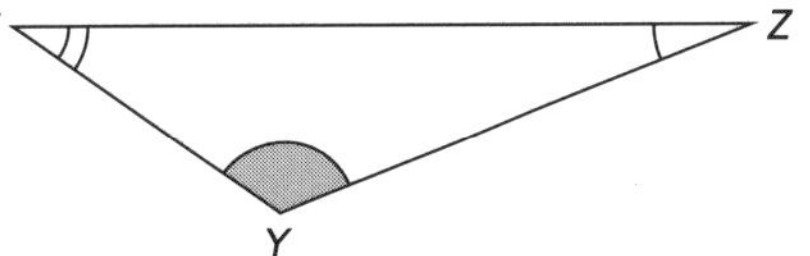


**i** $\angle YZX$ ____________ **ii** $\angle ZYX$ ____________

**iii** $\angle ZXY$ ____________

**f**

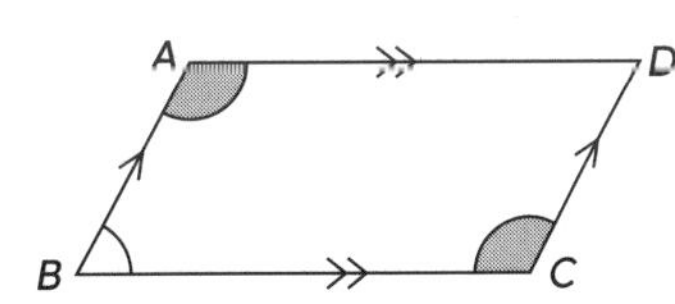


**i** $\angle BAD$ ____________ **ii** $\angle CBA$ ____________

**iii** $\angle DCB$ ____________

**g**

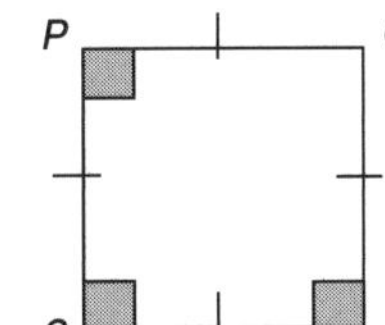


**i** $\angle QPS$ ____________

**ii** $\angle QRS$ ____________

**iii** $\angle RSP$ ____________

UNIT 6

## GEOMETRY

# Types of angles

**1** Draw two different examples of an:

**a** acute angle **b** obtuse angle

**c** right angle **d** reflex angle

**2**

| Angle type | Angle |
|---|---|
| Acute | $0° < a° < 90°$ |
| Right | 90° |
| Obtuse | $90° < a° < 180°$ |
| Straight | 180° |
| Reflex | $180° < a° < 360°$ |
| Revolution | 360° |

Use the table above to complete the table below:

| Angle | Angle type |
|---|---|
| 45° | |
| 60° | |
| 125° | |
| 90° | |
| 180° | |
| 165° | |
| 179° | |
| 270° | |
| 210° | |
| 360° | |
| 79° | |
| 190° | |

**3** How many degrees need to be added to 30° to make a right angle?

________ – 30 = ________°

**4** How many degrees need to be added to 45° to make a right angle?

________ – 45 = ________

**5** How many degrees need to be added to 60° to make a right angle?

**6** How many degrees need to be added to 53° to make a right angle?

**7** How many degrees need to be added to 120° to make a straight angle?

**8** How many degrees need to be added to 80° to make a straight angle?

**9** How many degrees need to be added to 90° to make a straight angle?

**10** How many degrees need to be added to 270° to make an angle of revolution?

**11** How many degrees need to be added to 180° to make an angle of revolution?

**GEOMETRY**

# Measuring angles

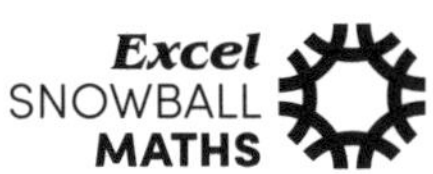


**1** Measure the size of each marked angle:

**Acute angles**

**a** ________°

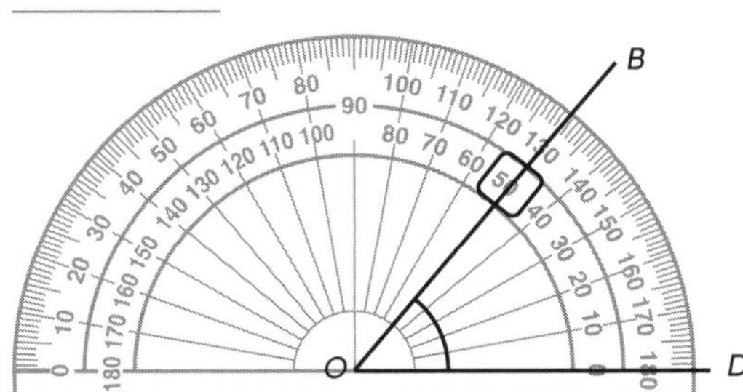


**b** ________°

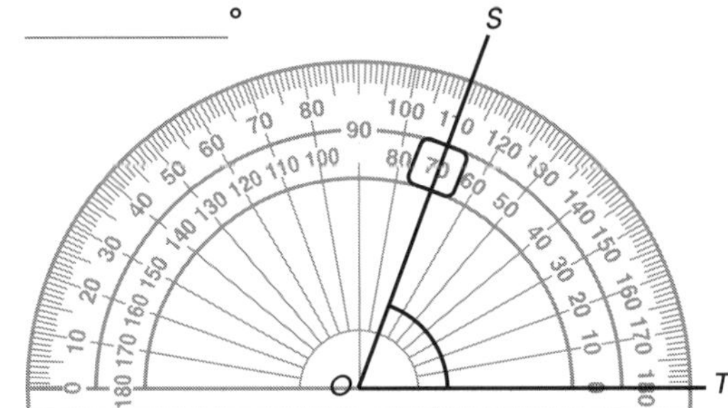


**c** ________

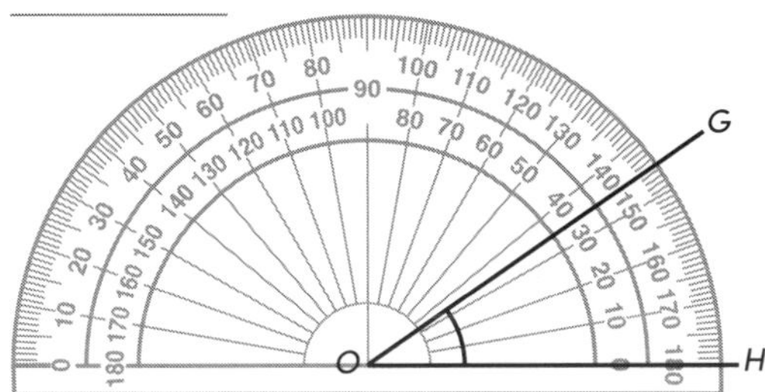


**d** ________

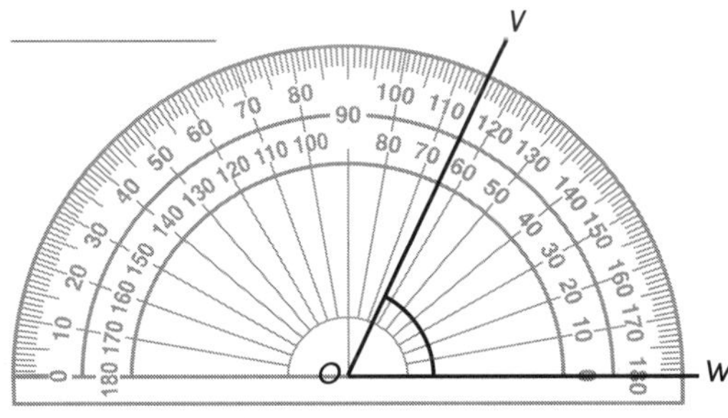


**e**

**f**

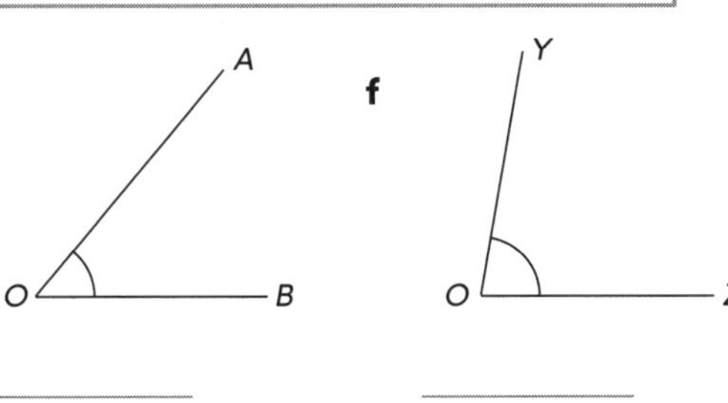


________ ________

**g**

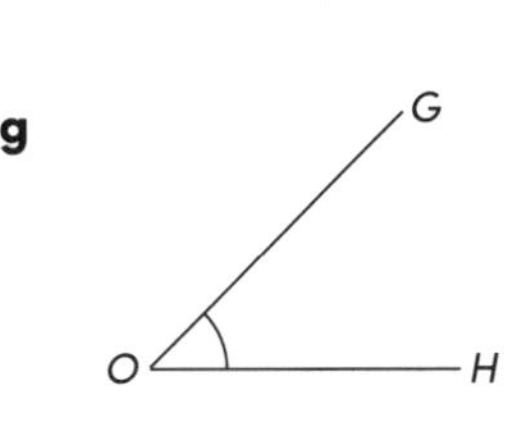


________

**Right angle**

**h**

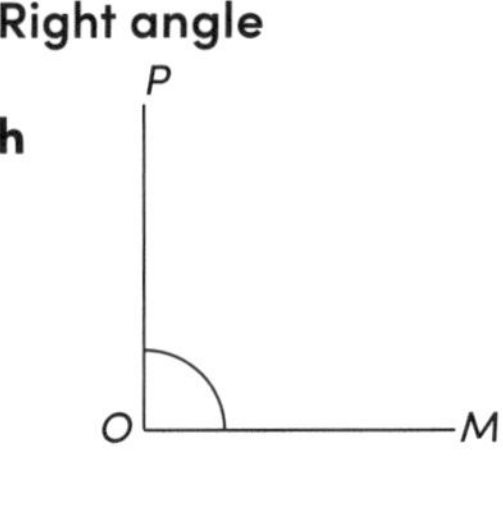


________

**Obtuse angles**

**i** ________

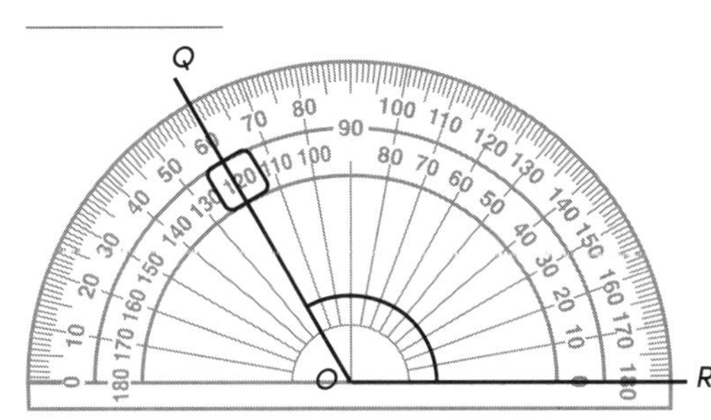


**j** ________

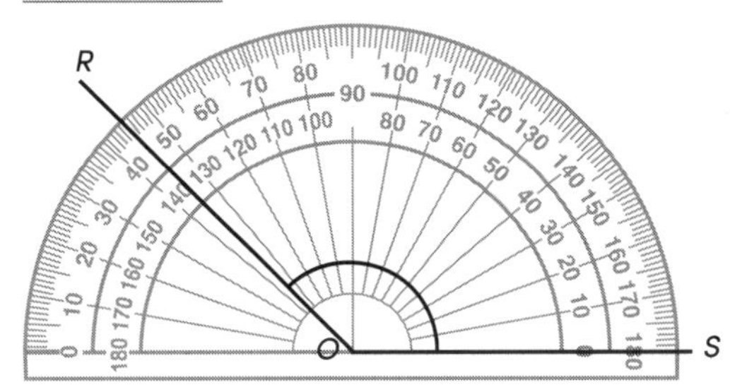


**k** ________

**l**

**m**

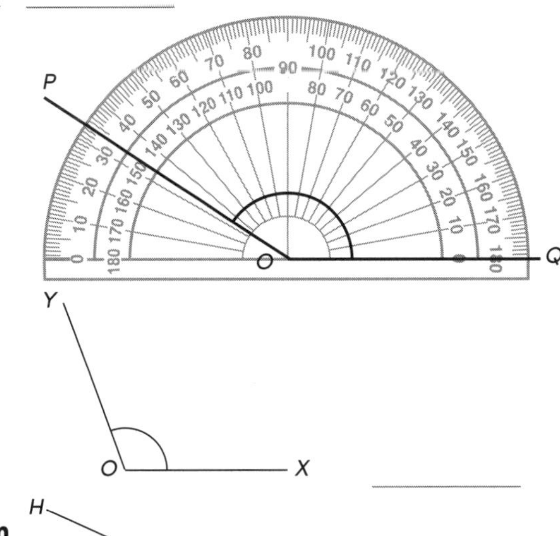


________

________

# UNIT 8

## GEOMETRY
# Measuring angles

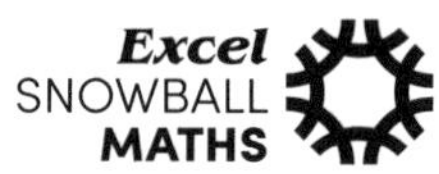


**1** Look at the top number and measure the size of each marked angle:

**a** ____________

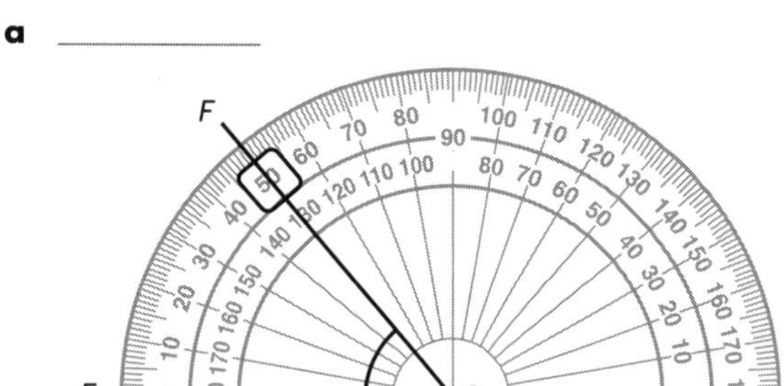

**b** ____________

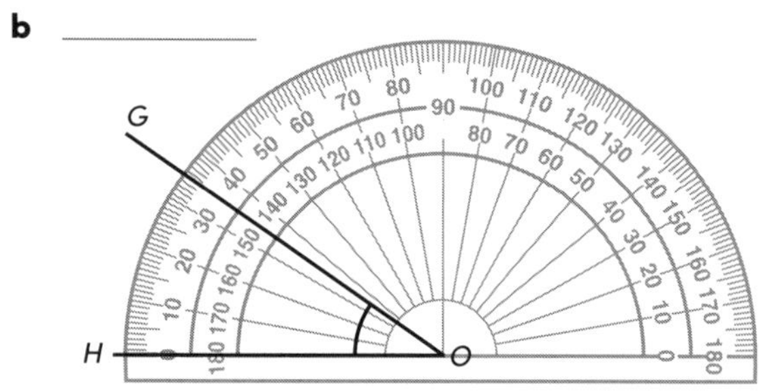

**c** **d**

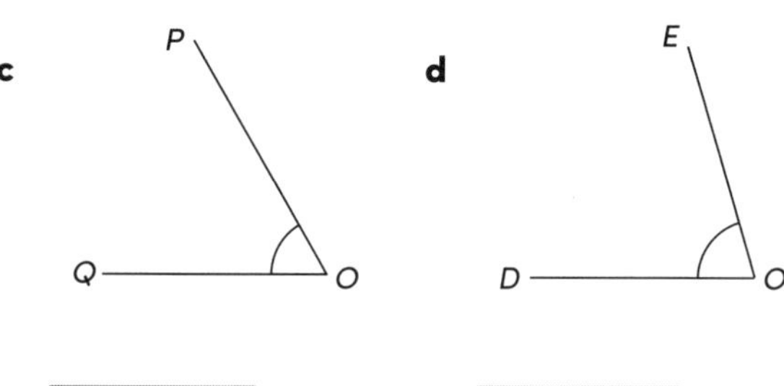

____________ ____________

**e** ____________

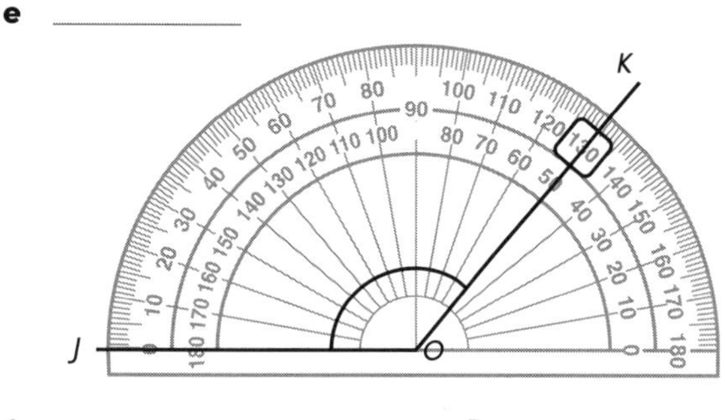

**f**

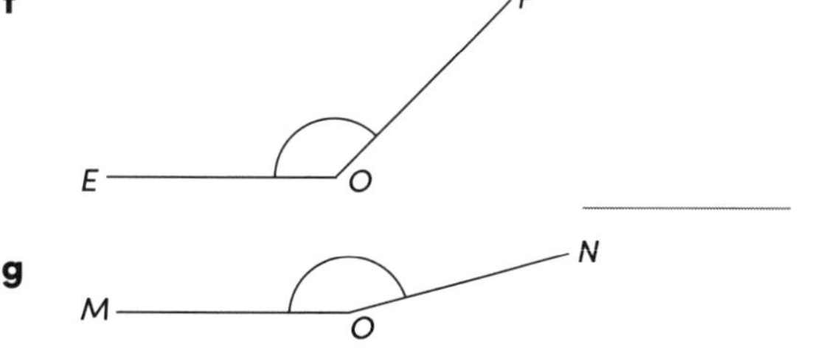

____________

**g**

____________

**2** Measure the size of each marked angle:

**a**

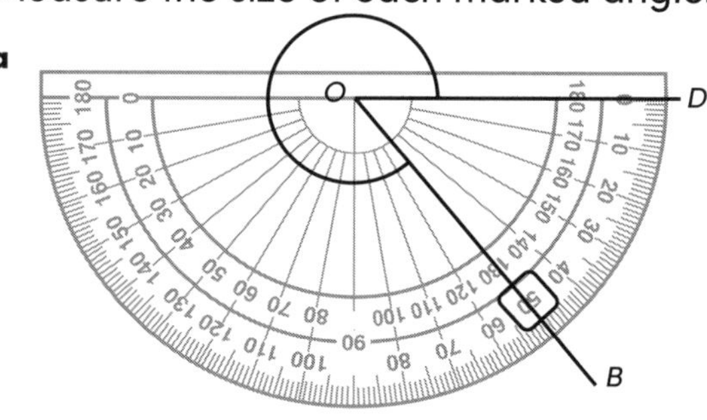

360 – ____________ = ____________

**b**

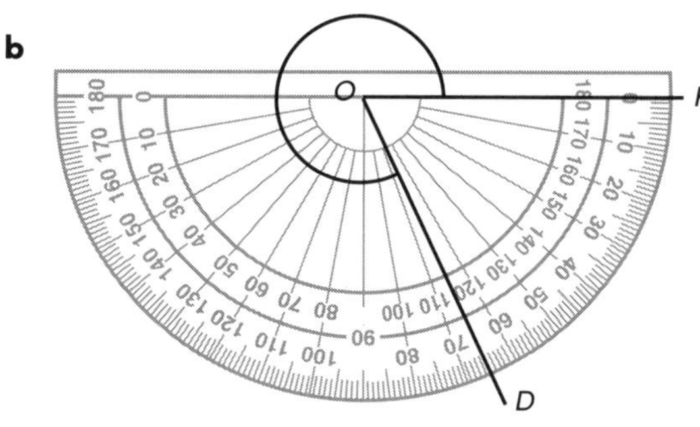

360 – ____________ = ____________

**c**

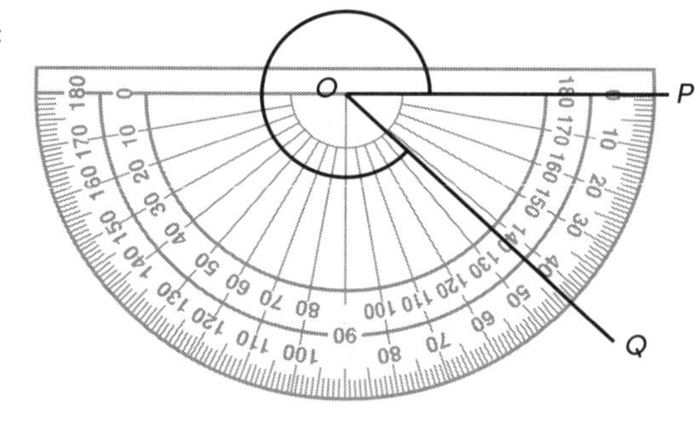

____________

**d** **e**

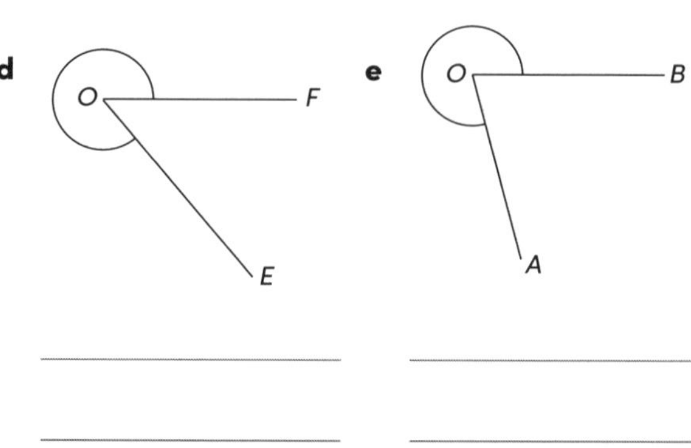

____________ ____________

____________ ____________

GEOMETRY
# Measuring angles

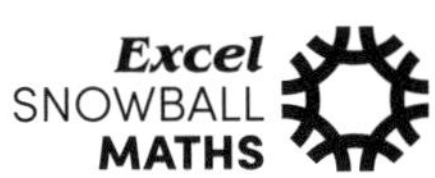


**1** Measure the following angles:

**a**

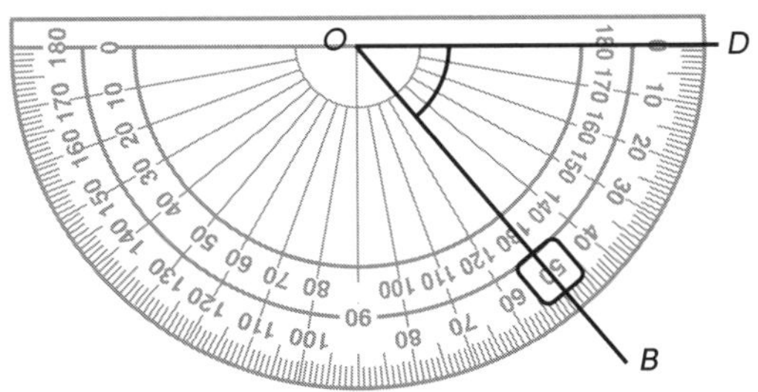


______

**b**

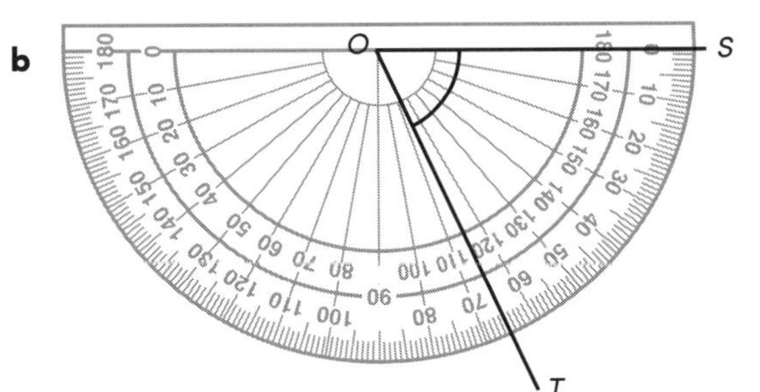


______

**c**

**d**

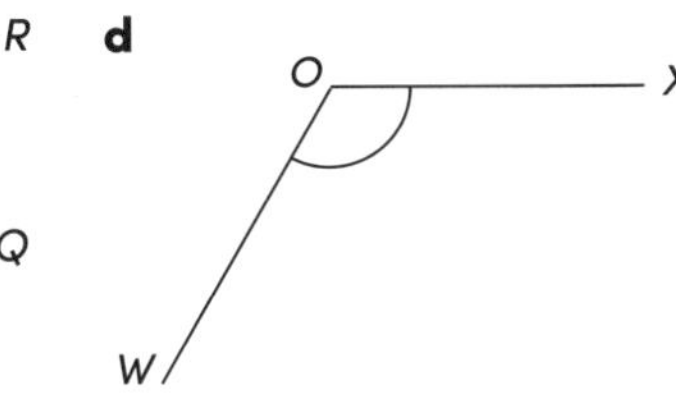


______ ______

**e**

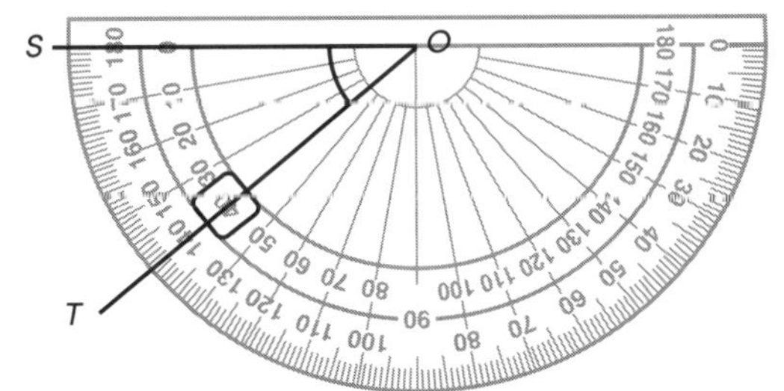


______

**f**

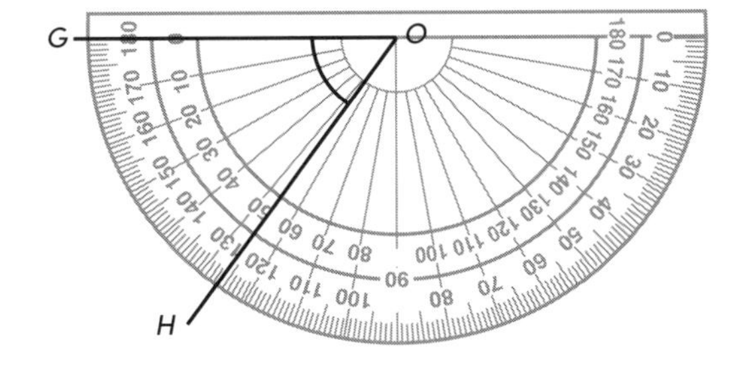


______

**g** **h**

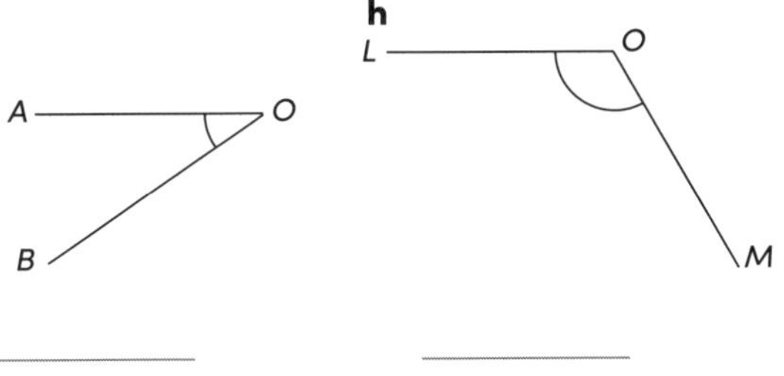


______ ______

**2** Measure the following angles:

**a**

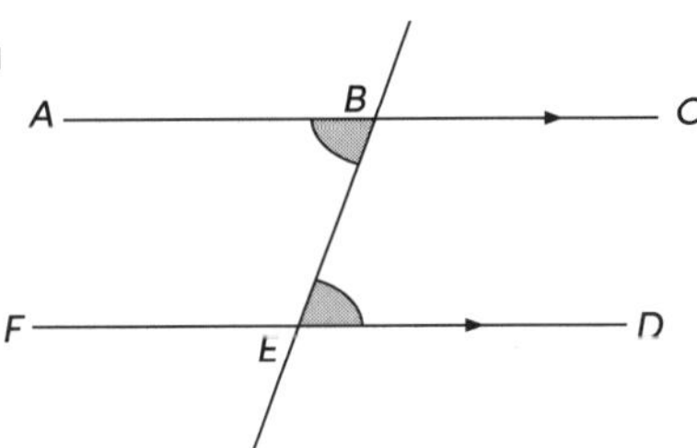


$\angle ABE =$ ______ ° $\angle BED =$ ______ °

**b**

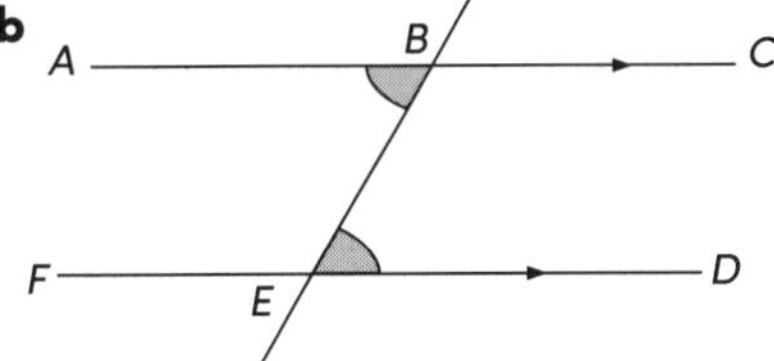


$\angle ABE =$ ______ ° $\angle BED =$ ______ °

**c**

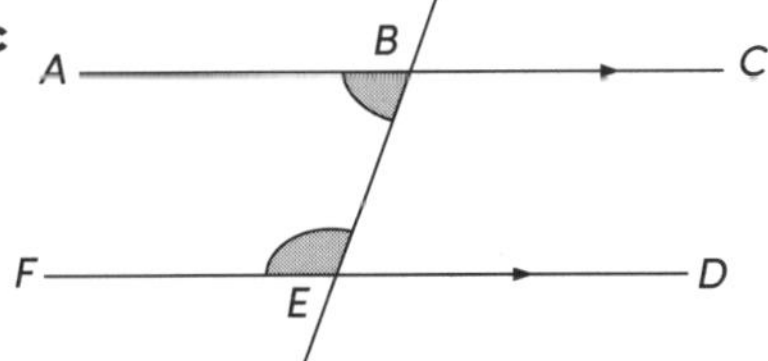


$\angle ABE =$ ______ $\angle FEB =$ ______

**d**

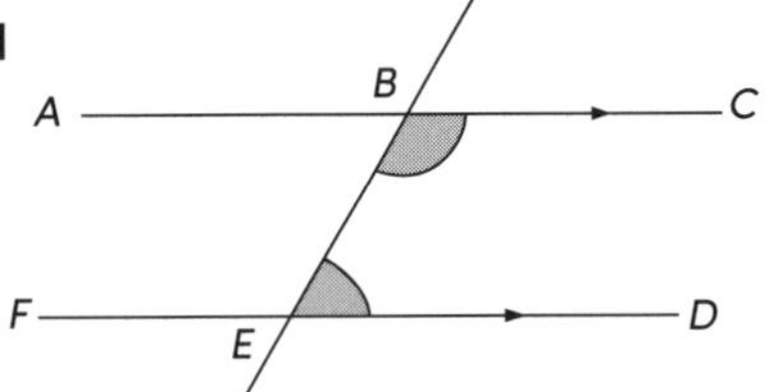


$\angle CBE =$ ______ $\angle BED =$ ______

## GEOMETRY
# Complementary and supplementary angles

**1** Find the complement of the following angles:

**a** 40° 90 – 40 = ________

**b** 52° 90 – 52 = ________

**c** 60° ________

**d** 45° ________

**e** 10° ________

**f** 30° ________

**2** Find the supplement of the following angles:

**a** 42° 180 – ______ = ______

**b** 64° 180 – ______ = ______

**c** 130° ________

**d** 115° ________

**e** 90° ________

**f** 120° ________

**3** Add each pair of angles. Are they complementary? Answer Yes or No.

**a** 30° and 60° ________

**b** 40° and 50° ________

**c** 45° and 60° ________

**d** 20° and 53° ________

**e** 12° and 78° ________

**f** 80° and 9° ________

**4** Add each pair of angles. Are they supplementary? Answer Yes or No.

**a** 130° and 50° ________

**b** 140° and 30° ________

**c** 60° and 120° ________

**d** 75° and 105° ________

**e** 90° and 95° ________

**5** Find the value of $x$ in each right angle:

**a**

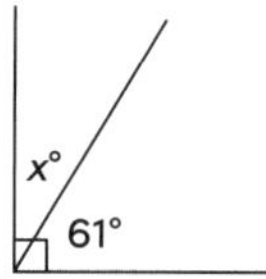


$x$ = 90 – 61
= ______°

**b**

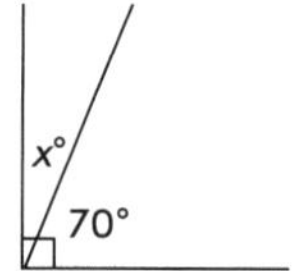


$x$ = 90 – 70
= ______°

**c**

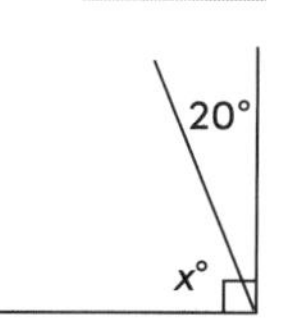


$x$ = ________
= ________

**d**

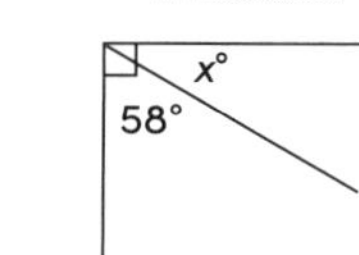


$x$ = ________
= ________

**e**

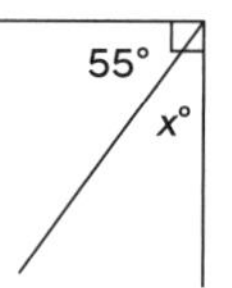


________
________

**f**

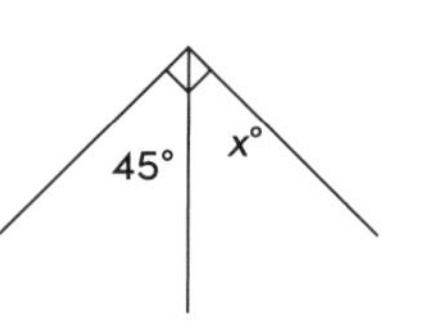


________
________

**g**

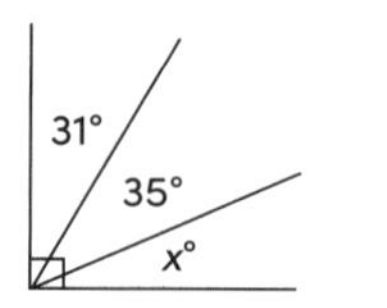


________
________

**h**

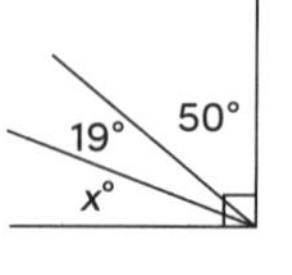


________
________

## GEOMETRY

# Complementary and supplementary angles

**1** Find the value of $a$ in each straight angle:

**a** 125° a°

$a =$ 180 $- 125$

$= \_\_\_\_°$

**b** 75° a°

$a =$ 180 $- 75$

$= \_\_\_\_°$

**c**

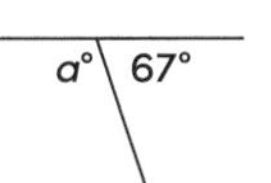


$a = \_\_\_\_ - \_\_\_\_$

$= \_\_\_\_°$

**d** 54° a°

$a = \_\_\_\_ - \_\_\_\_$

$= \_\_\_\_°$

**e**

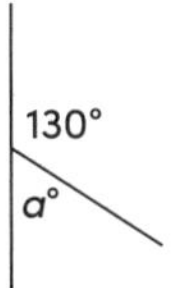


$a = \_\_\_\_$

$= \_\_\_\_$

**f**

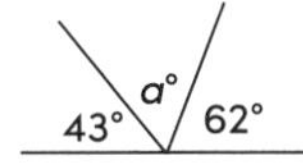


$a = \_\_\_\_$

$= \_\_\_\_$

**g**

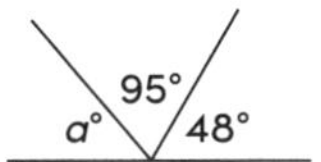


**h**

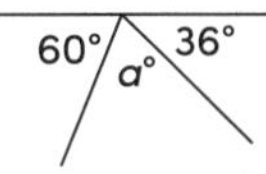


**i**

**j**

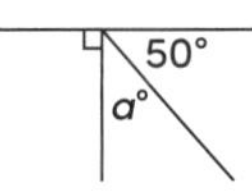

**k**

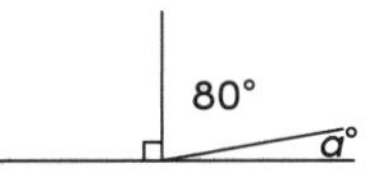


**l**

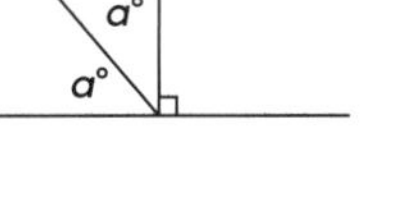


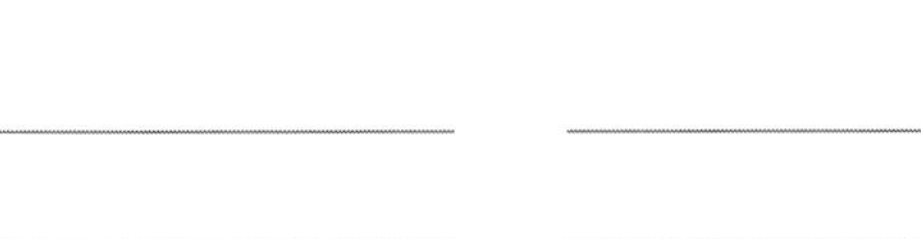

**2** Find the value of $x$ and $y$ in each straight angle:

**a**

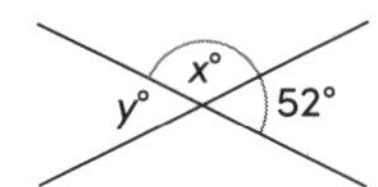


$x =$ 180 $- 52$

$= \_\_\_\_°$

$y =$ 180 $- \_\_\_\_$

$= \_\_\_\_°$

**b**

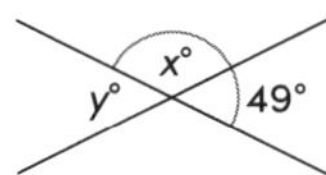


$x =$ 180 $- 49$

$= \_\_\_\_°$

$y =$ 180 $- \_\_\_\_$

$= \_\_\_\_°$

**c**

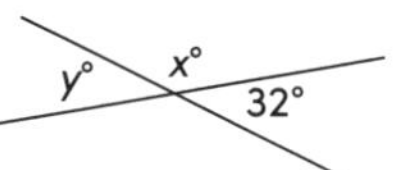


$x = \_\_\_\_$

$= \_\_\_\_$

$y = \_\_\_\_$

$= \_\_\_\_$

**d**

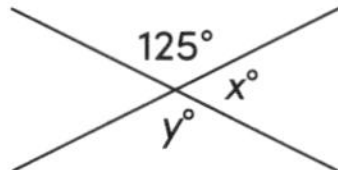


$x = \_\_\_\_$

$= \_\_\_\_$

$y = \_\_\_\_$

$= \_\_\_\_$

**e**

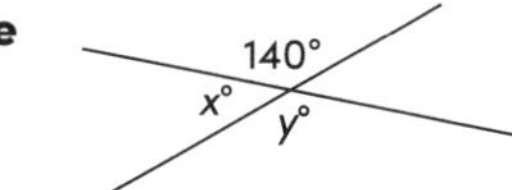


**f**

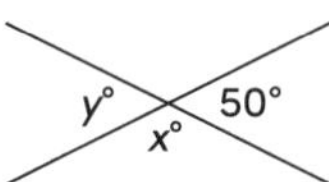


**g**

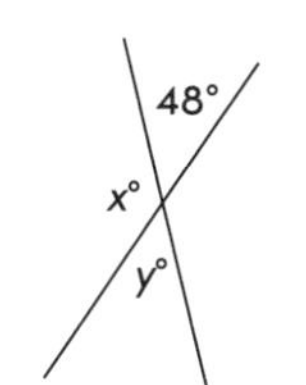


**h**

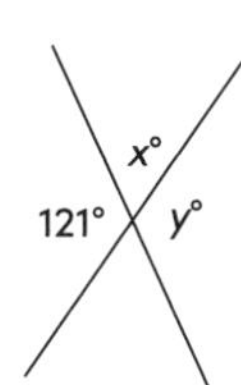

## GEOMETRY
# Angles at a point and vertically opposite angles

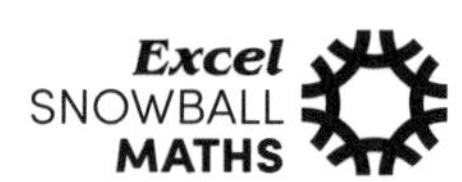


**1** Find the value of $x$ and $y$ in each diagram:

**a**
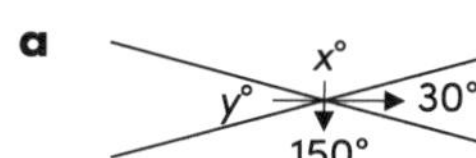


$x =$ ______°

$y =$ ______°

**b**

$x =$ ______°

$y =$ ______°

**c**
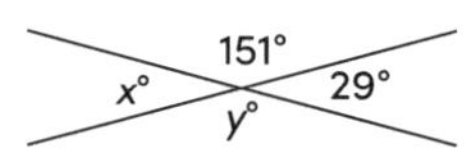


$x =$ ______

$y =$ ______

**d**
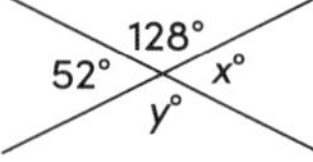


$x =$ ______

$y =$ ______

**e**
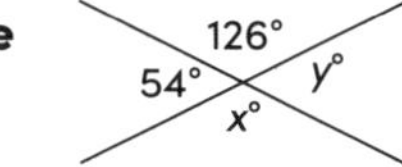


$x =$ ______

$y =$ ______

**f**
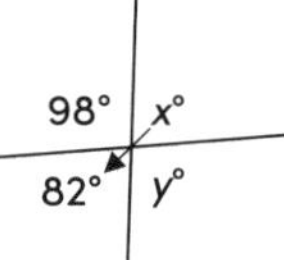


$x =$ ______

$y =$ ______

**g**
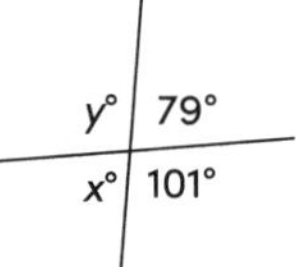


$x =$ ______

$y =$ ______

**h**
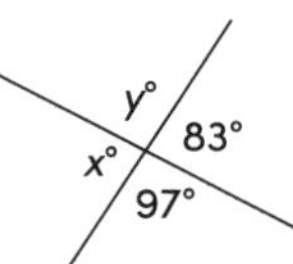


$x =$ ______

$y =$ ______

**i**
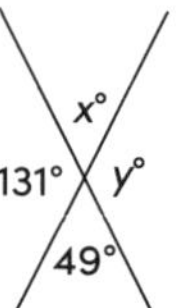


$x =$ ______

$y =$ ______

**2** Find the value of $x$ and $y$ in each diagram:

**a**
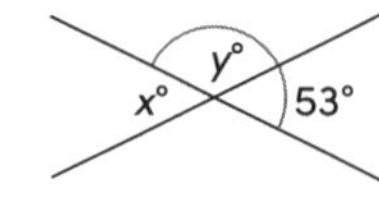


$x =$ ______°

$y =$ 180 – 53

= ______

**b**
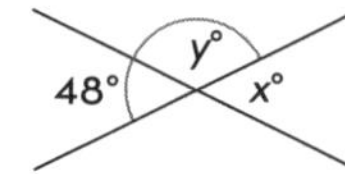


$x =$ ______°

$y =$ 180 – 48

= ______

**c**
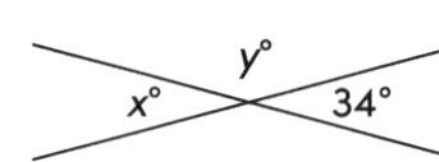


$x =$ ______

$y =$ ______

= ______

**d**
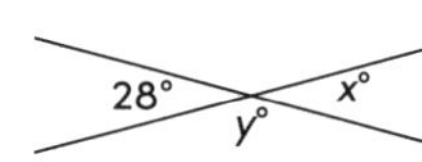


$x =$ ______

$y =$ ______

= ______

**e**

x°
y°
147°

______

______

______

**f**

x°
y°
139°

______

______

______

**g**

x°
y°
33°

______

______

______

**h**
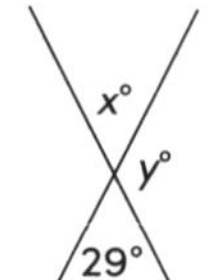


______

______

______

**GEOMETRY**

# Angles at a point and vertically opposite angles

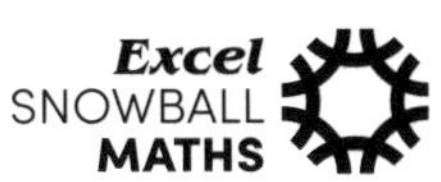


**1** Find the value of *b* in each diagram:

**a**

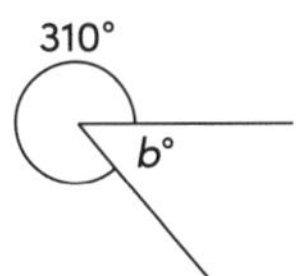


$b =$ 360 $- 310$

$=$ ______ °

**b**

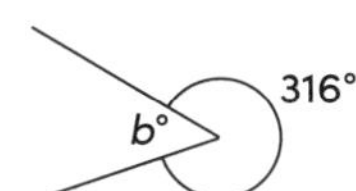


$b =$ 360 $- 316$

$=$ ______ °

**c**

$b =$ ______________

$=$ ______

**d**

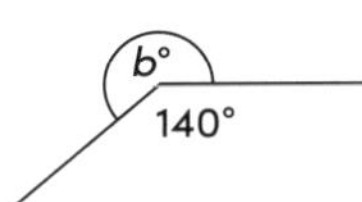


$b =$ ______________

$=$ ______

**e**

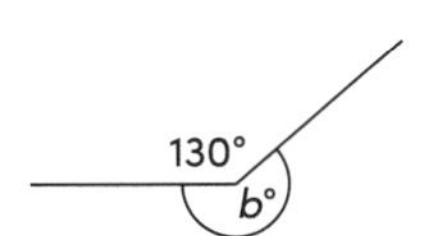


______________

______________

**f**

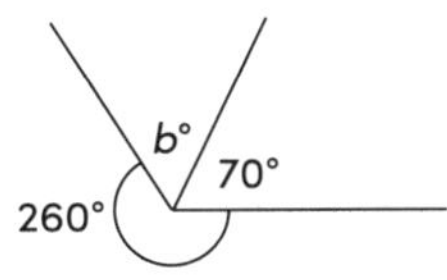


______________

______________

**g**

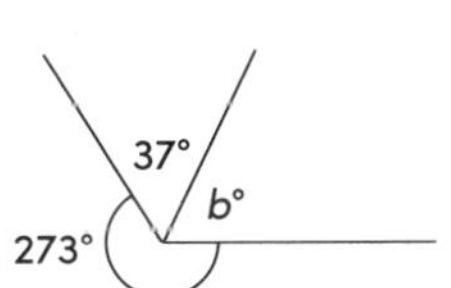


______________

______________

**h**

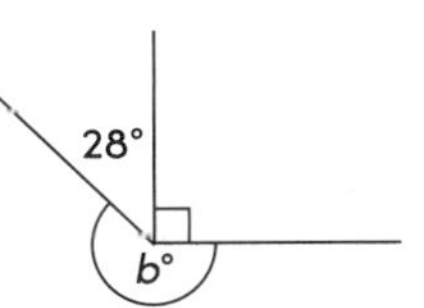


______________

______________

**i**

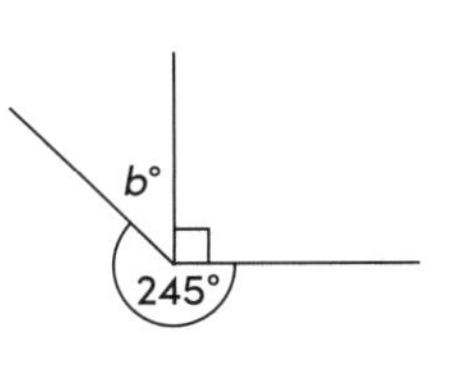


______________

______________

**j**

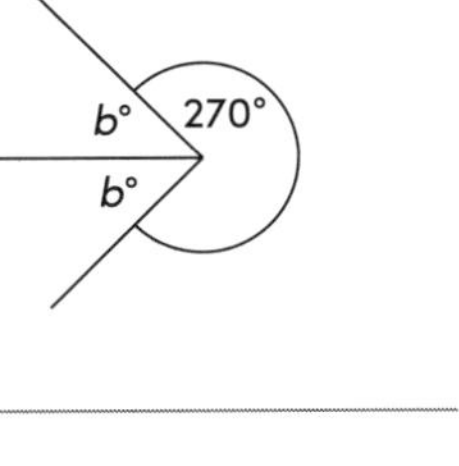


______________

______________

**2** Find the value of *a* in each diagram:

**a**

a°
107° 164°

$a =$ 360 $-$ ______ $-$ ______

$=$ ______________

**b**

169° a°
82°

$a =$ 360 $-$ ______ $-$ ______

$=$ ______________

**c**

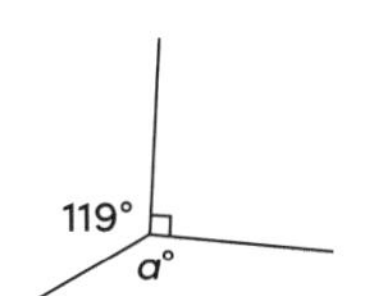


$a =$ ______________

$=$ ______

**d**

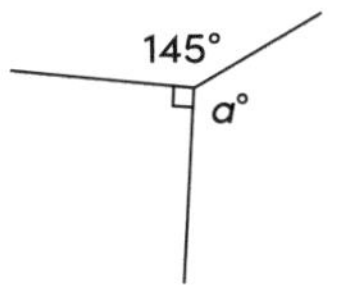


$a =$ ______________

$=$ ______

**e**

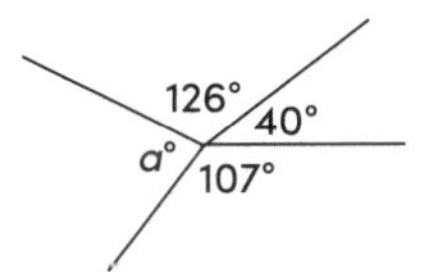


______________

______________

**f**

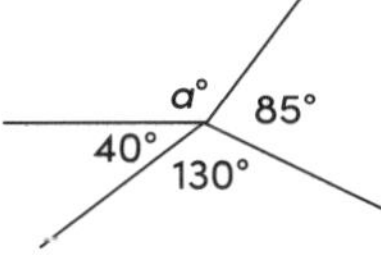

______________

______________

**g**

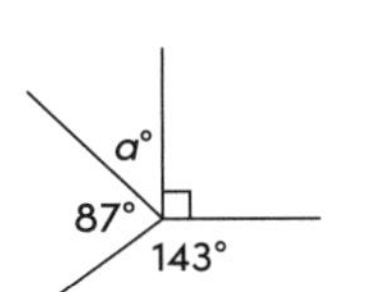


______________

______________

**h**

34°
a°
142°

______________

______________

**GEOMETRY**

# Corresponding angles

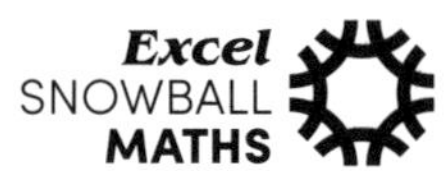


1 Find the value of each pronumeral:

a
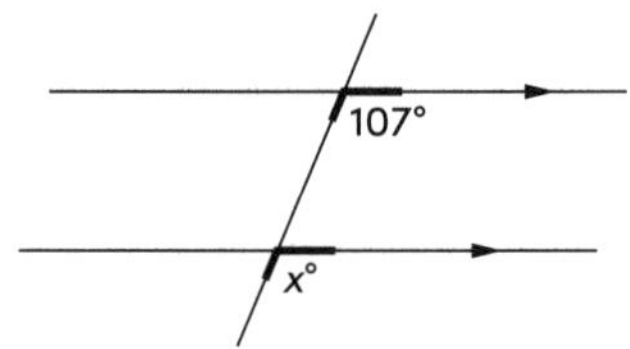


$x =$ ______

b
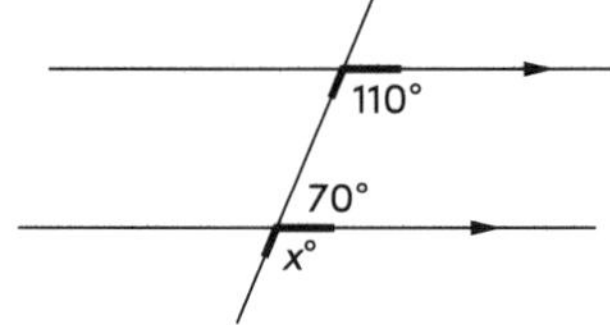


$x =$ ______

c
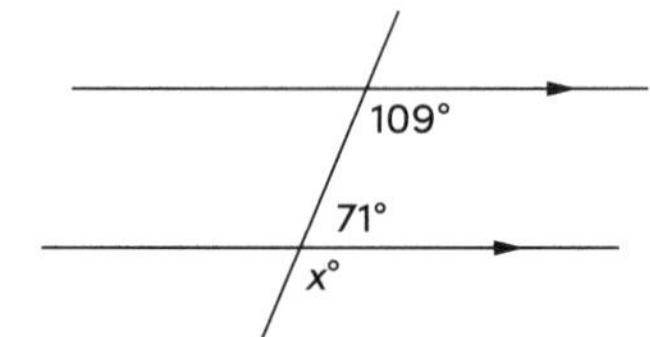


$x =$ ______

d
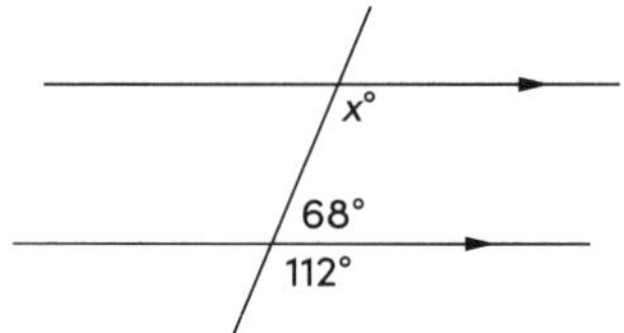


$x =$ ______

e
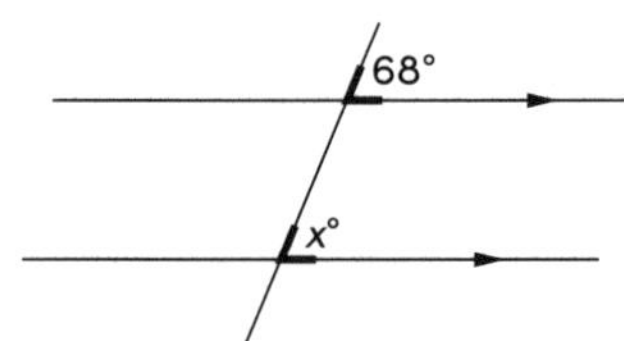


$x =$ ______

f
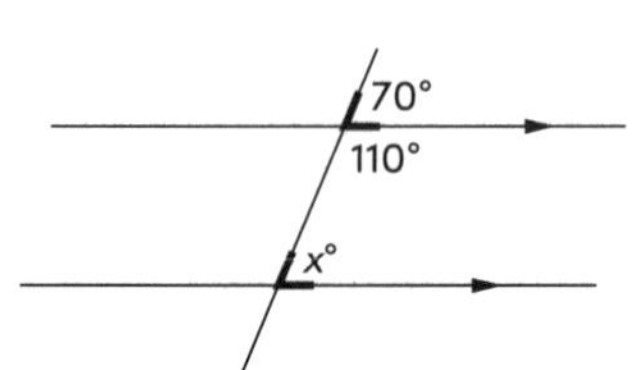


$x =$ ______

g
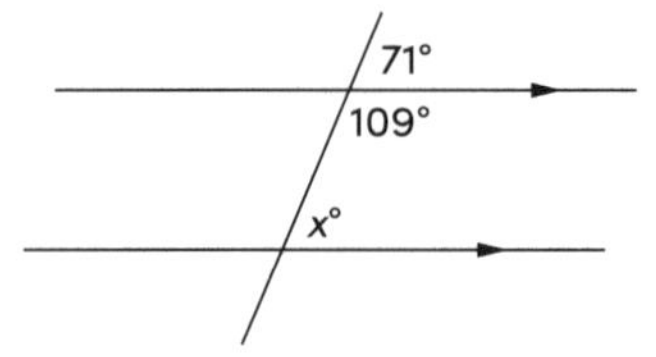


$x =$ ______

h
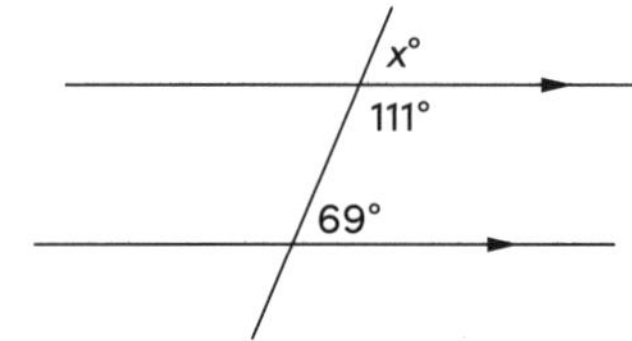


$x =$ ______

i
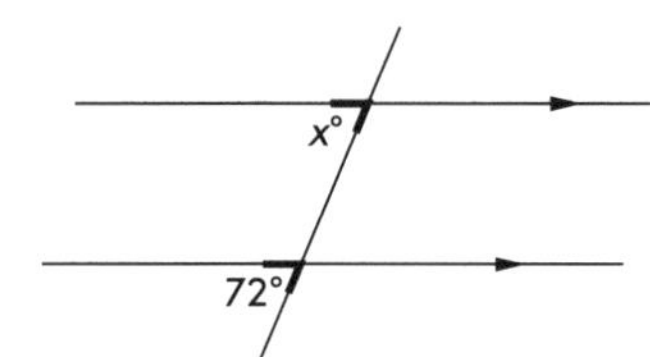


$x =$ ______

j
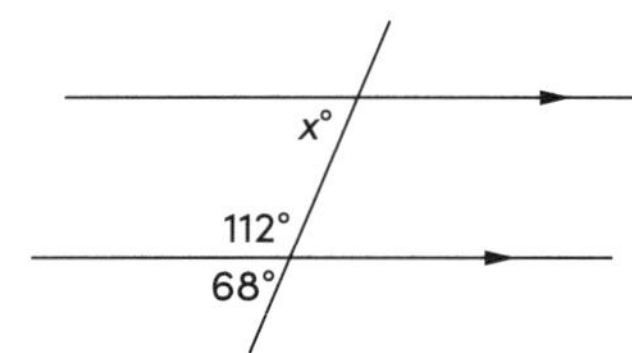


$x =$ ______

k
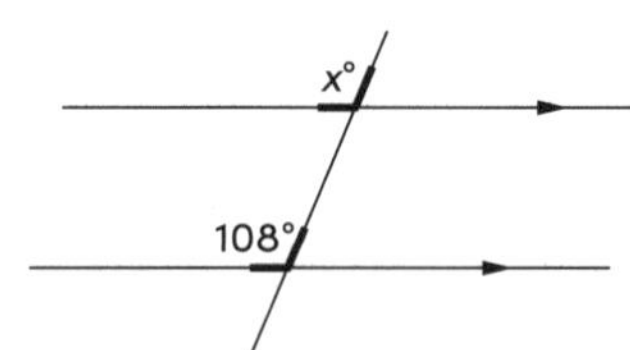


$x =$ ______

l
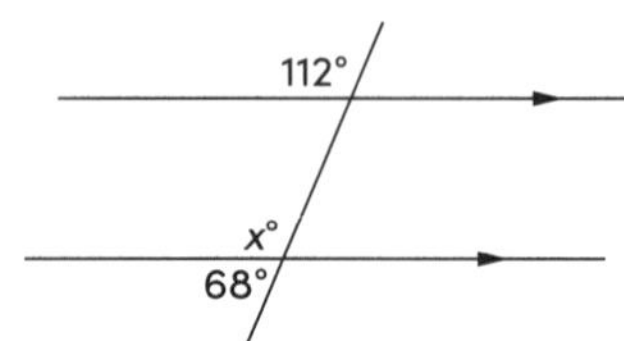


$x =$ ______

UNIT 15

# GEOMETRY
## Corresponding angles

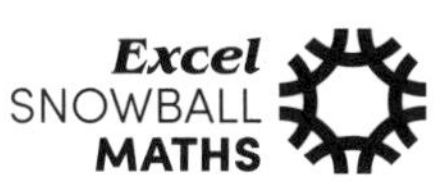


**1** Find the value of each pronumeral:

**a**

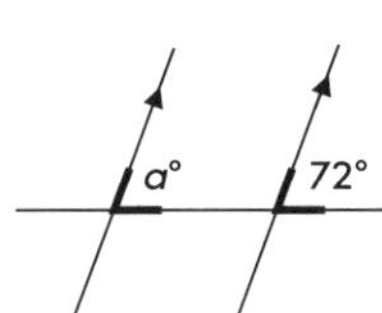


$a$ = ______

**b**

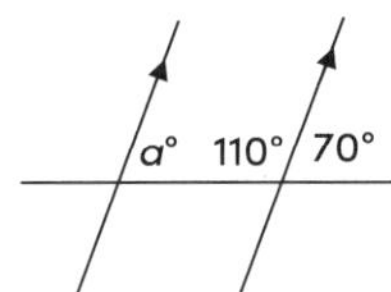


$a$ = ______

**c**

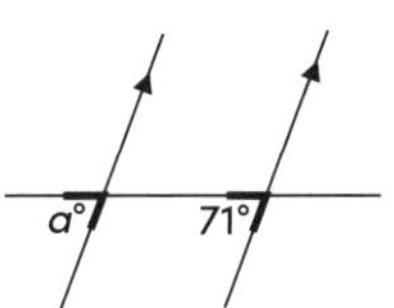


$a$ = ______

**d**

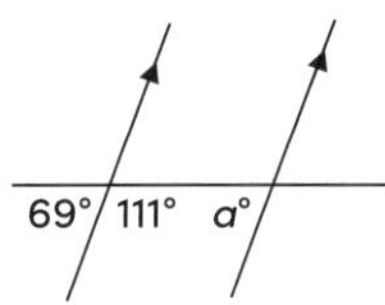


$a$ = ______

**2** Find the value of each pronumeral:

**a**

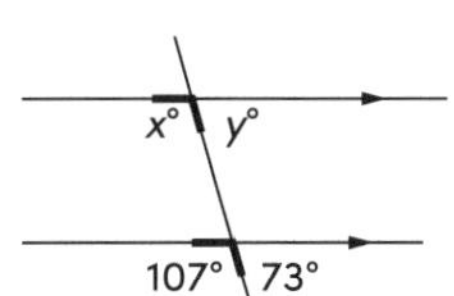


$x$ = ______

$y$ = ______

**b**

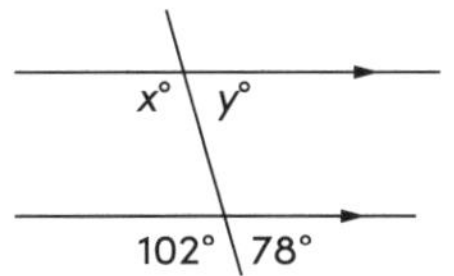


$x$ = ______

$y$ = ______

**c**

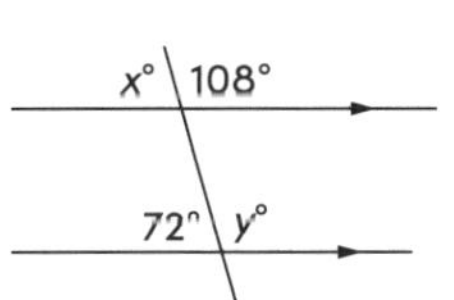


$x$ = ______

$y$ = ______

**d**

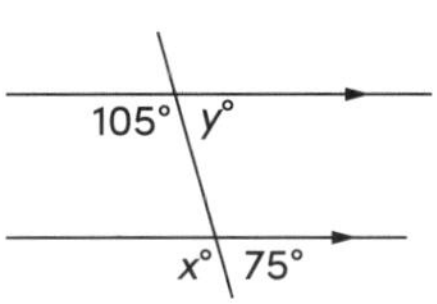


$x$ = ______

$y$ = ______

**3** Find the value of each pronumeral:

**a**

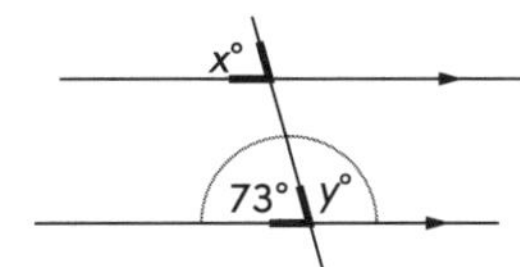


$x$ = ______

$y$ = 180 − 73

= ______

**b**

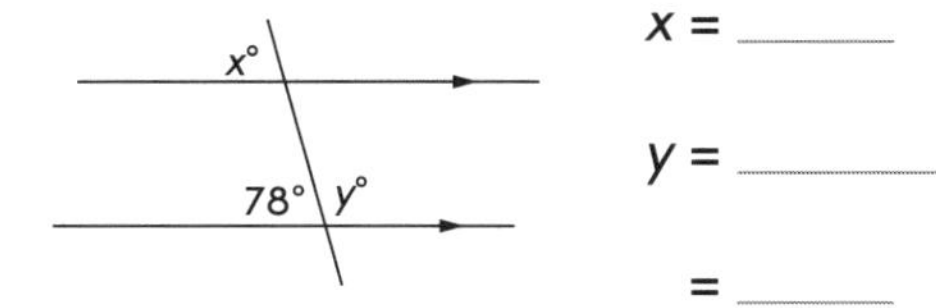


$x$ = ______

$y$ = ______

= ______

**c**

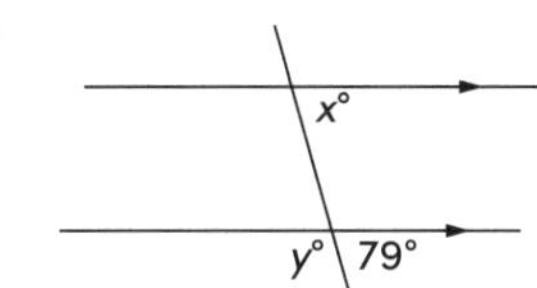


$x$ = ______

$y$ = ______

= ______

**d**

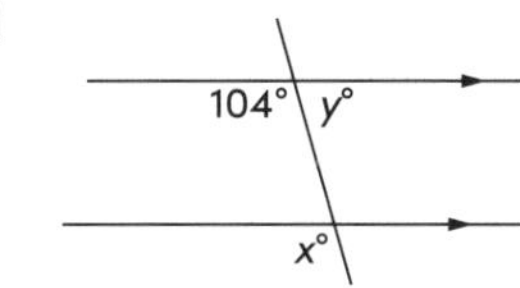


$x$ = ______

$y$ = ______

= ______

**e**

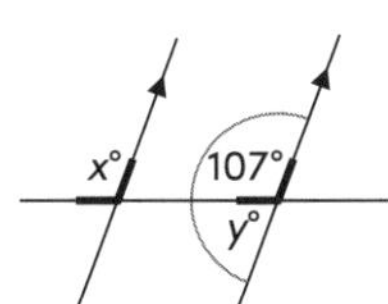


$x$ = ______

$y$ = ______

= ______

**f**

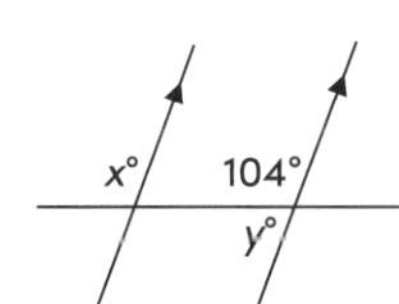


$x$ = ______

$y$ = ______

= ______

**g**

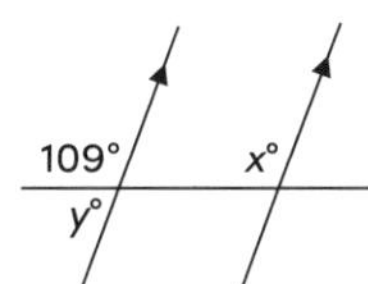


______

______

______

**h**

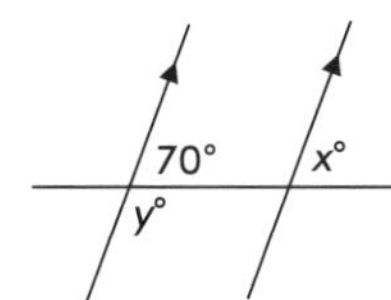


______

______

______

UNIT 16

GEOMETRY

# Corresponding angles

**1** Find the value of each pronumeral:

**a**

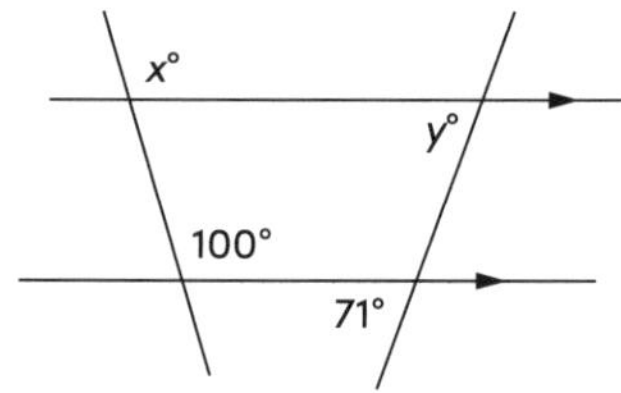


$x =$ ______

$y =$ ______

**b**

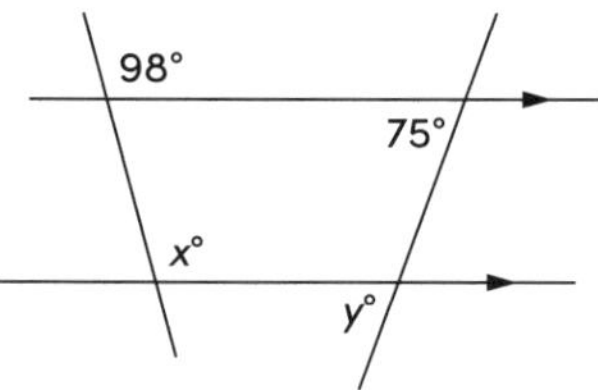


$x =$ ______

$y =$ ______

**c**

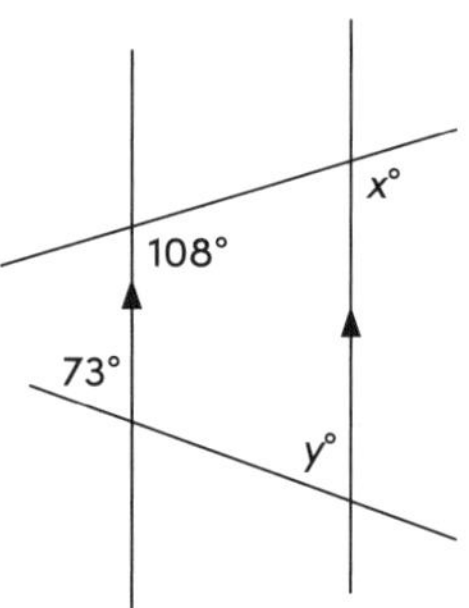


$x =$ ______

$y =$ ______

**d**

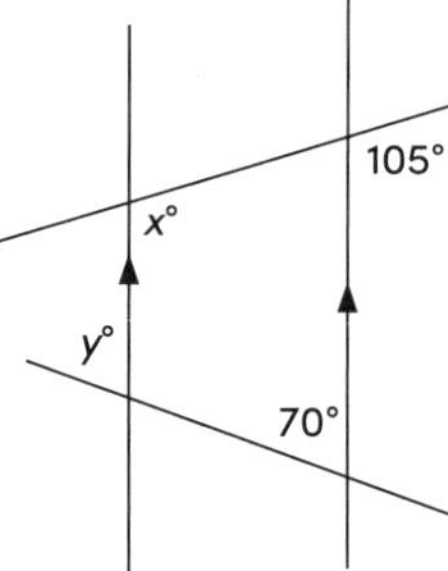


$x =$ ______

$y =$ ______

**e**

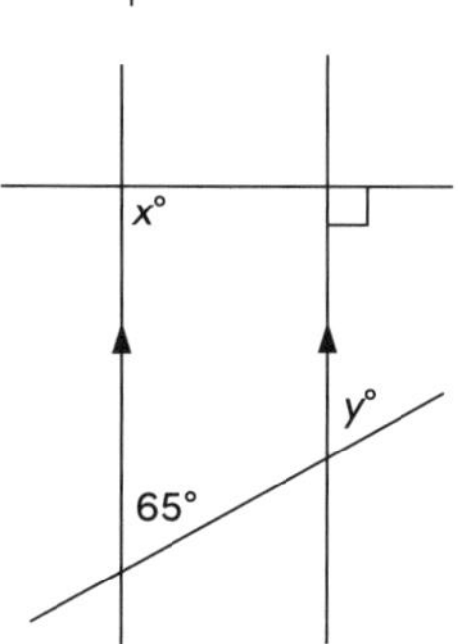


$x =$ ______

$y =$ ______

**2** Find the value of the pronumerals:

**a**

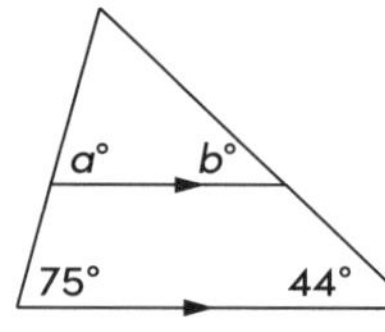


$a =$ ______

$b =$ ______

**b**

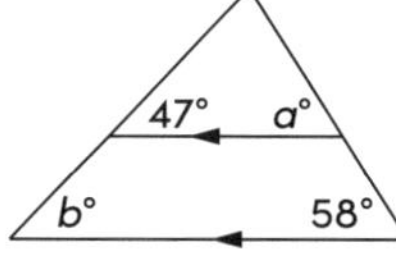


$a =$ ______

$b =$ ______

**c**

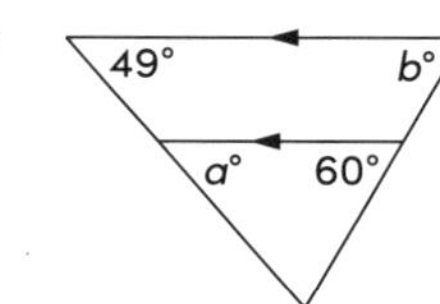


$a =$ ______

$b =$ ______

**d**

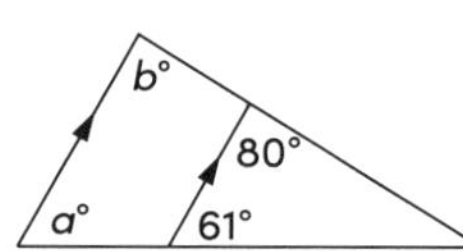


$a =$ ______

$b =$ ______

**e**

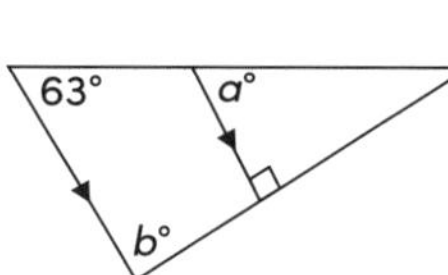


$a =$ ______

$b =$ ______

**f**

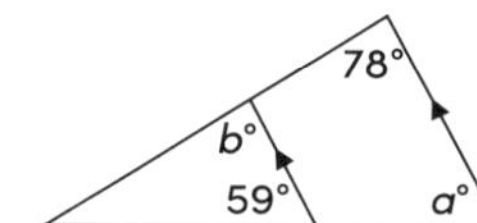


$a =$ ______

$b =$ ______

UNIT 17

## GEOMETRY
# Alternate angles

**1** Find the value of each pronumeral:

**a**

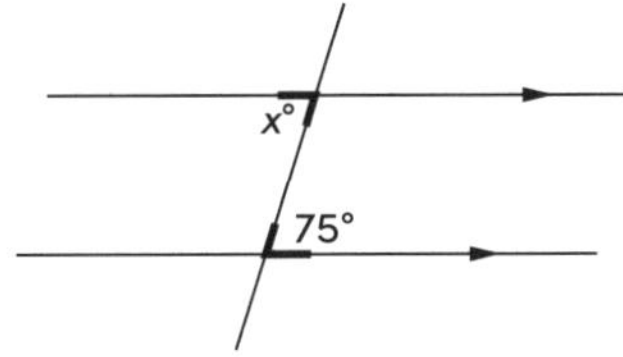


$x =$ ______

**b**

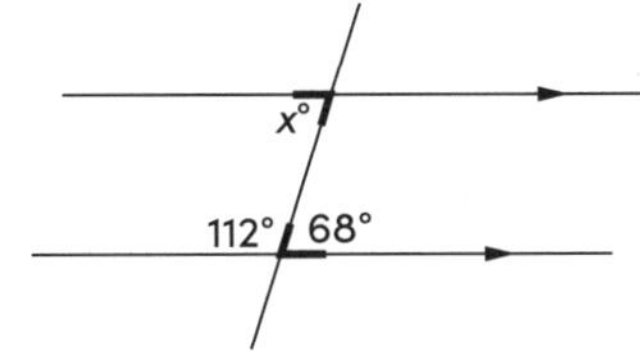


$x =$ ______

**c**

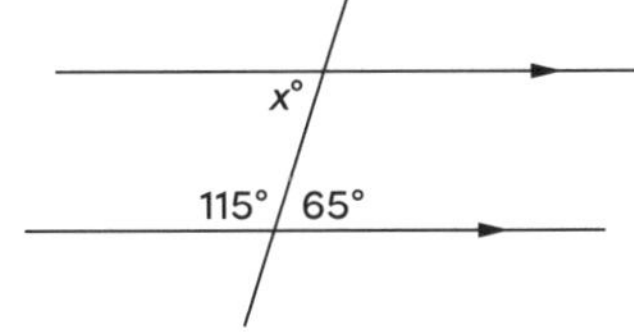


$x =$ ______

**d**

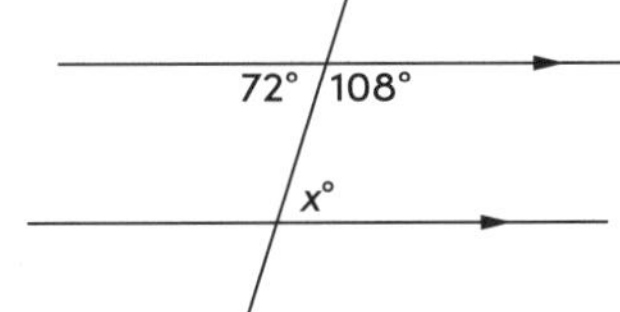


$x =$ ______

**e**

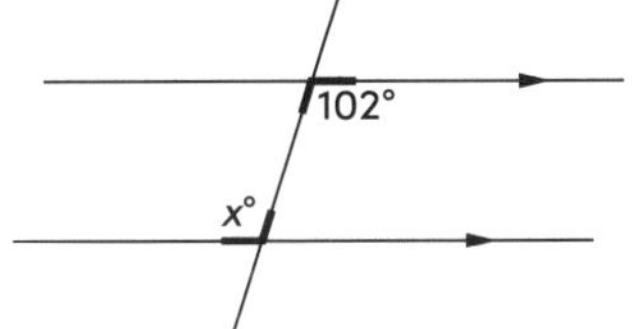


$x =$ ______

**f**

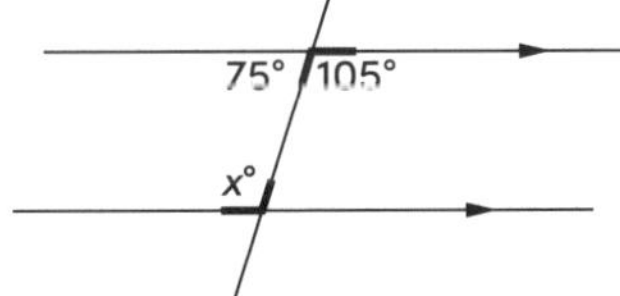


$x =$ ______

**g**

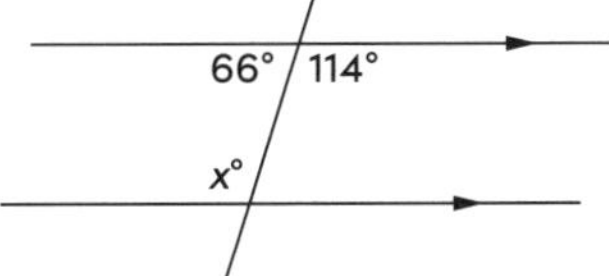


$x =$ ______

**h**

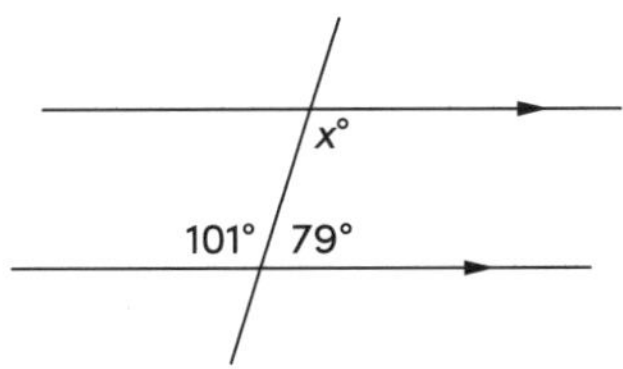


$x =$ ______

**i**

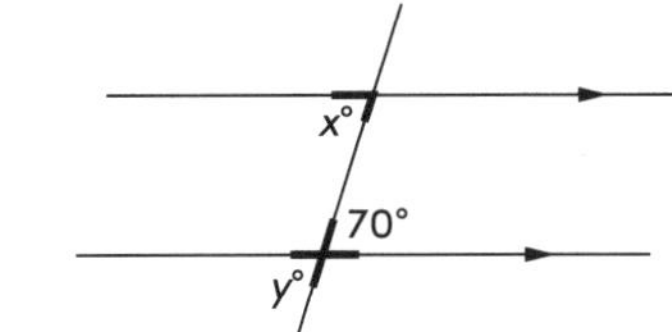


$x =$ ______

$y =$ ______

**j**

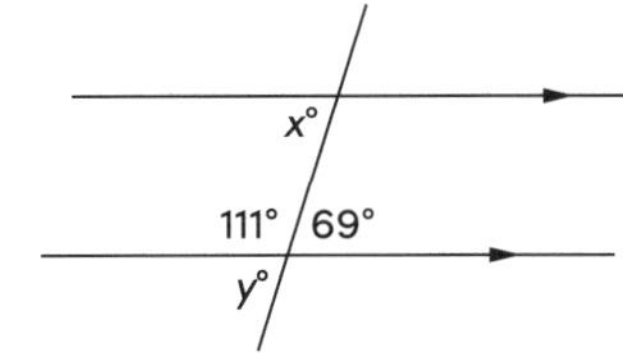


$x =$ ______

$y =$ ______

**k**

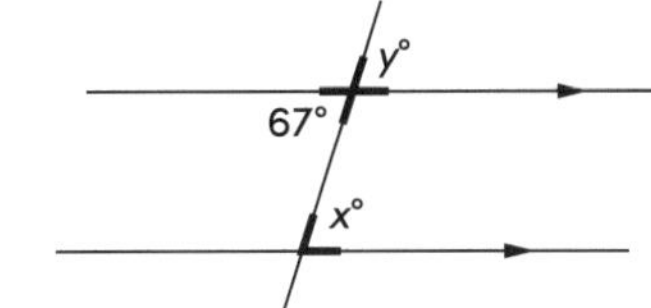


$x =$ ______

$y =$ ______

**l**

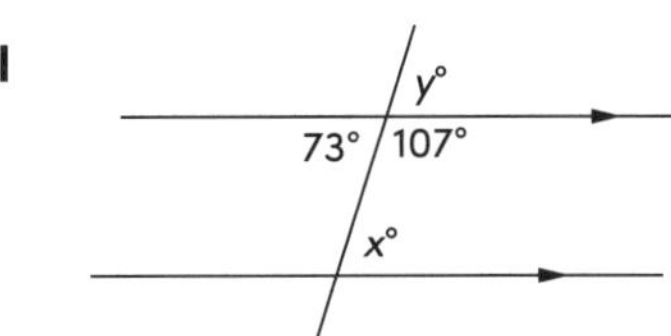


$x =$ ______

$y =$ ______

**m**

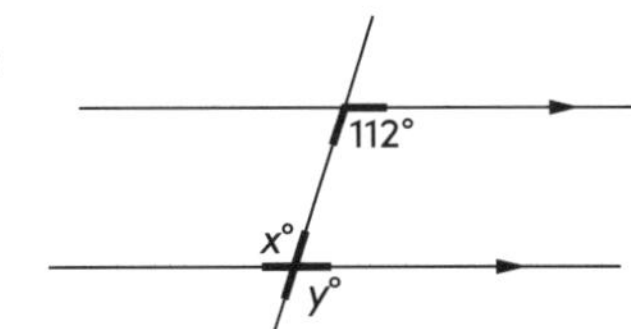


$x =$ ______

$y =$ ______

**n**

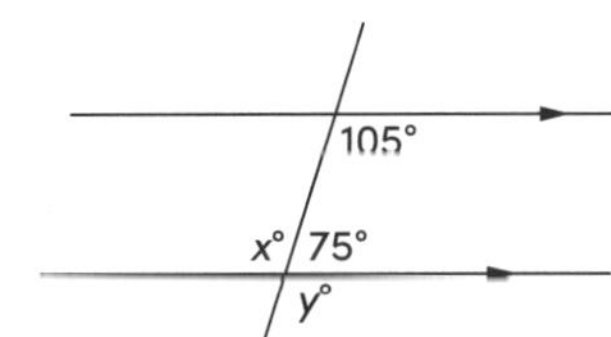


$x =$ ______

$y =$ ______

**o**

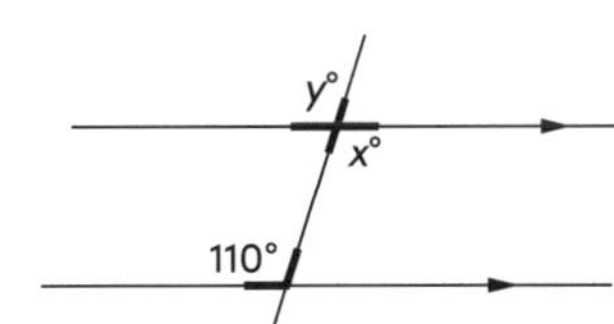


$x =$ ______

$y =$ ______

**p**

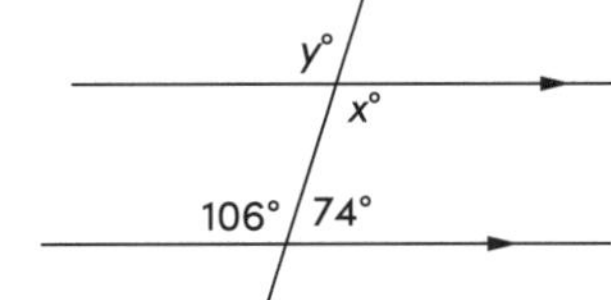


$x =$ ______

$y =$ ______

UNIT 18

# GEOMETRY
## Alternate angles

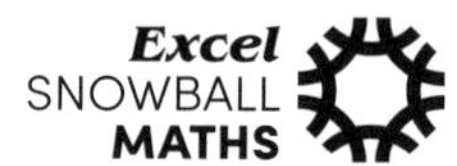


**1** Find the value of each pronumeral:

**a**

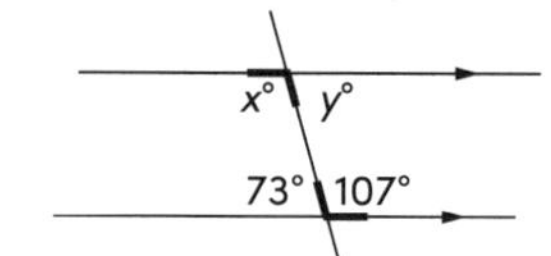


$x =$ ______

$y =$ ______

**b**

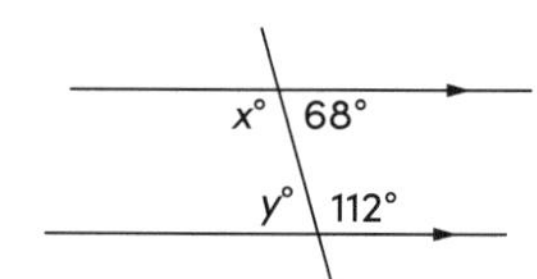


$x =$ ______

$y =$ ______

**c**

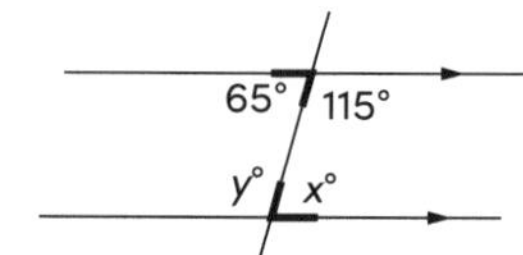


$x =$ ______

$y =$ ______

**d**

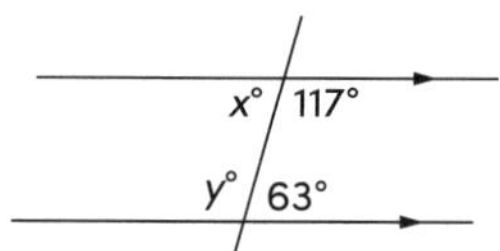


$x =$ ______

$y =$ ______

**2** Find the value of each pronumeral:

**a**

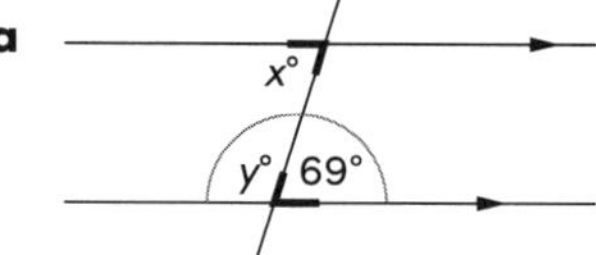


$x =$ ______

$y =$ 180 $- 69$

$=$ ______

**b**

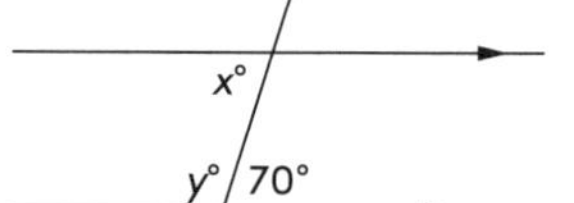


$x =$ ______

$y =$ ______

$=$ ______

**c**

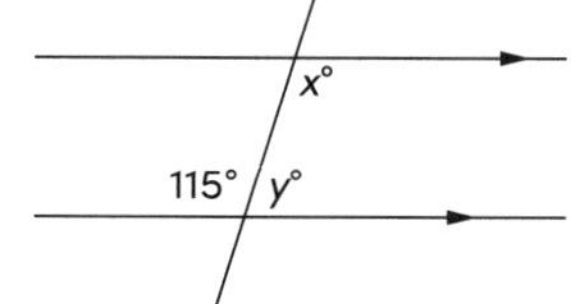


$x =$ ______

$y =$ ______

$=$ ______

**d**

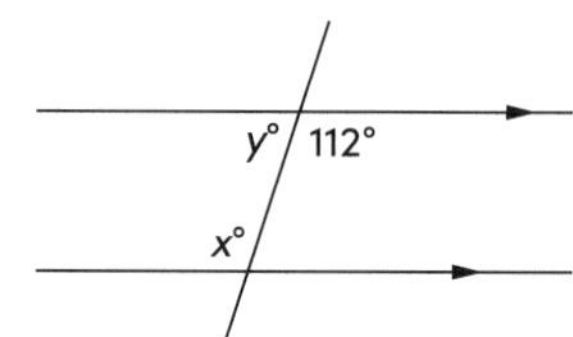


$x =$ ______

$y =$ ______

$=$ ______

**e**

$x =$ ______

$y =$ ______

$=$ ______

**f**

$x =$ ______

$y =$ ______

$=$ ______

**g**

$x =$ ______

$y =$ ______

$=$ ______

**h**

$x =$ ______

$y =$ ______

$=$ ______

**i**

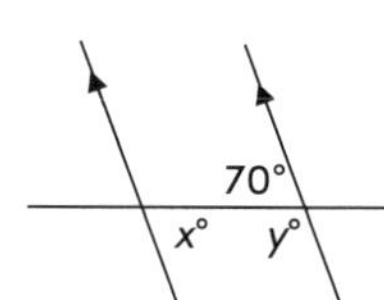


______

______

______

**j**

______

______

______

**k**

______

______

______

UNIT 19

## GEOMETRY
# Alternate angles

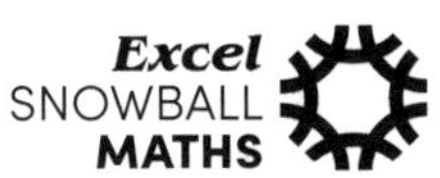

1 Find the value of each pronumeral:

a
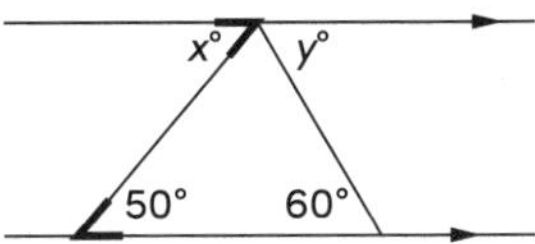


$x =$ ______

$y =$ ______

b
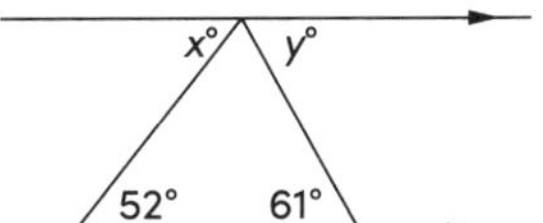


$x =$ ______

$y =$ ______

c
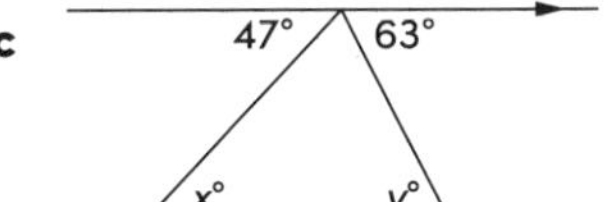


$x =$ ______

$y =$ ______

d
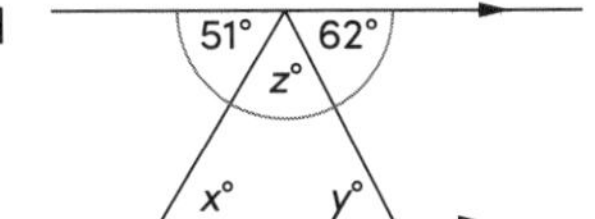


$x =$ ______

$y =$ ______

$z =$ ______

$=$ ______

e
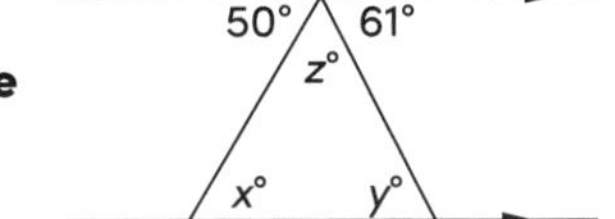


$x =$ ______

$y =$ ______

$z =$ ______

$=$ ______

f
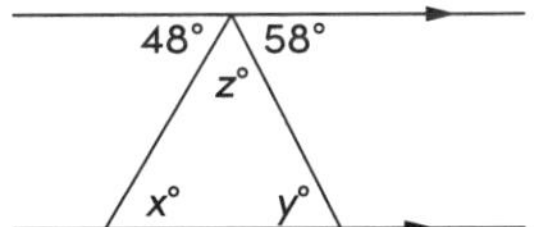


______

______

______

______

g
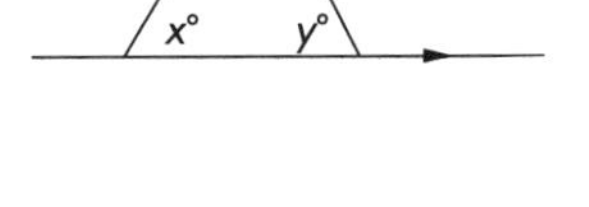
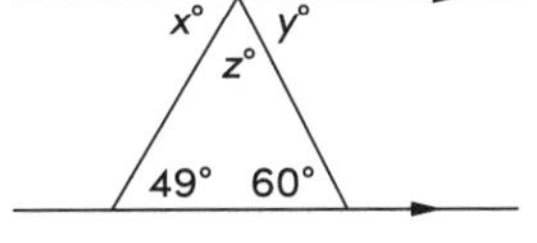


______

______

______

______

2 Find the value of each pronumeral:

a
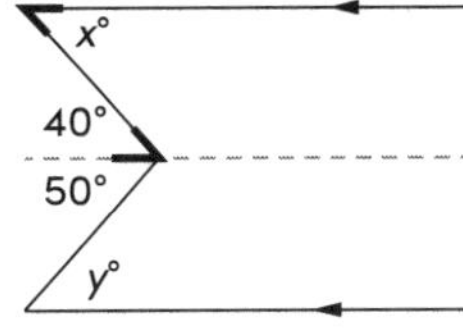


$x =$ ______

$y =$ ______

b
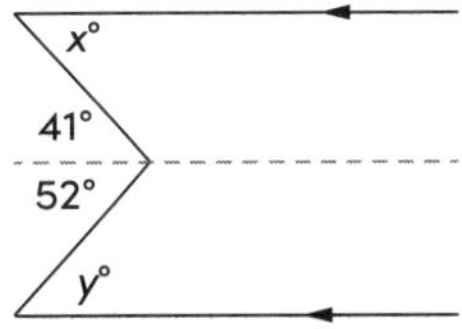


$x =$ ______

$y =$ ______

c
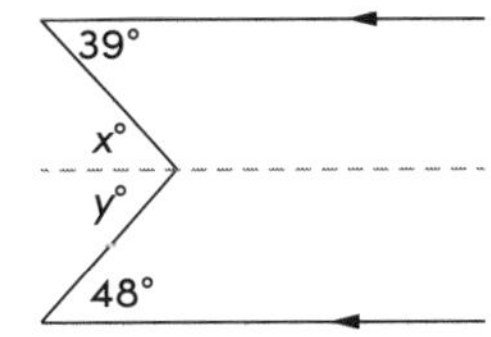


$x =$ ______

$y =$ ______

d
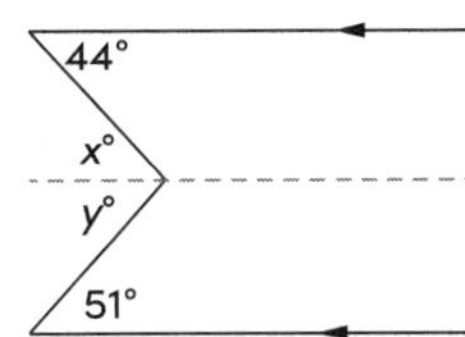


$x =$ ______

$y =$ ______

e
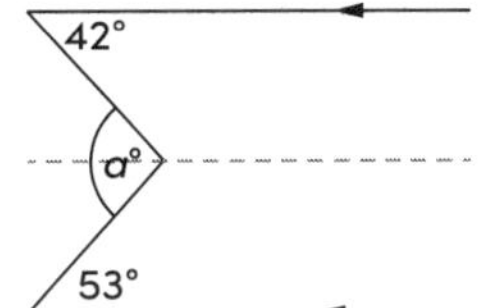


$a =$ ______

$=$ ______

f
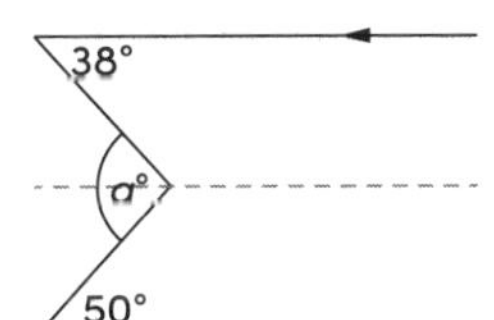


$a =$ ______

$=$ ______

g
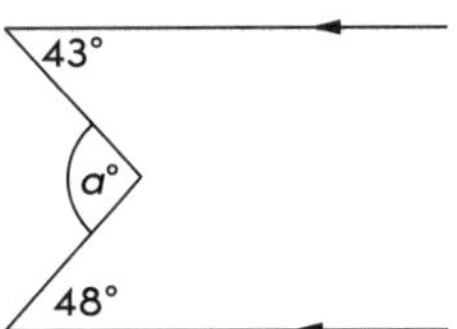


______

______

h
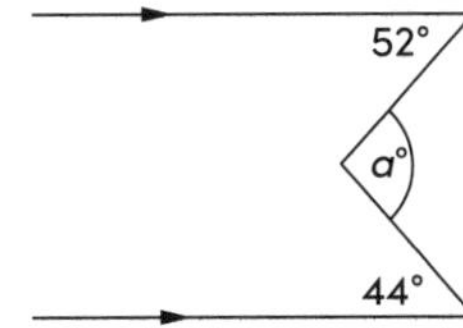


______

______

GEOMETRY

# Co-interior angles

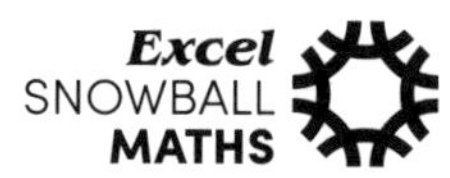


**1** Find the value of each pronumeral:

**a**

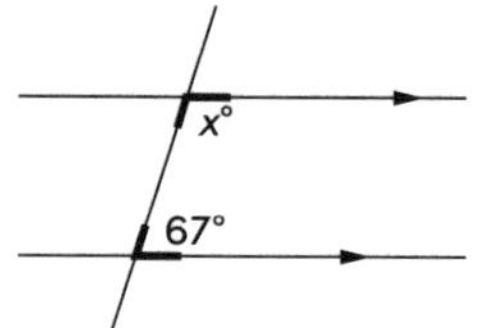


$x =$ 180 − 67

= ______

**b**

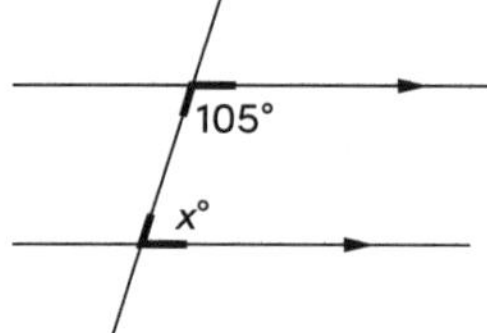


$x =$ 180 − 105

= ______

**c**

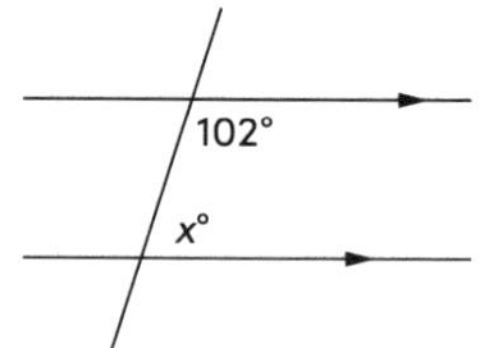


$x =$ ______

= ______

**d**

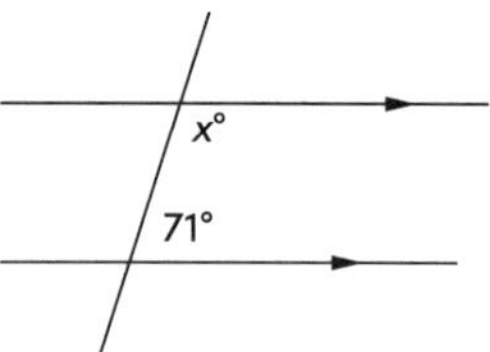


$x =$ ______

= ______

**e**

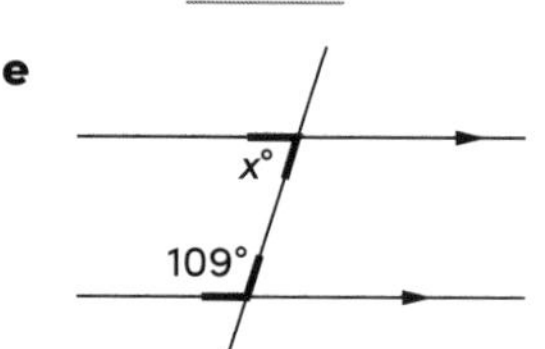


$x =$ ______

= ______

**f**

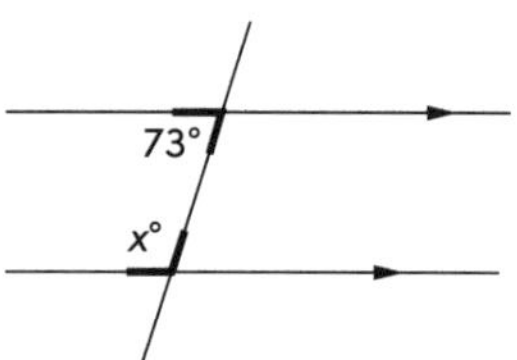


$x =$ ______

= ______

**g**

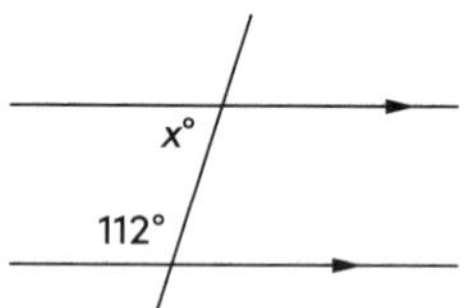


$x =$ ______

= ______

**h**

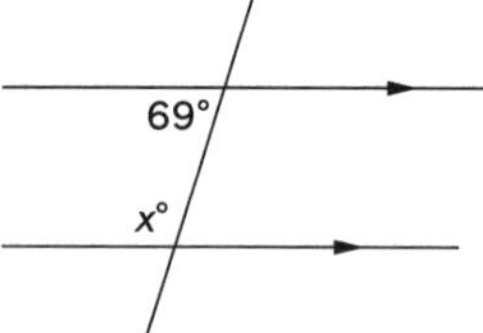


$x =$ ______

= ______

**2** Find the value of each pronumeral:

**a**

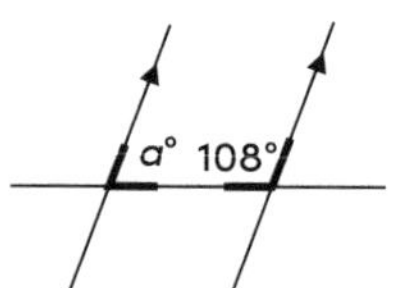


$a =$ ______

= ______

**b**

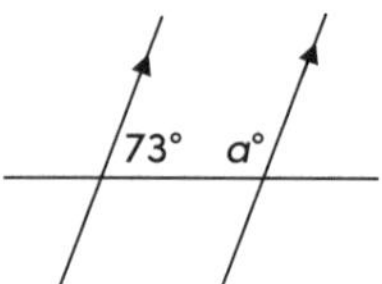


______

______

**c**

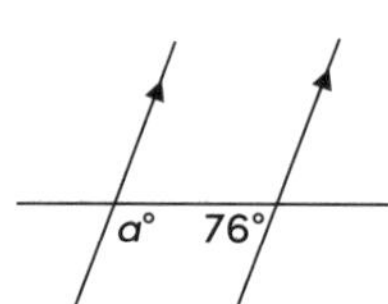


______

______

**d**

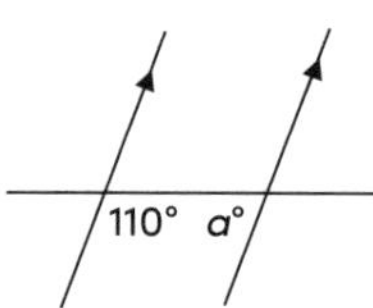


______

______

**3** Find the value of each pronumeral:

**a**

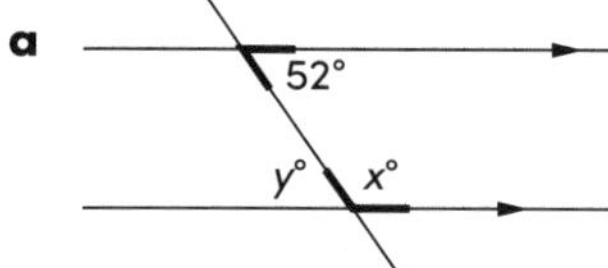


$x =$ ______

______

$y =$ ______

**b**

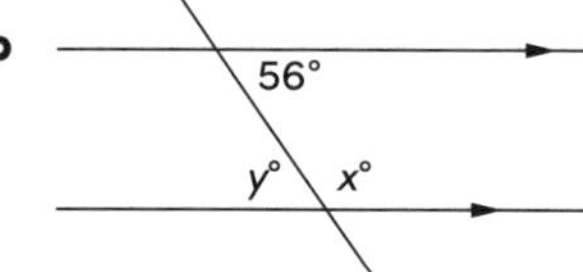


$x =$ ______

______

$y =$ ______

**c**

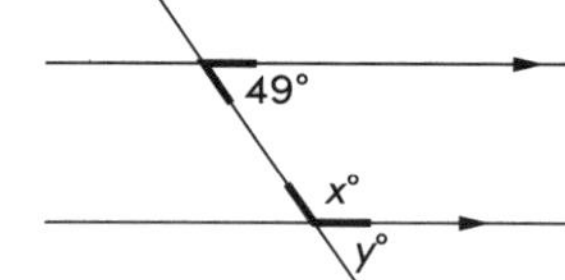


$x =$ ______

______

$y =$ ______

**d**

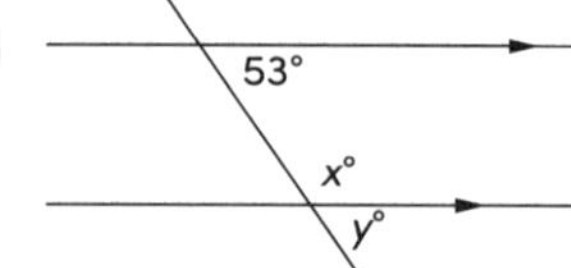


$x =$ ______

______

$y =$ ______

**e**

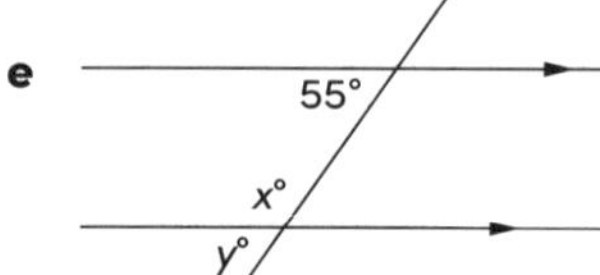


$x =$ ______

______

$y =$ ______

UNIT 21

## GEOMETRY

# Co-interior angles

**1** Find the value of each pronumeral:

**a**

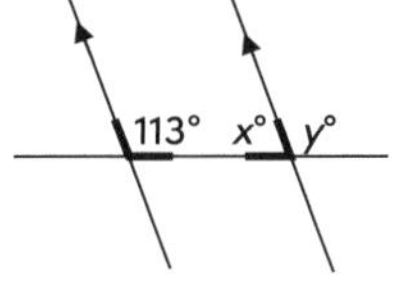


**b**

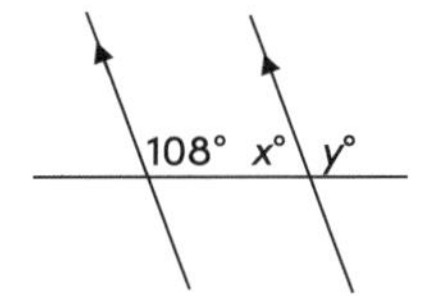


**c**

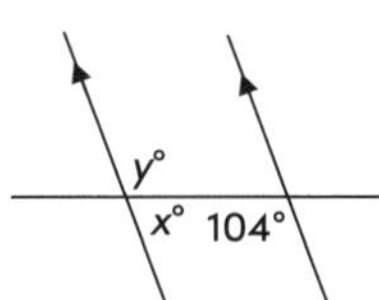


**d**

y°
x°
104°

**2** Find the value of each pronumeral:

**a** x° y° 64° 116°

$x =$ 180 – ______

= ______

$y =$ 180 – ______

= ______

**b** x° y° 58° 122°

$x =$ 180 – ______

= ______

$y =$ 180 – ______

= ______

**c** 120° 60° x° y°

$x =$ ______

= ______

$y =$ ______

= ______

**d** 118° x° y° 118°

$x =$ ______

= ______

$y =$ ______

= ______

**3** Find the value of each pronumeral:

**a**

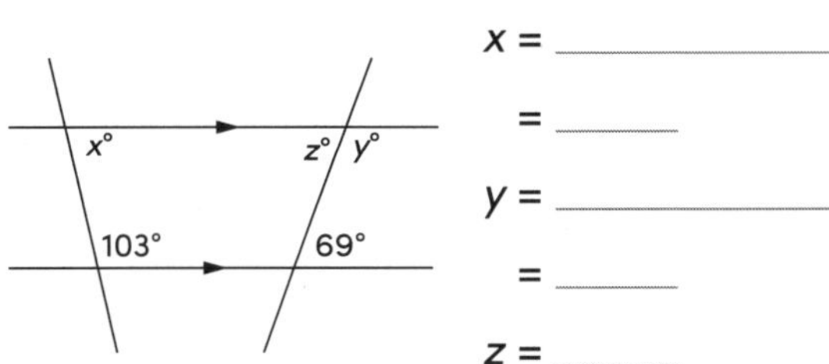


$x =$ ______

= ______

$y =$ ______

= ______

$z =$ ______

**b**

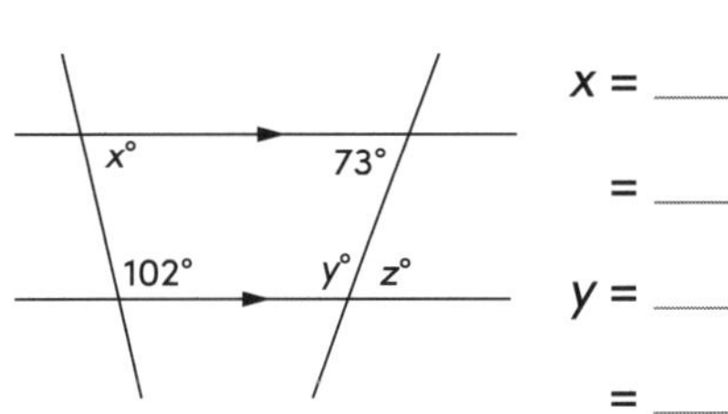


$x =$ ______

= ______

$y =$ ______

= ______

$z =$ ______

**c**

109°
x°
z°
y°
74°

**d**

y°
105°
z°
82°
x°

**e**

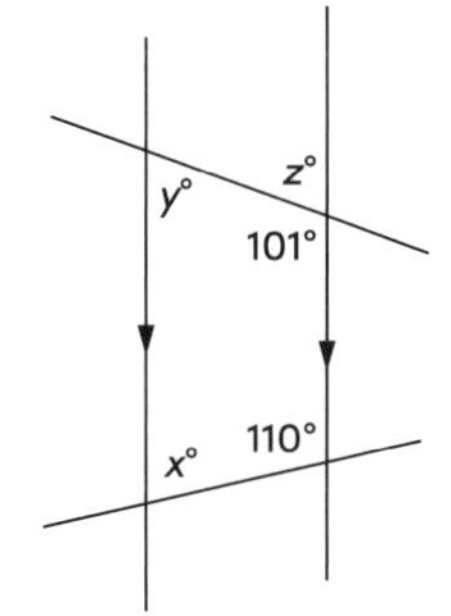

UNIT 22

## GEOMETRY

# Angles on parallel lines

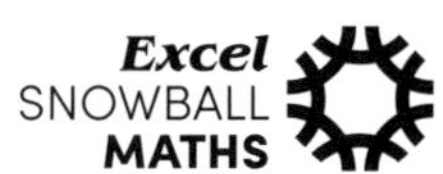

**1**

| Corresponding | Co-interior |
| --- | --- |
| Alternate | |

By choosing from the box above, name the type of angles on each pair of parallel lines:

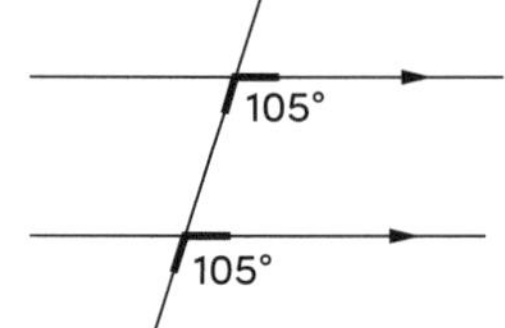


......................................

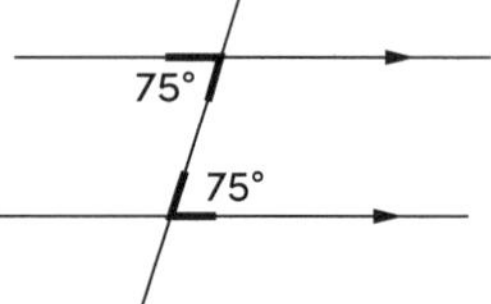


......................................

**c**

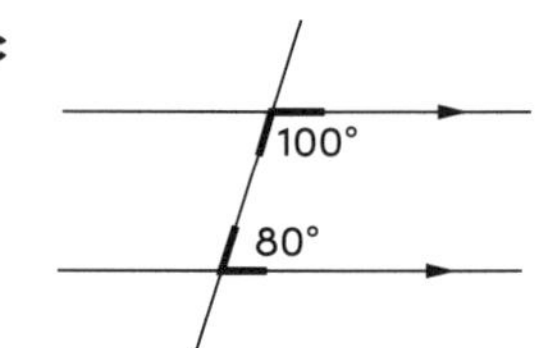


......................................

**d**

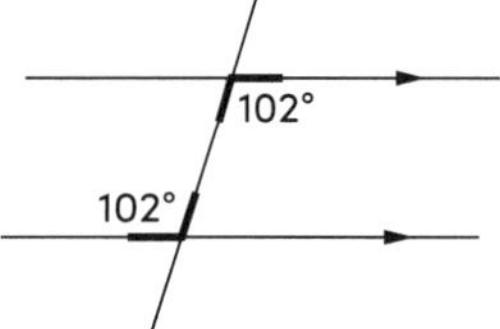


......................................

**e**

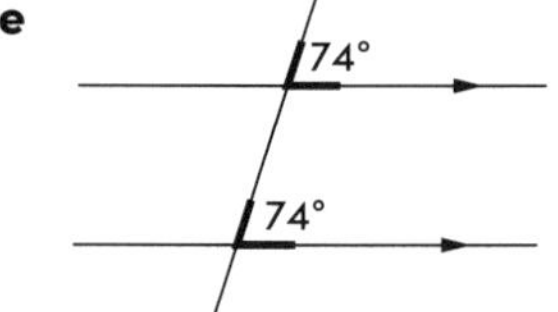


......................................

**f**

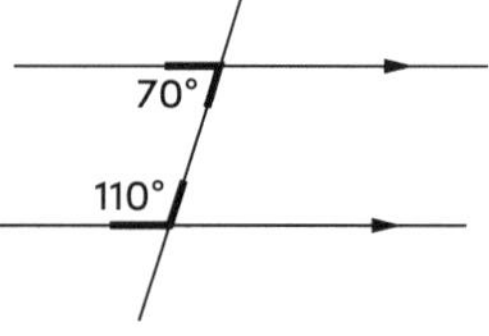


......................................

**g**

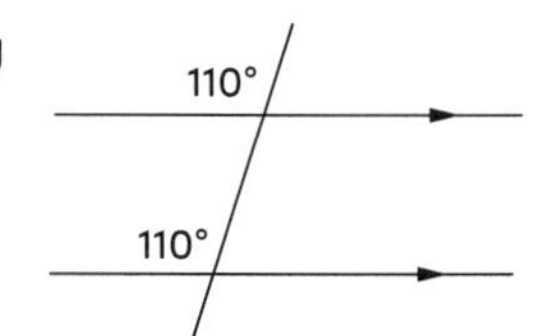


......................................

**h**

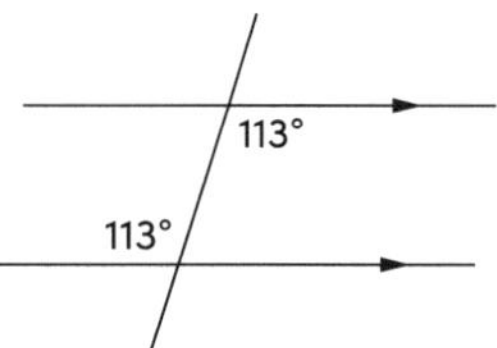


......................................

**i**

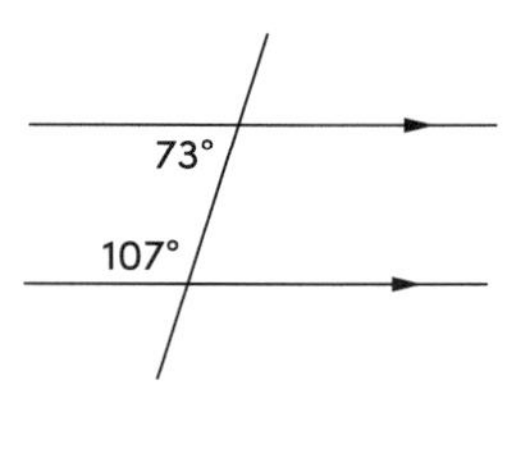


......................................

**j**

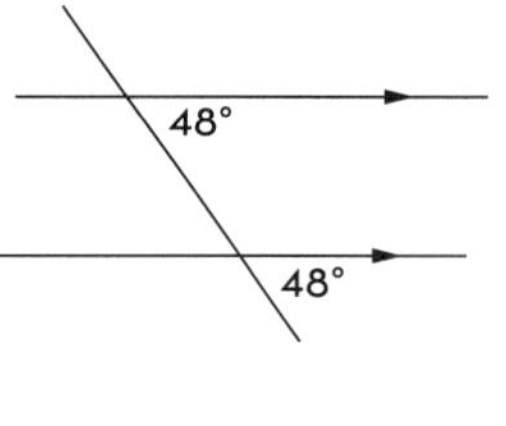


......................................

**2** Name the type of angles on each pair of parallel lines:

**a**

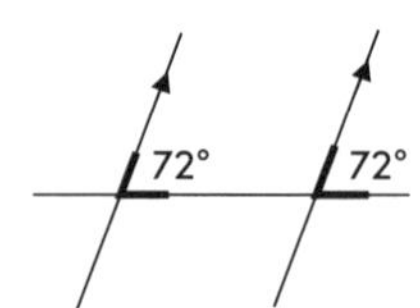


......................................

**b**

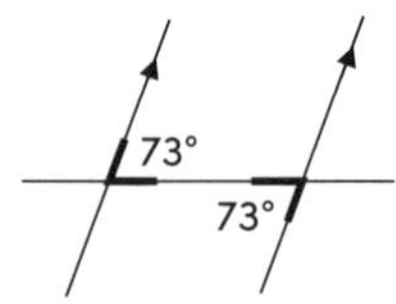


......................................

**c**

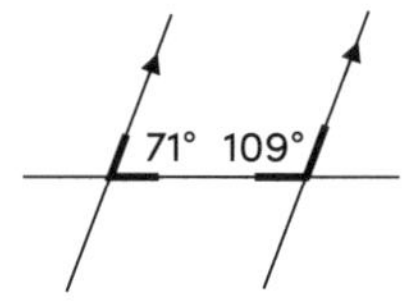


......................................

**d**

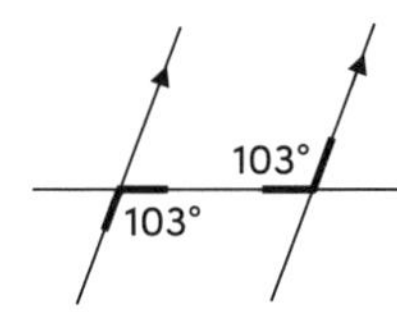


......................................

**e**

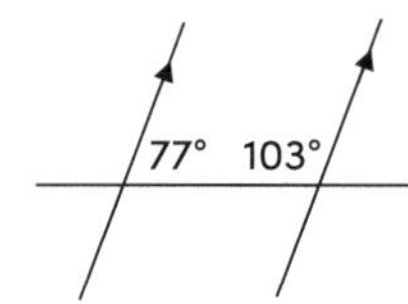


......................................

**f**

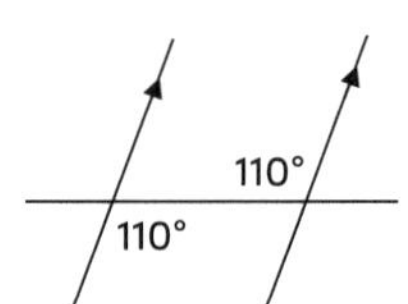


......................................

**g**

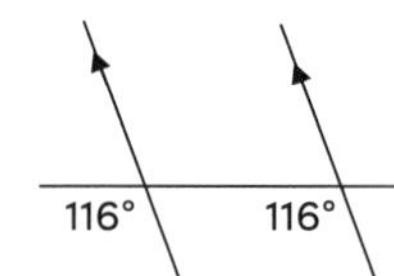


......................................

**h**

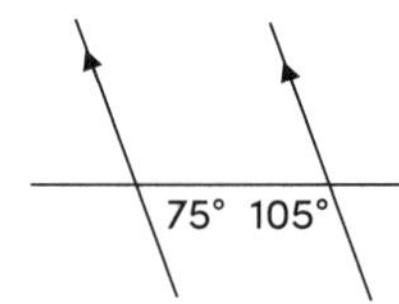


......................................

**i**

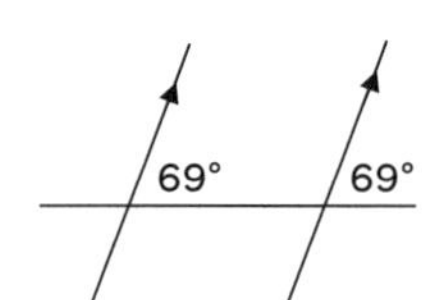


......................................

**j**

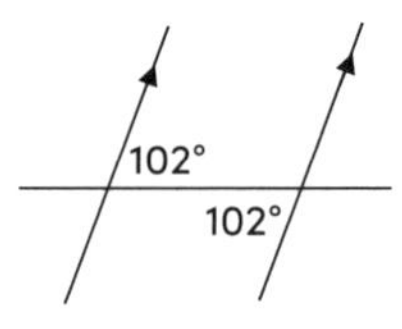


......................................

**k**

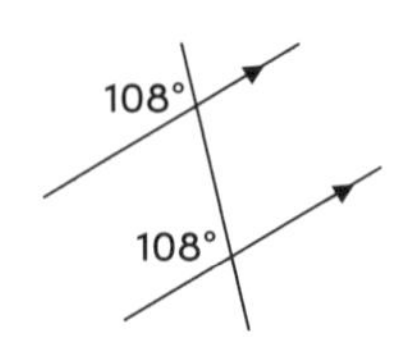


......................................

**l**

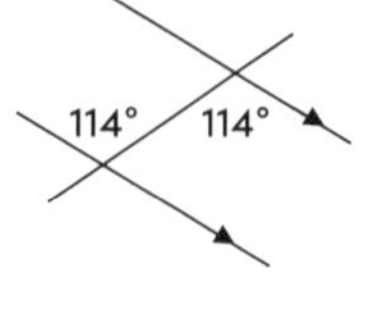


......................................

**GEOMETRY**

# Unknown angles on parallel lines

**1** Find the unknown angles, giving reasons:

**a**

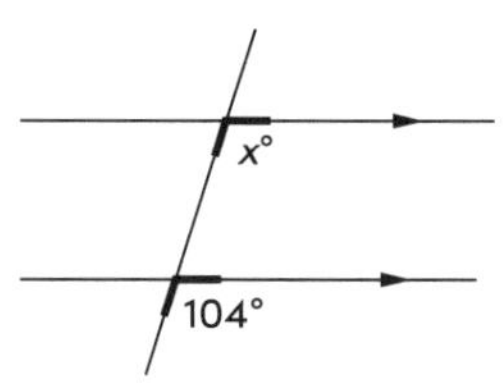


$x =$ ______

Circle one:
Corresponding
Alternate
Co-interior

**b**

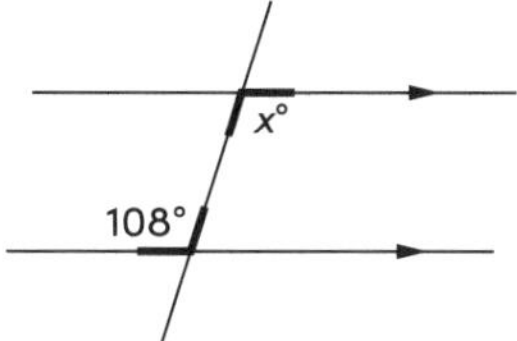


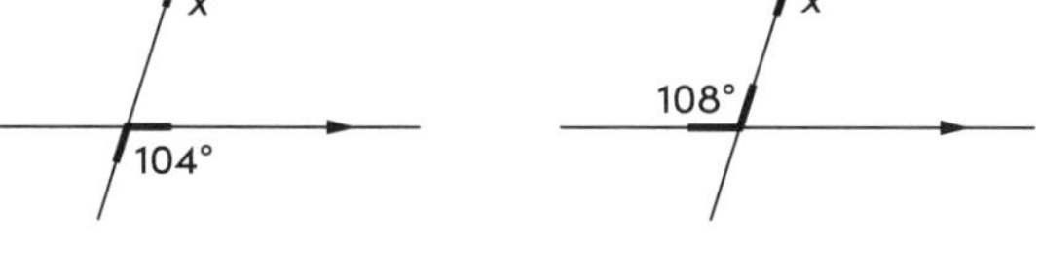

$x =$ ______

Circle one:
Corresponding
Alternate
Co-interior

**c**

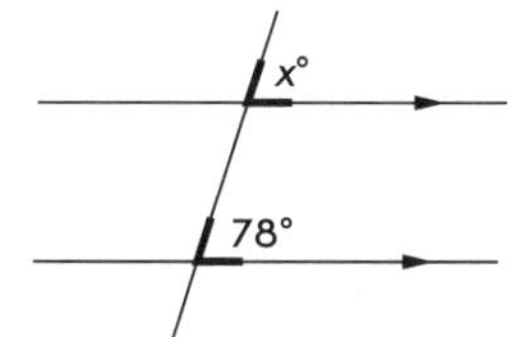


$x =$ ______

Circle one:
Corresponding
Alternate
Co-interior

**d**

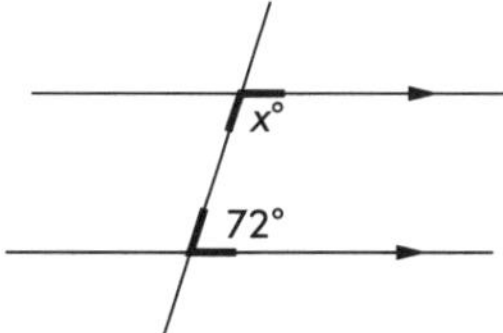


$x =$ 180 $- 72$

$=$ ______

Circle one:
Corresponding
Alternate
Co-interior

**e**

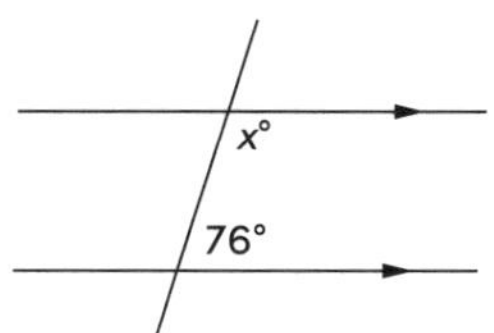


$x =$ ______

$=$ ______

______

**f**

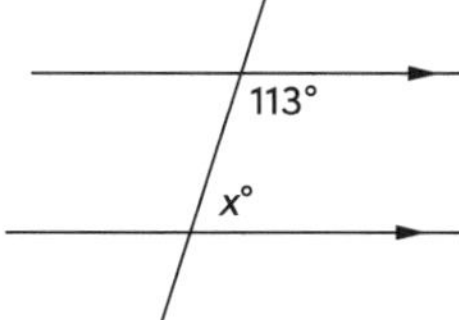


$x =$ ______

$=$ ______

______

**g**

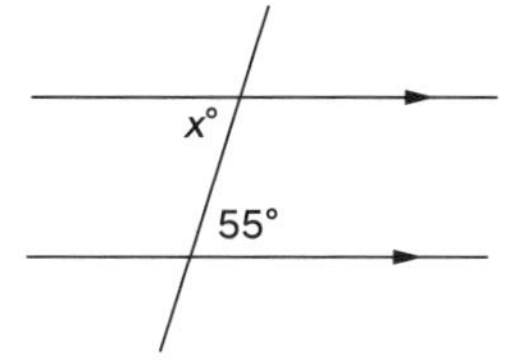


$x =$ ______

______

**h**

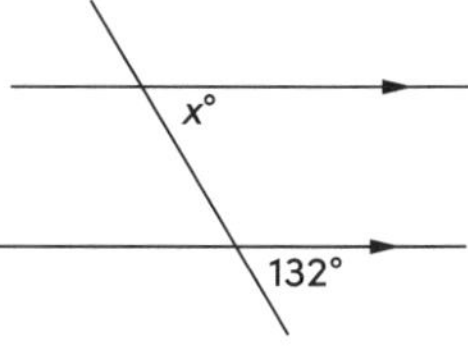


$x =$ ______

______

**2** Find the unknown angles, giving the correct reasons:

**a**

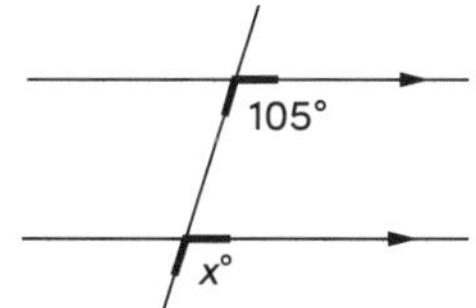


$x =$ ______

______

**b**

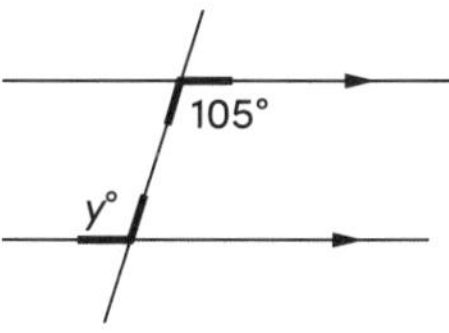


$y =$ ______

______

**c**

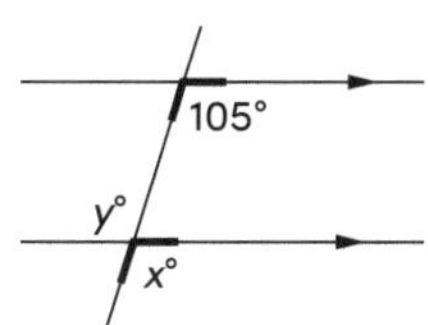


$x =$ ______

______

$y =$ ______

______

**d**

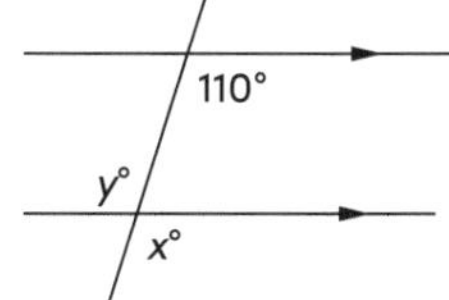


$x =$ ______

______

$y =$ ______

______

**e**

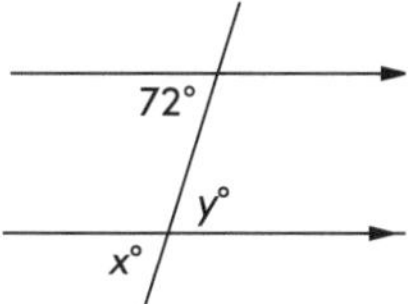


$x =$ ______

______

$y =$ ______

______

**f**

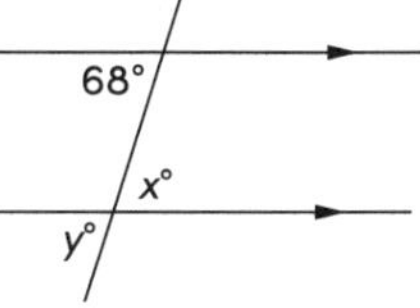


$x =$ ______

______

$y =$ ______

______

GEOMETRY

# Unknown angles on parallel lines

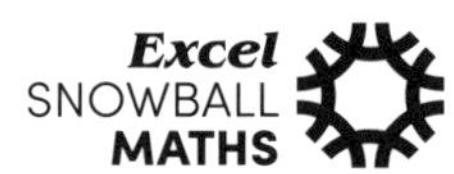

**1** Find the unknown angles, giving the correct reasons:

**a**

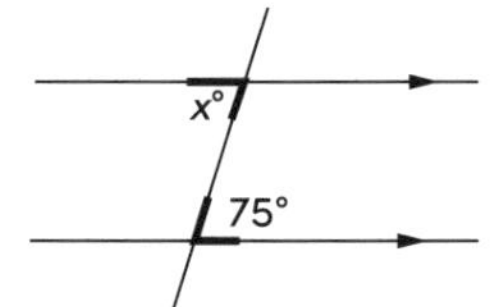


$x =$ ______

______

**b**

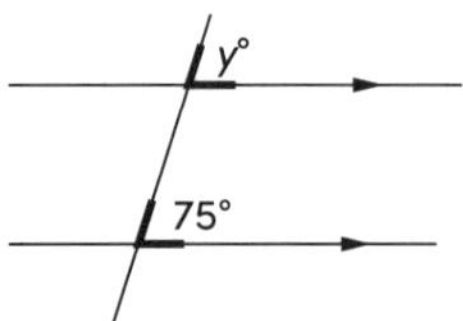


$y =$ ______

______

**c**

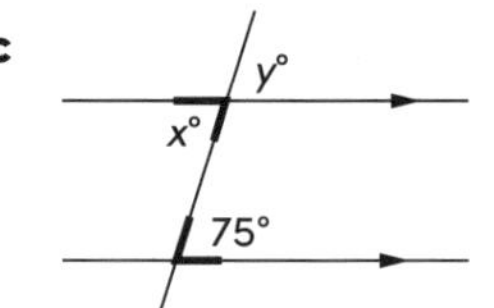


$x =$ ______

______

$y =$ ______

______

**d**

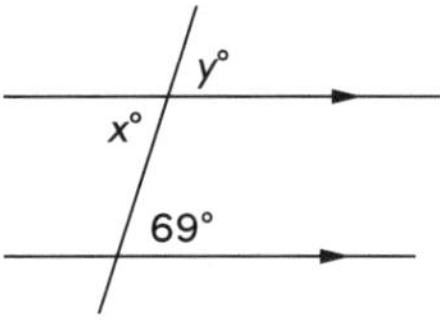


$x =$ ______

______

$y =$ ______

______

**e**

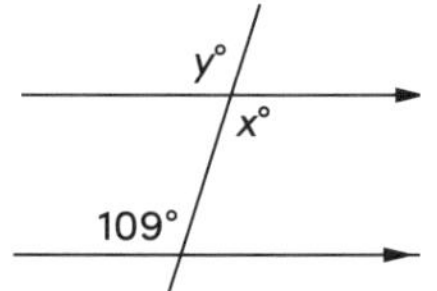


$x =$ ______

______

$y =$ ______

______

**f**

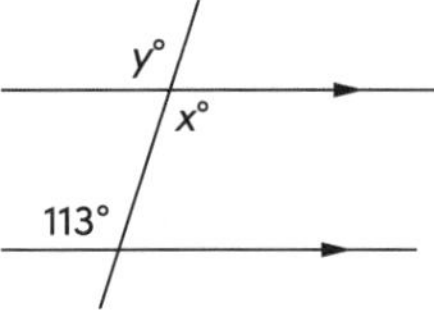


$x =$ ______

______

$y =$ ______

______

**g**

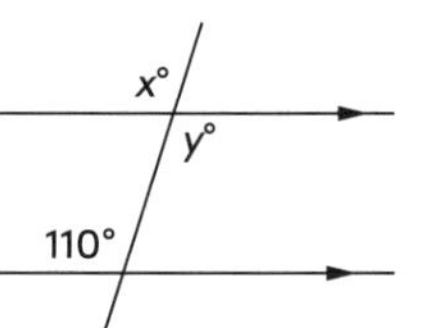


$x =$ ______

______

$y =$ ______

______

**h**

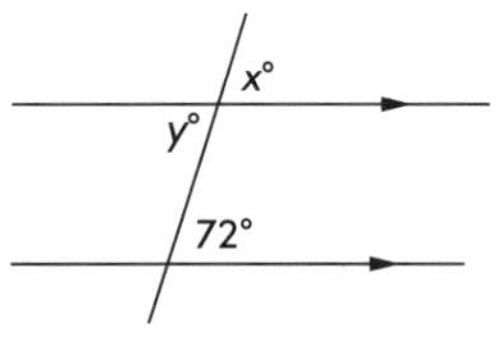


$x =$ ______

______

$y =$ ______

______

**2** Find the unknown angles, giving the correct reasons:

**a**

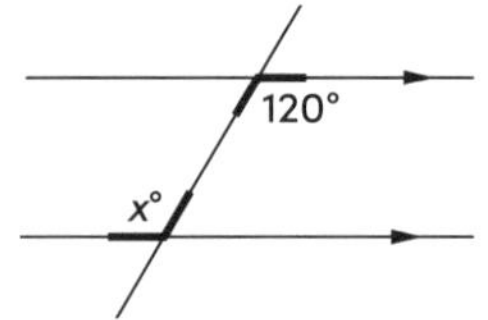


$x =$ ______

______

**b**

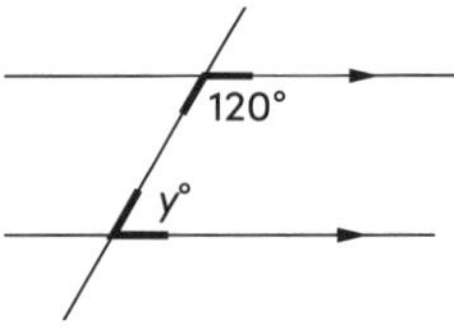


$y =$ ______ $- 120$

$=$ ______

______

**c**

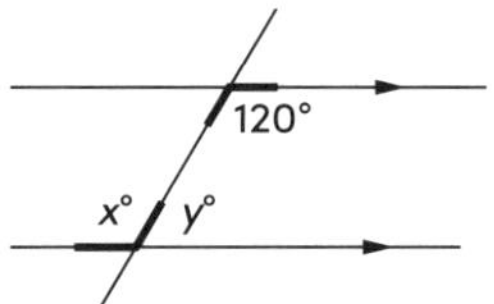


$x =$ ______

______

$y =$ ______

$=$ ______

______

**d**

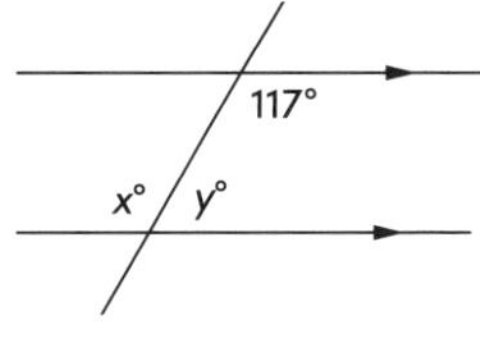


$x =$ ______

______

$y =$ ______

$=$ ______

______

**e**

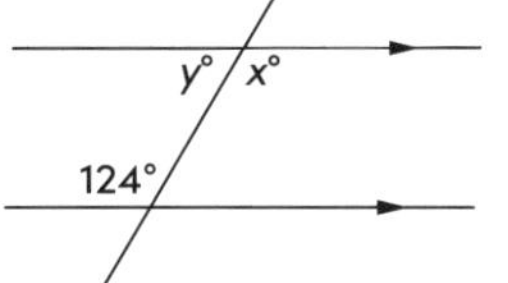


______

______

______

______

______

**f**

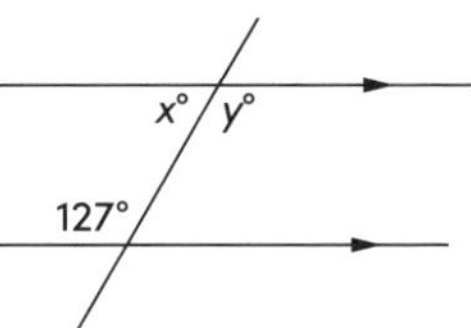


______

______

______

______

______

## GEOMETRY

# Unknown angles on parallel lines

**1** Find the unknown angles, giving the correct reasons:

**a**

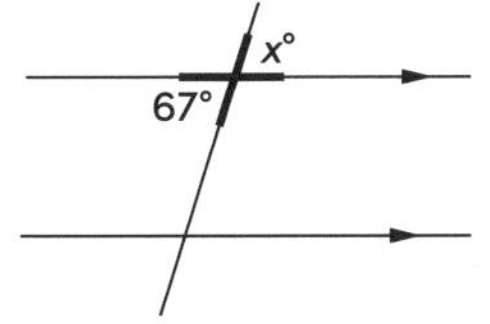


$x =$ ______

(Vertically opposite angles are equal)

**b**

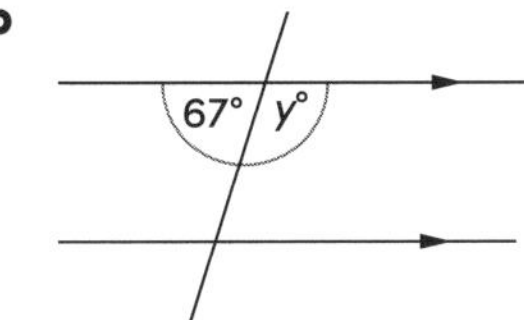


$y =$ ______ $- 67$

$=$ ______

(Straight angle equals 180°)

**c**

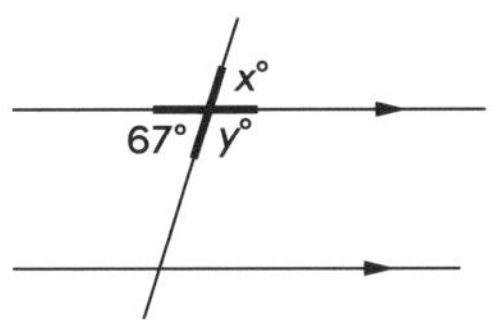


$x =$ ______

______

$y =$ ______

$=$ ______

______

**d**

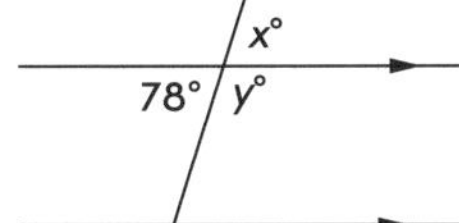


$x =$ ______

______

$y =$ ______

$=$ ______

______

**e**

**f**

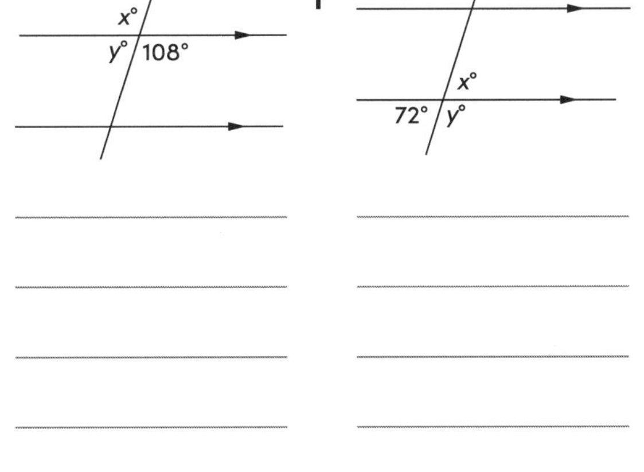


______

______

______

______

______

**2** Find the unknown angles, giving the correct reasons:

**a**

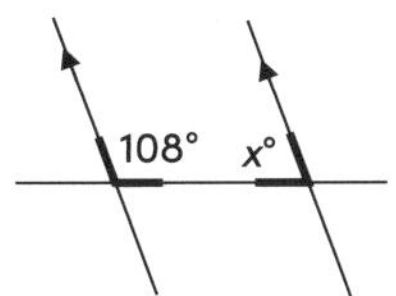


$x =$ ______ $- 108$

$=$ ______

______

**b**

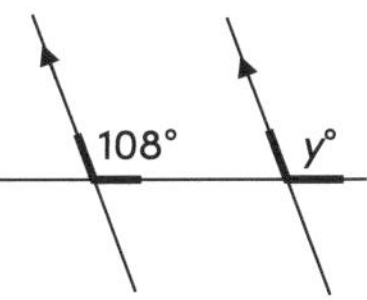


$y =$ ______

______

**c**

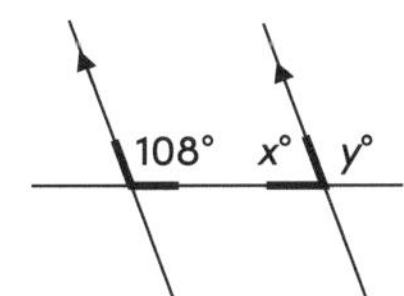


$x =$ ______

$=$ ______

______

$y =$ ______

______

**d**

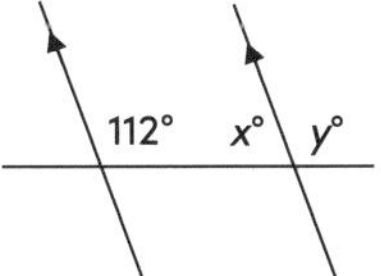


$x =$ ______

$=$ ______

______

$y =$ ______

______

**e**

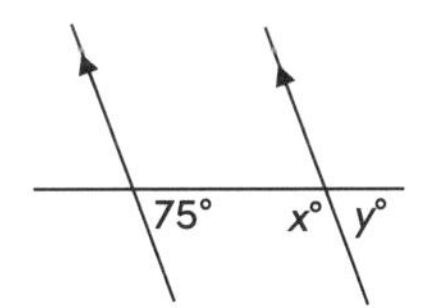


______

______

______

______

______

**f**

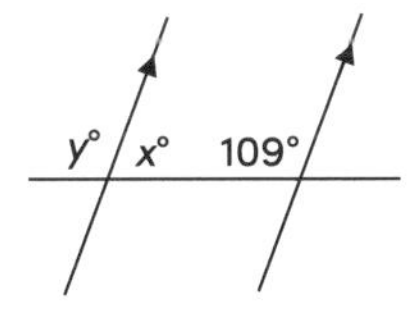


______

______

______

______

______

## GEOMETRY
# Classifying triangles

**1** State whether each triangle is **equilateral** (all 3 sides are the same), **isosceles** (2 sides are the same) or **scalene** (no sides are the same):

**a**
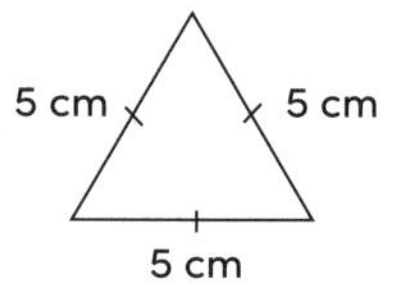


**b**
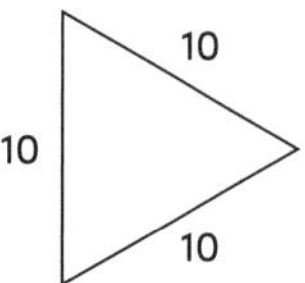


**c**
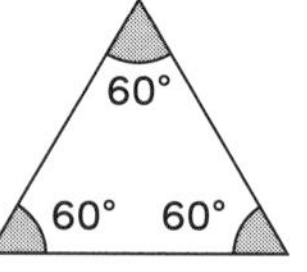


**d**
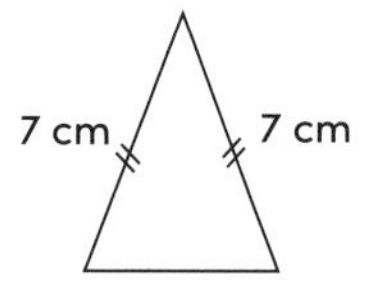


**e**
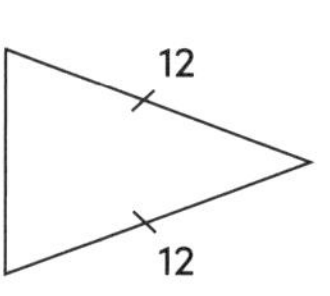


**f**
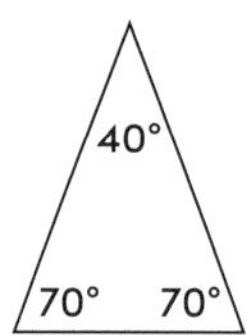


**g**
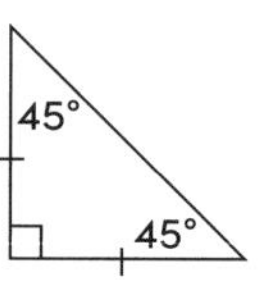


**h**
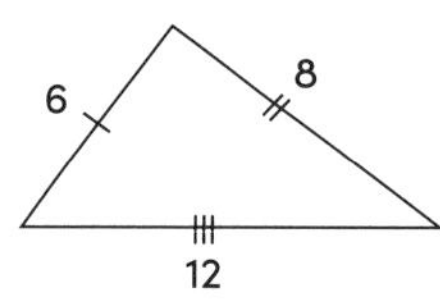


**i**
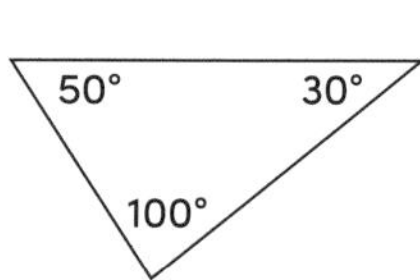


**j**
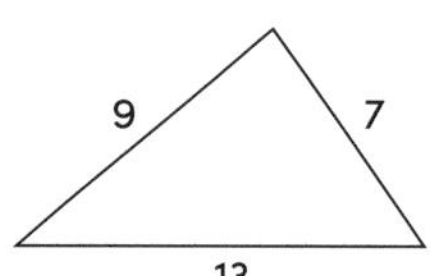


**k**
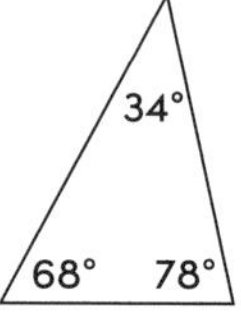


**l**
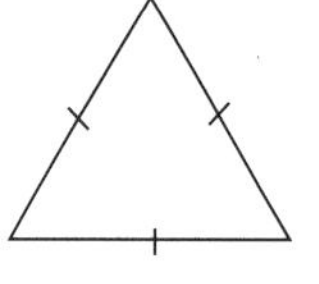

**m**
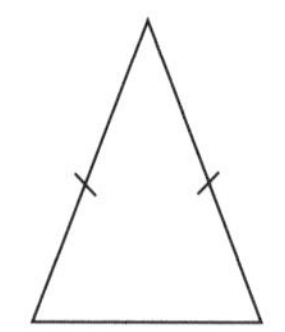

**n**
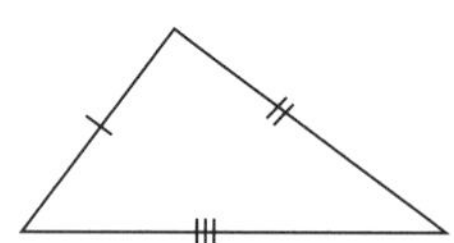

**o**
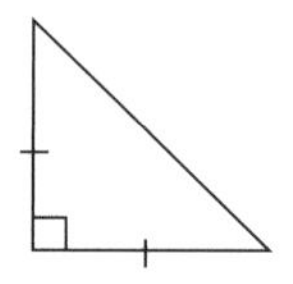

**p**
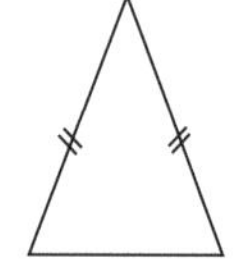

**q**
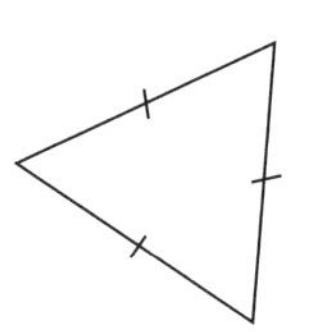

**r**
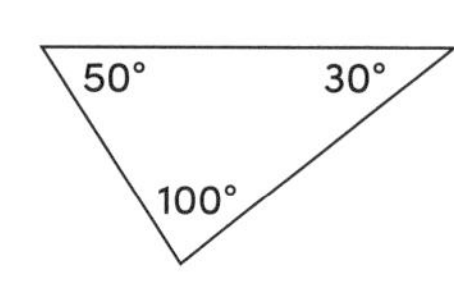


**2** State whether each triangle is **acute-angled**, **right-angled** or **obtuse-angled**:

**a**
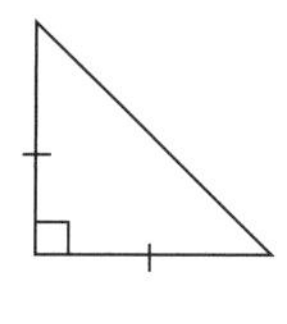

**b**
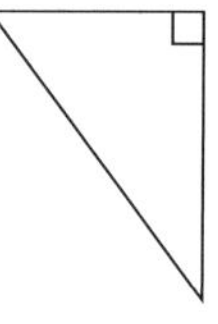

**c**
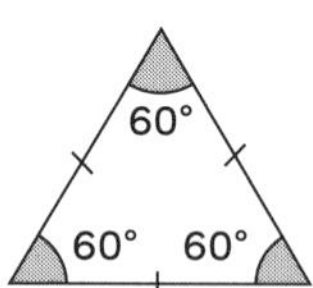


**d**
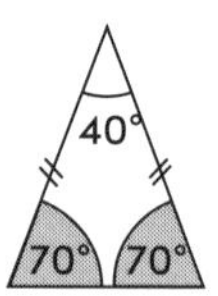


**e**
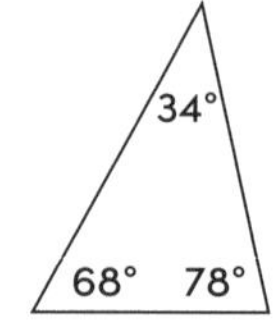


**f**
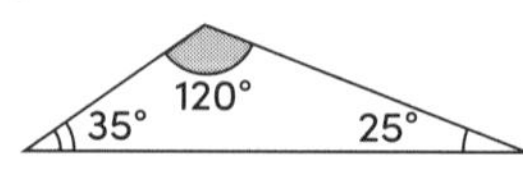

UNIT 27

# GEOMETRY
## Classifying triangles

**1** State whether each triangle is acute-angled, right-angled or obtuse-angled:

**a**

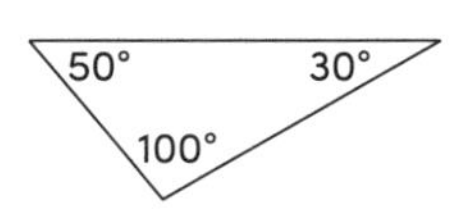


______________

**b**

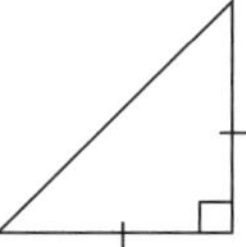

______________

**c**

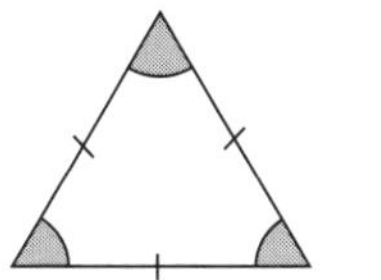

______________

**d**

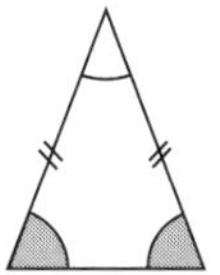

______________

**e**

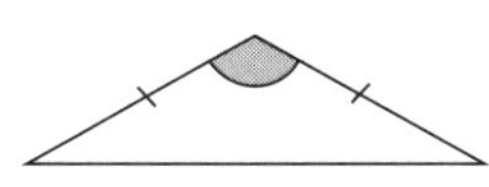

______________

**f**

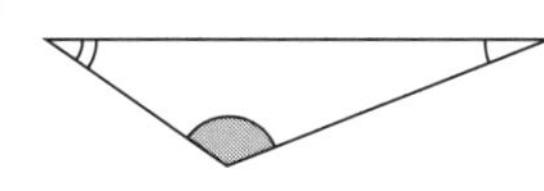

______________

Complete each sentence to classify the triangle by sides and angle. Choose the correct option from each box.

**2 a**

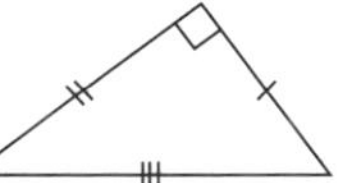

| Equilateral | Isosceles | Scalene |
|---|---|---|
| Acute | Right | Obtuse |

The triangle is ______________ and ______________ angled.

**b**

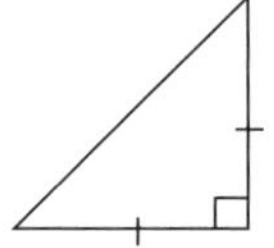

| Equilateral | Isosceles | Scalene |
|---|---|---|
| Acute | Right | Obtuse |

The triangle is ______________ and ______________ angled.

**c**

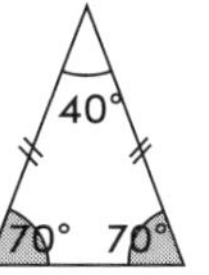


| Equilateral | Isosceles | Scalene |
|---|---|---|
| Acute | Right | Obtuse |

The triangle is ______________ and ______________ angled.

**d**

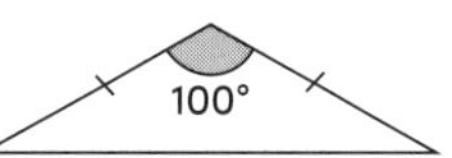


| Equilateral | Isosceles | Scalene |
|---|---|---|
| Acute | Right | Obtuse |

The triangle is ________________________

________________________

**e**

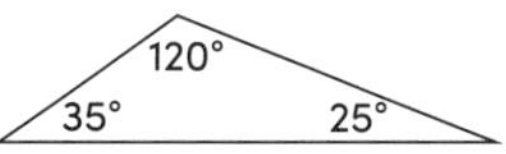


| Equilateral | Isosceles | Scalene |
|---|---|---|
| Acute | Right | Obtuse |

The triangle is ________________________

________________________

**f**

| Equilateral | Isosceles | Scalene |
|---|---|---|
| Acute | Right | Obtuse |

The triangle is ________________________

________________________

UNIT 28

**GEOMETRY**

# Classifying triangles

**1** Complete each sentence to classify the triangle by sides and angle. Choose the correct option from each box.

**a**

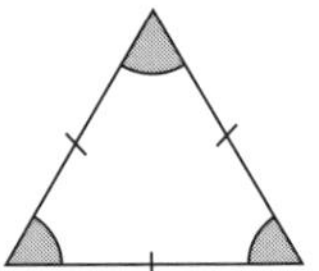

| Equilateral | Isosceles | Scalene |
|---|---|---|
| Acute | Right | Obtuse |

The triangle is ______________________

______________________

**b**

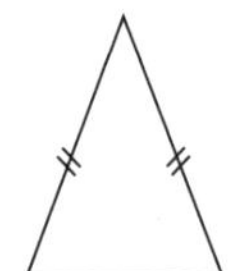

| Equilateral | Isosceles | Scalene |
|---|---|---|
| Acute | Right | Obtuse |

The triangle is ______________________

______________________

**c**

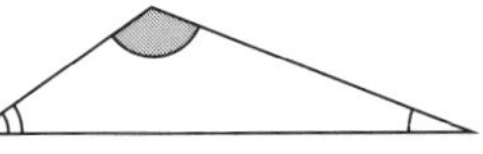

| Equilateral | Isosceles | Scalene |
|---|---|---|
| Acute | Right | Obtuse |

The triangle is ______________________

______________________

**d**

| Equilateral | Isosceles | Scalene |
|---|---|---|
| Acute | Right | Obtuse |

The triangle is ______________________

______________________

**2** Draw an example of the following:

**a** equilateral triangle

**b** isosceles triangle

**c** scalene triangle

**d** right-angled triangle

**e** acute-angled triangle

**f** obtuse-angled triangle

**g** isosceles, right-angled triangle

**h** scalene, right-angled triangle

**i** isosceles, acute-angled triangle

**j** isosceles, obtuse-angled triangle

**k** scalene, acute-angled triangle

**l** right-angled, isosceles triangle

UNIT 29

GEOMETRY

# Angle sum of triangles

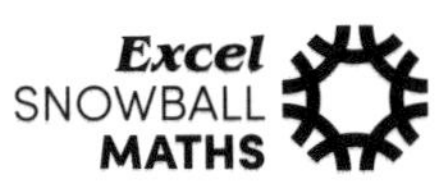


**1** Find the value of each pronumeral:

**a**

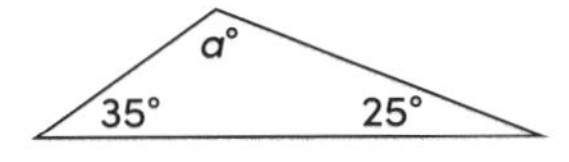


$a$ = 180 – ______ – ______

= ______

**b**

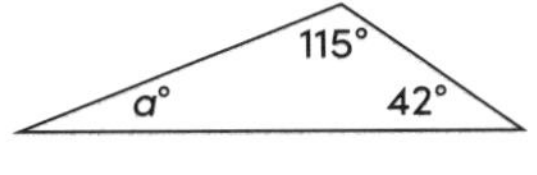


$a$ = 180 – ______ – ______

= ______

**c**

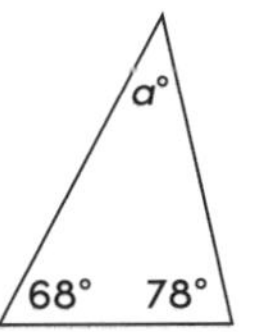


$a$ = ______ – ______ – ______

= ______

**d**

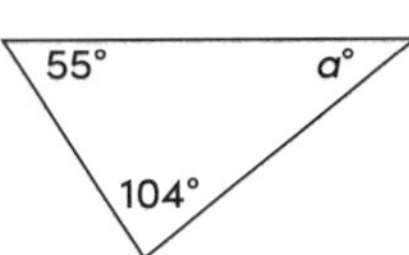


$a$ = ______ – ______ – ______

= ______

**e**

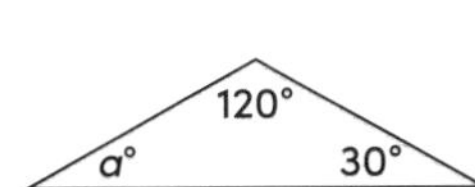


$a$ = ______

= ______

**f**

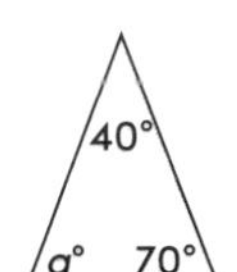


$a$ = ______

= ______

**g**

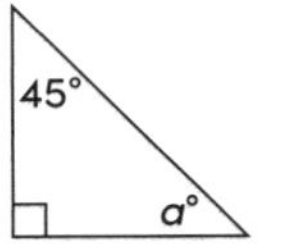


______

______

**h**

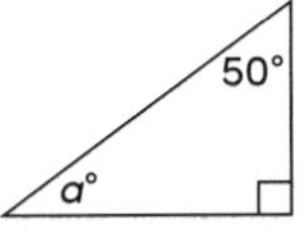


______

______

**2** Find the value of each pronumeral:

**a**

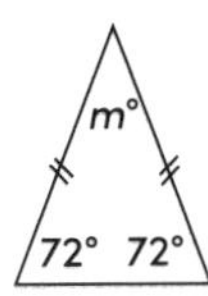


$m$ = 180 – ______ – ______

= ______

**b**

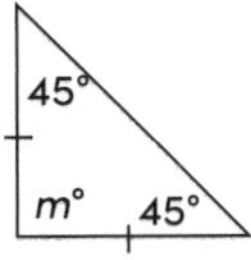


$m$ = 180 – ______ – ______

= ______

**c**

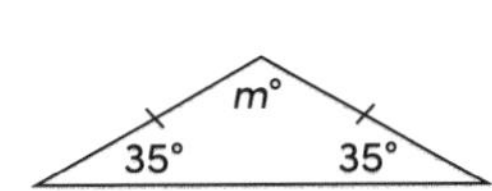


$m$ = ______

= ______

**d**

$m$ = ______

= ______

**e**

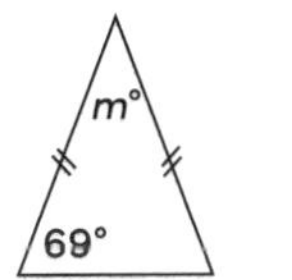


$m$ = ______

= ______

**f**

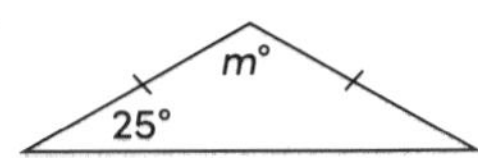


$m$ = ______

= ______

**g**

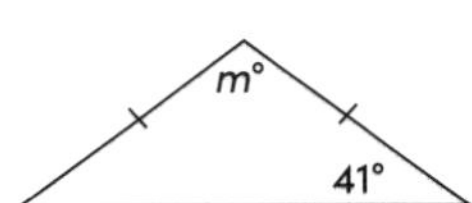


______

______

**h**

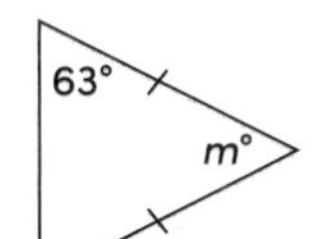


______

______

**i**

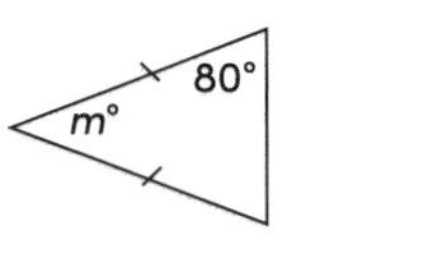


______

______

**j**

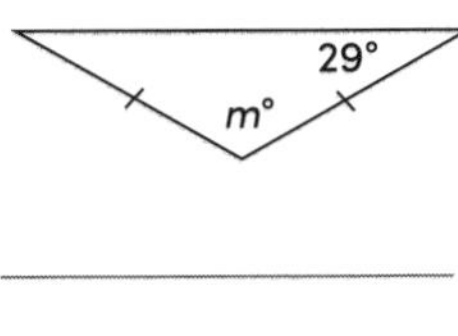


______

______

UNIT 30

GEOMETRY

# Angle sum of triangles

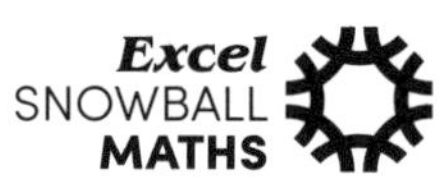


**1** Find the value of each pronumeral:

**a**

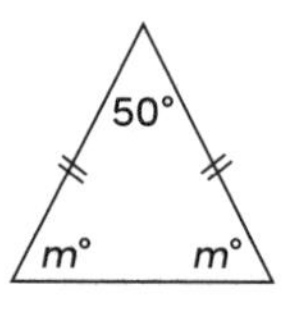


$2m =$ 180 − ____

$=$ ____

$m = \frac{\square}{2}$

$=$ ____

**b**

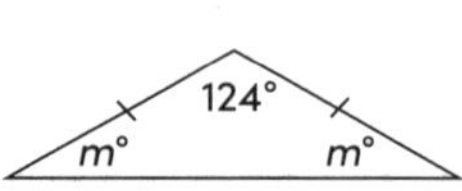


$2m =$ 180 − ____

$=$ ____

$m = \frac{\square}{2}$

$=$ ____

**c**

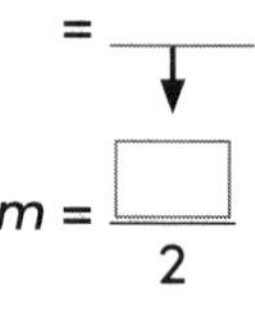


$2m =$ ____

$=$ ____

$m = \frac{\square}{\square}$

$=$ ____

**d**

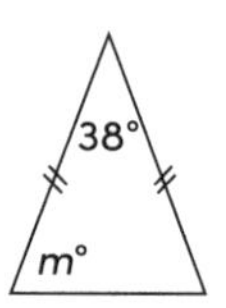


$2m =$ ____

$=$ ____

$m = \frac{\square}{\square}$

$=$ ____

**e**

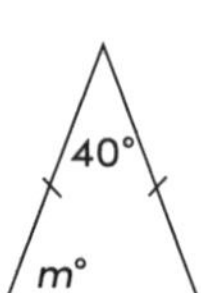


____

____

____

____

**f**

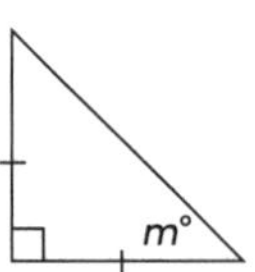


____

____

____

____

**2** Find the value of each pronumeral:

**a**

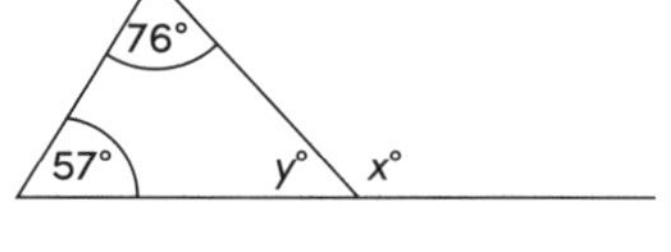


$x =$ ____ + ____

$=$ ____

$y =$ 180 − ____

$=$ ____

**a**

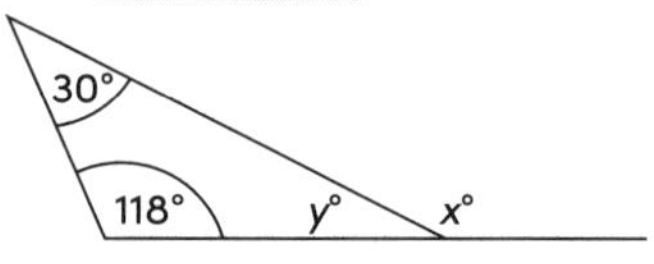


$x =$ ____ + ____

$=$ ____

$y =$ 180 − ____

$=$ ____

**c**

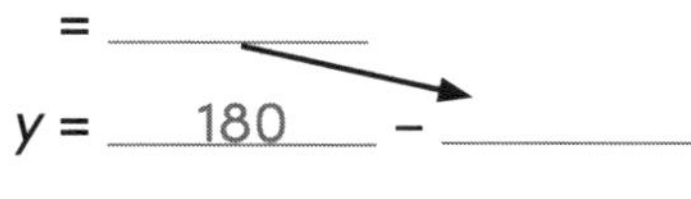


$x =$ ____

$=$ ____

$y =$ ____

$=$ ____

**d**

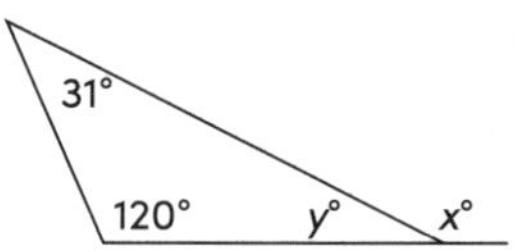
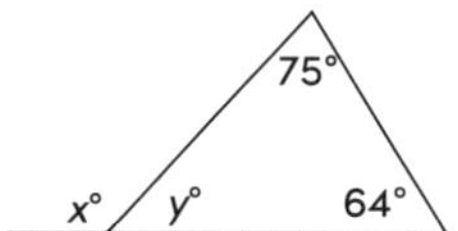


$x =$ ____

$=$ ____

$y =$ ____

$=$ ____

**e**

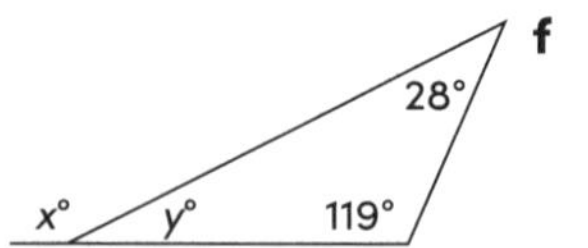


____

____

____

____

**f**

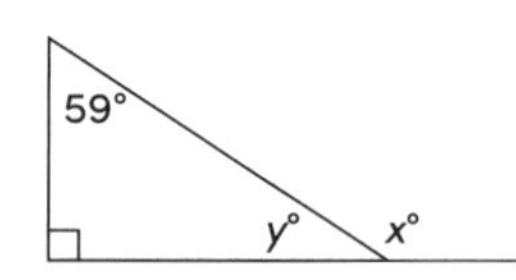


____

____

____

____

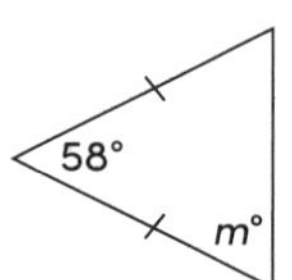

GEOMETRY

# Properties of quadrilaterals

equal | 90° | bisect

**1** Complete the sentence for each square:

**a**

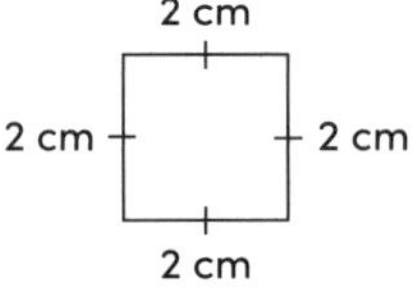


All sides of a square are ______________.

**b**

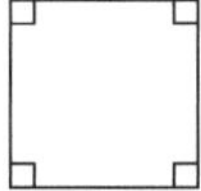

All angles of a square are ______________°.

**c**

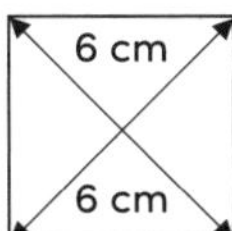


Diagonals of a square are ______________.

**d**

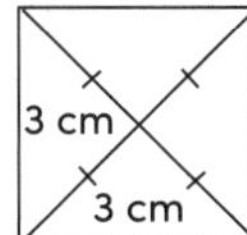


Diagonals of a square ______________ each other.

**e**

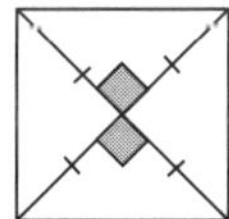

Diagonals of a square bisect each other at ______________°.

**f**

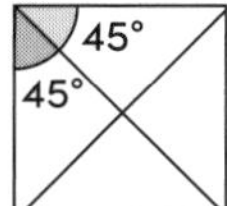


Diagonals ______________ the angles of a square.

**2** Write all the properties of a square:

**3** Complete the sentence for each rectangle:

**a**

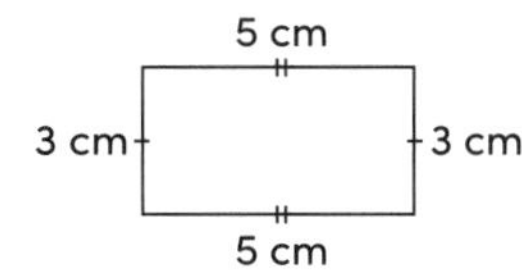


Opposite sides are ______________.

**b**

All angles of a rectangle are ______________.

**c**

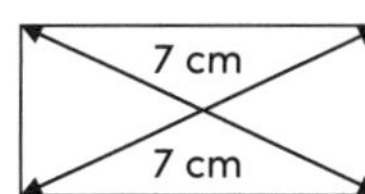


Diagonals of a rectangle are ______________.

**d**

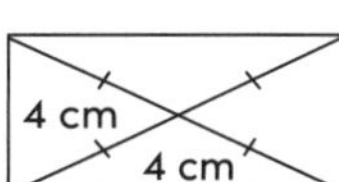


Diagonals of a rectangle ______________ each other.

GEOMETRY

# Properties of quadrilaterals

**1** Write all the properties of a rectangle:

_______________

_______________

_______________

_______________

_______________

_______________

_______________

_______________

equal | 90° | bisect | parallel

**2** Complete the sentence for each rhombus:

**a** 8 cm, 8 cm, 8 cm, 8 cm

All sides of a rhombus are ___________.

**b**

Opposite sides of a rhombus are

_______________.

**c** 70°, 70°

Opposite angles of a rhombus are

_______________.

**d** 5 cm, 5 cm

Diagonals of a rhombus

_______________ each other.

**e**

Diagonals of a rhombus bisect each other at _______________.

**f** 50°, 50°

Diagonals _______________ the angles of a rhombus.

**3** Write all the properties of a rhombus:

_______________

_______________

_______________

_______________

_______________

_______________

_______________

_______________

_______________

UNIT 33

**GEOMETRY**

# Properties of quadrilaterals

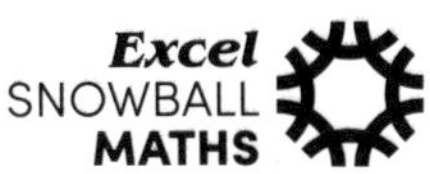


equal | bisect | parallel

**1** Complete the sentence for each parallelogram:

**a**

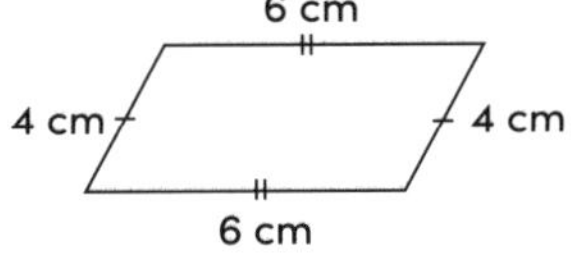


Opposite sides are ______________.

**b**

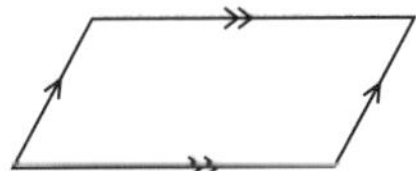

Opposite sides of a parallelogram are

______________.

**c**

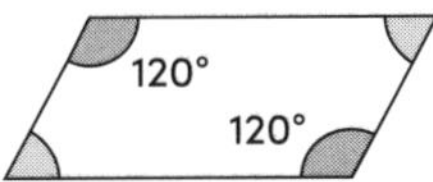


Opposite angles of a parallelogram are

______________.

**d**

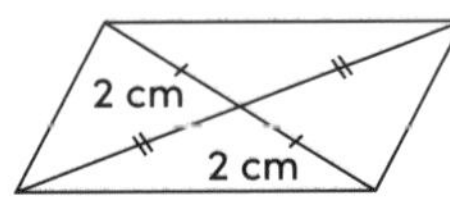


Diagonals of a parallelogram

______________ each other.

**2** Write all the properties of a parallelogram:

______________

______________

______________

______________

______________

equal | 90°

**3** Complete the sentence for each kite:

**a**

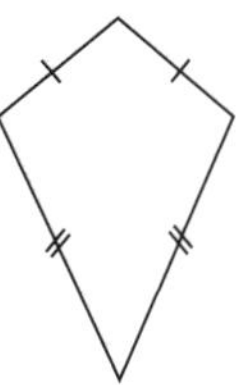

Two pairs of adjacent sides are

______________.

**b**

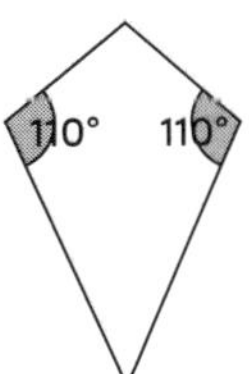


One pair of opposite angles are

______________.

**c**

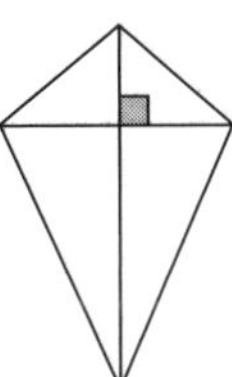

Diagonals intersect at

______________.

**4** Write all the properties of a kite:

______________

______________

______________

______________

______________

UNIT 34

GEOMETRY

# Properties of quadrilaterals

1 Complete the sentence for the trapezium:

One pair of sides are ______________.

2 Write three properties of a square:

3 Write three properties of a rectangle:

4 Write three properties of a rhombus:

5 Write three properties of a parallelogram:

6 True or False?

a All four sides of a square are equal. ________

b All four sides of a rectangle are equal. ________

c All four angles of a rectangle are 90°. ________

d All four angles of a rhombus are 90°. ________

e Opposite angles are equal in a parallelogram. ________

f Diagonals bisect at 90° in a rhombus. ________

g Diagonals bisect at right angles in a rectangle. ________

UNIT 35

## GEOMETRY

# Angle sum of quadrilaterals

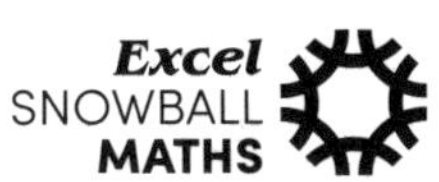


**1** Find the value of each pronumeral:

**a**

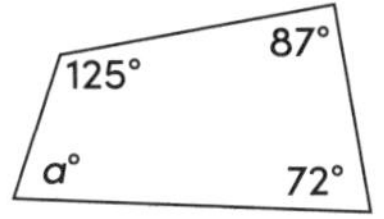


$a$ = 360 – ______ – ______ – ______

= ________ °

**b**

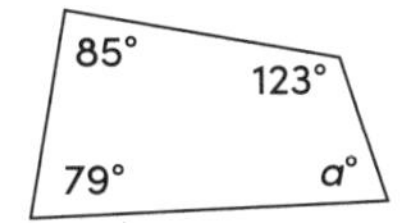


$a$ = 360 – ______ – ______ – ______

= ________ °

**c**

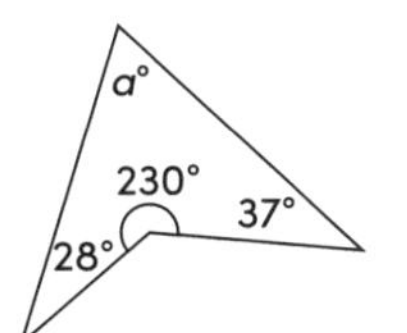


$a$ = ________

= ________

**d**

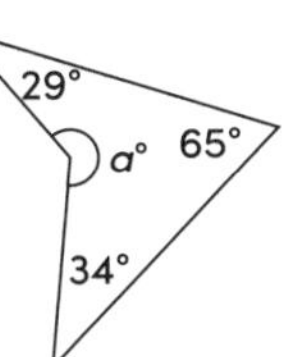


$a$ = ________

= ________

**e**

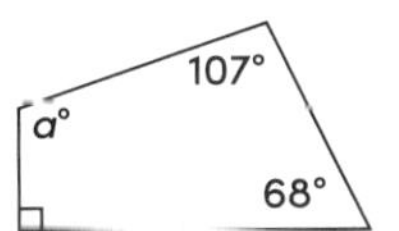


________

________

**f**

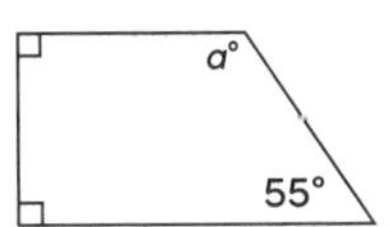


________

________

**g**

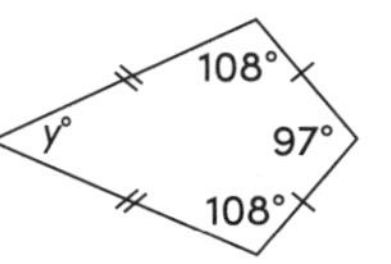


________

________

**h**

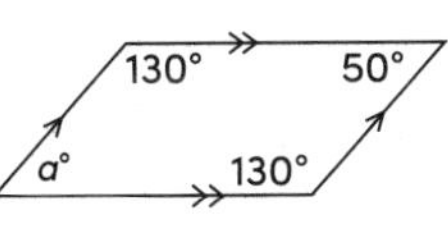


________

________

**2** Find the value of each pronumeral:

**a**

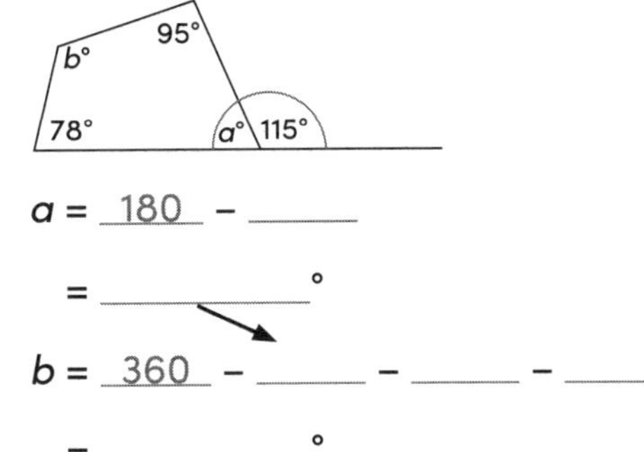


$a$ = 180 – ______

= ________ °

$b$ = 360 – ______ – ______ – ______

= ________ °

**b**

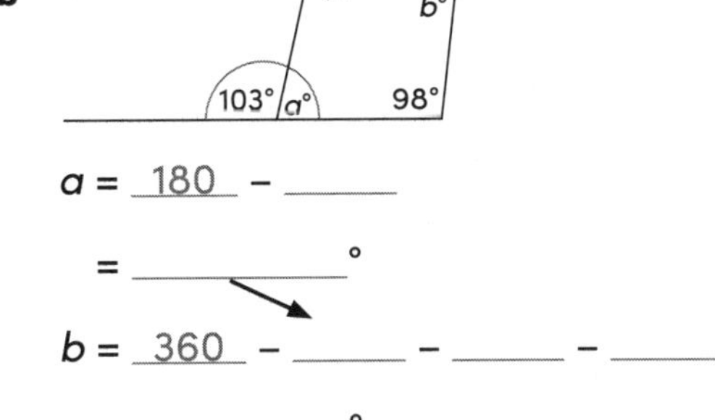


$a$ = 180 – ______

= ________ °

$b$ = 360 – ______ – ______ – ______

= ________ °

**c**

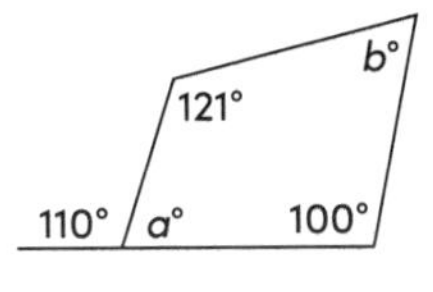


$a$ = ________

= ________

$b$ = ________

= ________

**d**

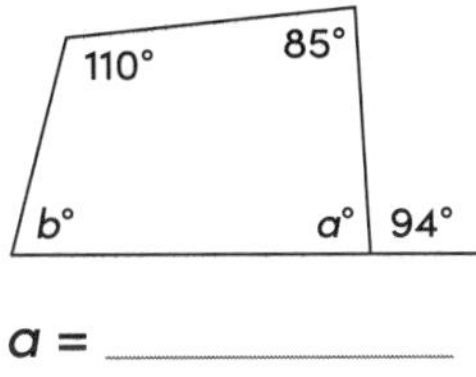


$a$ = ________

= ________

$b$ = ________

= ________

**e**

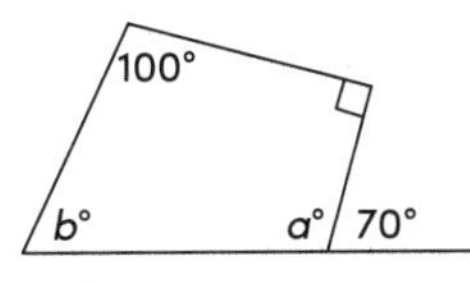


________

________

________

________

**f**

117°

80°

b°

78°

________

________

________

________

# Geometry



**1** Name the following marked angles:

**a**

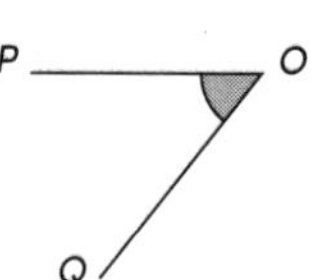


**b**

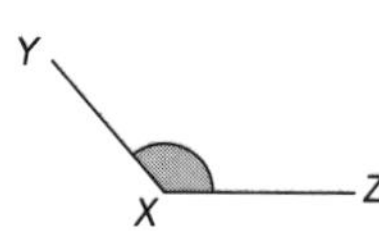


**c**

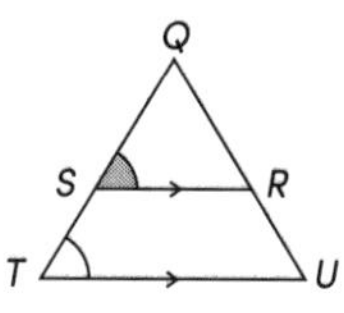


**d**

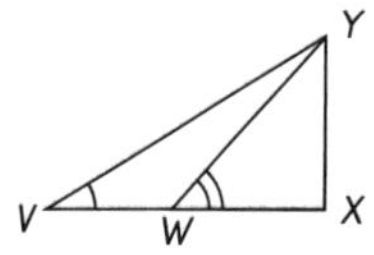


**2** Name each type of angle:

**a**

**b**

**c**

**d**

**e**

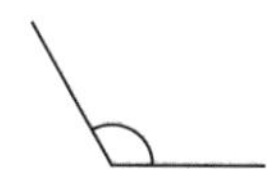

**f**

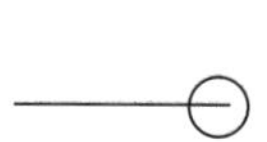

**3** Using a protractor, measure the following marked angles:

**a**

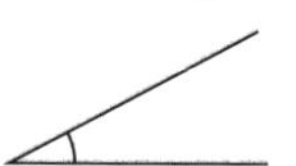

**b**

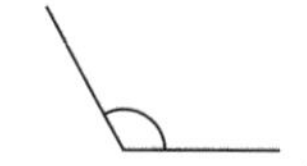

**4** What is the complement of 50°? ________

**5** What is the supplement of 110°? ________

**6** Find the value of each unknown angle:

**a**

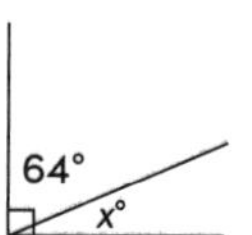


**b**

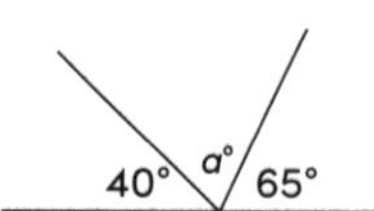


**c**

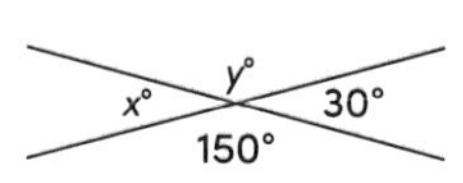


**d**

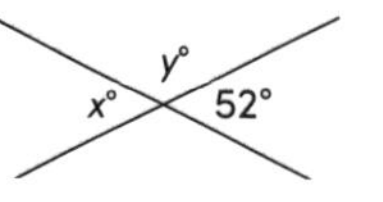


**e**

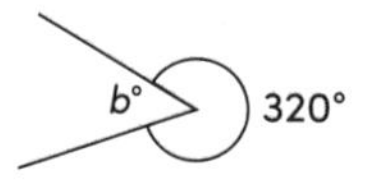


**f**

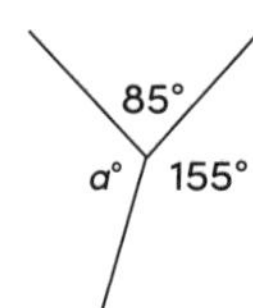


**7** Find each unknown angle, giving reasons:

**a**

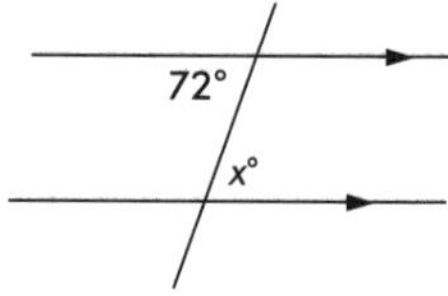


**b**

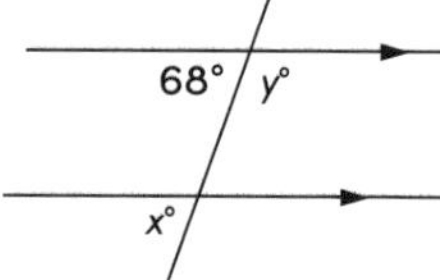


**c**

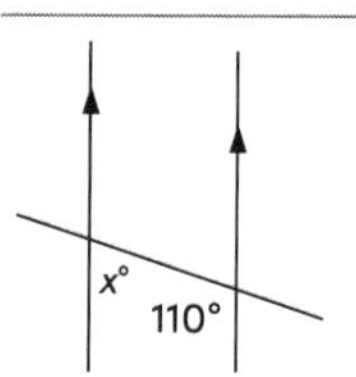


**d**

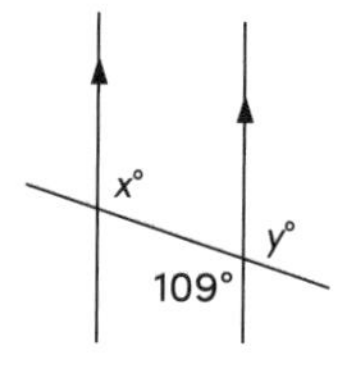


**8** What type of triangle is shown?

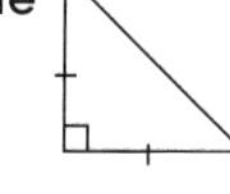

**9** Find the unknown angles for the following:

**a**

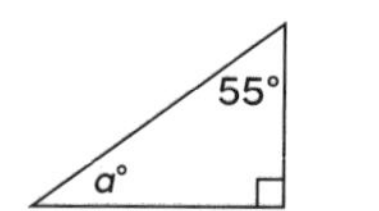


**b**

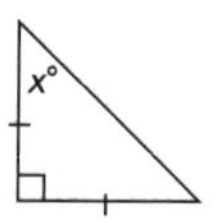


**c**

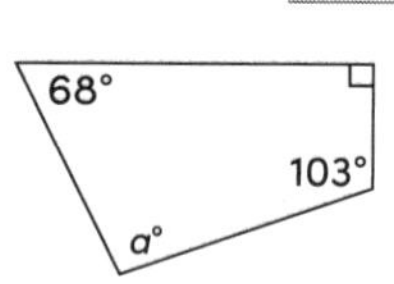


**d**

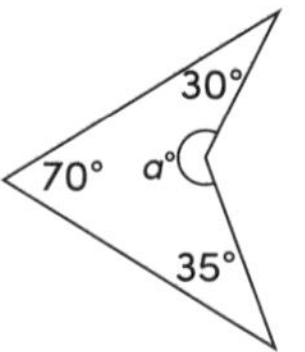


**e**

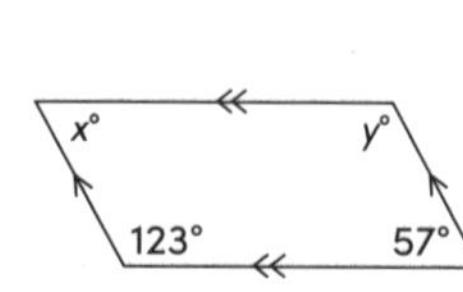


**f**

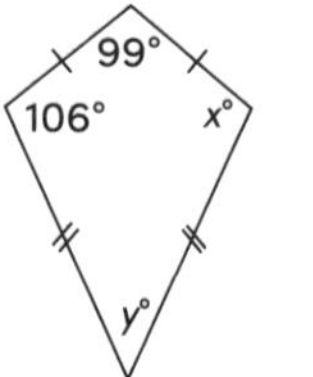


**10** Give two properties of a parallelogram:

________________

________________

## GRAPHS AND DATA

# Drawing frequency histograms and polygons

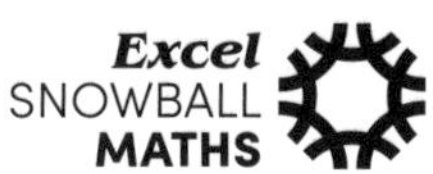


**1** Complete the frequency histogram and polygon from the frequency table below:

| Score ($x$) | Frequency ($f$) |
|---|---|
| 1 | 4 |
| 2 | 8 |
| 3 | 12 |
| 4 | 5 |

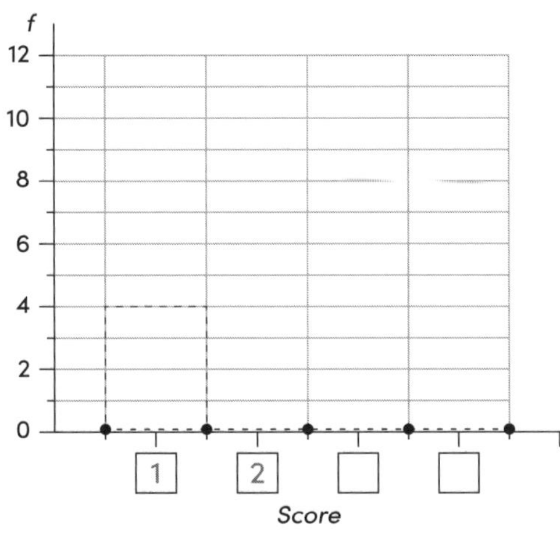


**2** Complete:

| Score ($x$) | Frequency ($f$) |
|---|---|
| 5 | 6 |
| 6 | 5 |
| 7 | 12 |
| 8 | 10 |

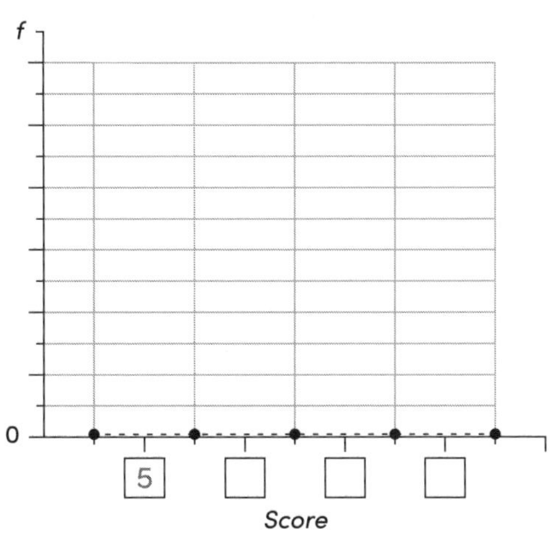


**3** Complete:

| Score ($x$) | Frequency ($f$) |
|---|---|
| 10 | 3 |
| 20 | 8 |
| 30 | 11 |
| 40 | 6 |

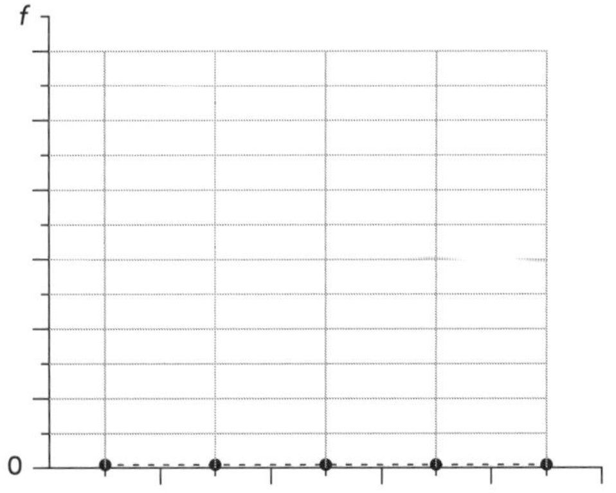


**4** Complete:

| Score ($x$) | Frequency ($f$) |
|---|---|
| 4 | 5 |
| 5 | 10 |
| 6 | 20 |
| 7 | 30 |
| 8 | 23 |

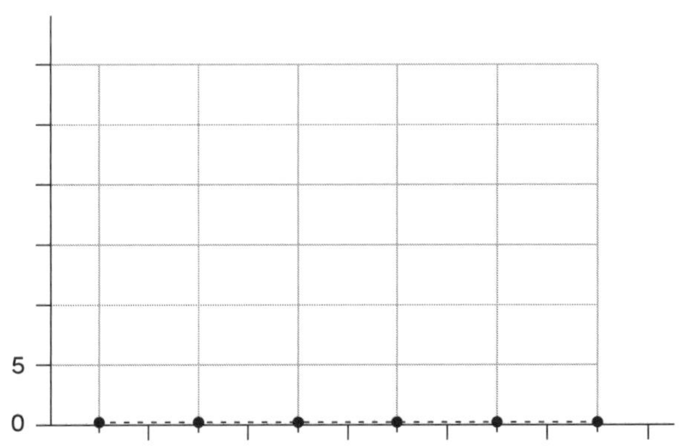

## GRAPHS AND DATA

# Drawing frequency histograms and polygons

**1** Complete:

| Score ($x$) | Frequency ($f$) |
|---|---|
| 10 | 15 |
| 11 | 25 |
| 12 | 4 |
| 13 | 10 |
| 14 | 27 |

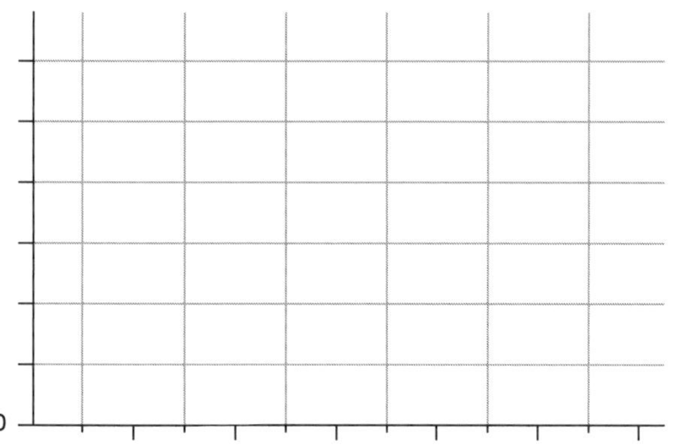


**2** Complete:

| Score ($x$) | Frequency ($f$) |
|---|---|
| 3 | 13 |
| 5 | 29 |
| 7 | 15 |
| 9 | 10 |
| 11 | 6 |

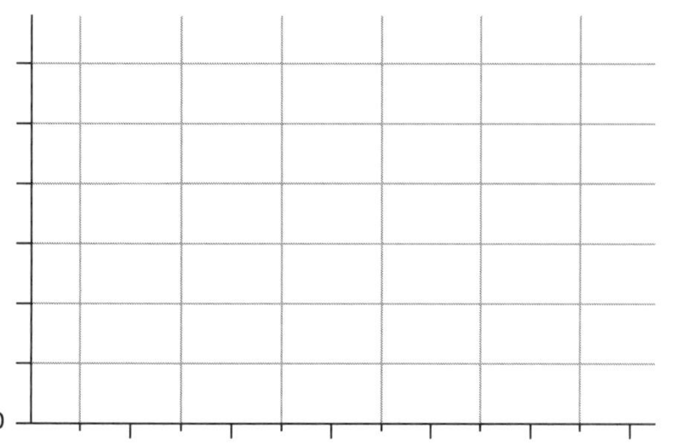


**3** Complete:

| Score ($x$) | Frequency ($f$) |
|---|---|
| 7 | 10 |
| 8 | 20 |
| 9 | 60 |
| 10 | 45 |
| 11 | 30 |

**4** The scores below show the ages of people doing bowling and arcade activities. Represent this data as a frequency histogram and polygon:

15 17 17 18 15 15 16 16 19 19
15 15 17 17 17 17 16 18 19 15
17 18 15 16 16 19 18 16 17 17

| Score ($x$) | Frequency ($f$) |
|---|---|
| | |
| | |
| | |
| | |
| | |

GRAPHS AND DATA

# Drawing stem-and-leaf plots

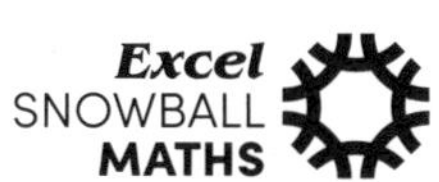


**1** Draw stem-and-leaf plots for the following sets of scores:

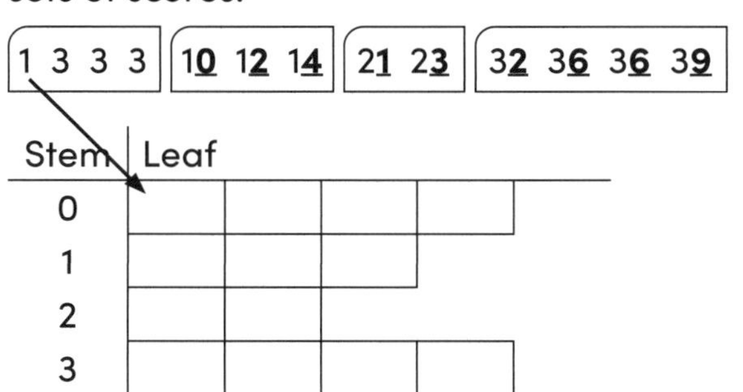


**2** 3 4 7 12 14 16 16 21 23 32 35 38

| Stem | Leaf |
|---|---|
| 0 | |
| 1 | |
| 2 | |
| 3 | |

**3** 5 6 6 7 11 13 20 24 26 34 35 38 39

| Stem | Leaf |
|---|---|
| 0 | |
| 1 | |
| 2 | |
| 3 | |

**4** 1 5 6 7 10 15 18 20 24 26 31 33 33 42 47

| Stem | Leaf |
|---|---|
| 0 | |
| 1 | |
| 2 | |
| 3 | |
| 4 | |

**5** 3 5 5 12 13 18 26 26 30 32 32 35 43 44 44 46

| Stem | Leaf |
|---|---|
| 0 | |
| 1 | |
| 2 | |
| 3 | |
| 4 | |

**6** 1 2 2 4 12 16 16 20 22 24 30 33 42 42 47 47

| Stem | Leaf |
|---|---|
| 0 | |
| 1 | |
| 2 | |
| 3 | |
| 4 | |

**7** 3 5 16 17 17 24 25 26 32 37 39 43 46 47 49

| Stem | Leaf |
|---|---|
| | |
| | |
| | |
| | |
| | |

**8** 2 4 6 9 12 12 13 24 28 32 36 36 42 43 47 48

| Stem | Leaf |
|---|---|
| | |
| | |
| | |
| | |
| | |

## GRAPHS AND DATA

# Drawing stem-and-leaf plots

Draw stem-and-leaf plots for the following sets of scores:

**1** 2 5 6 10 11 18 20 21 22 24 27 32 35 36 43 44 44 50 51 53 53

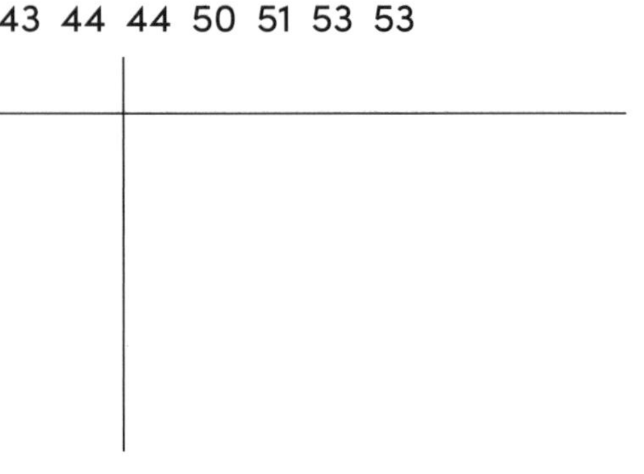

**2** 2 2 2 4 13 16 17 20 24 26 28 32 37 41 45 50 51 52 56

**3** 60 62 63 64 73 78 82 86 93 93 94 103 105 106 107

| Stem | Leaf |
|---|---|
| 6 | |
| 7 | |
| 8 | |
| 9 | |
| 10 | |

**4** 62 64 73 74 75 85 86 88 89 89 96 96 97 98 101 103 103

**5** 103 105 105 107 112 112 115 116 124 125 127 132 136 139 139 141 144 145

| Stem | Leaf |
|---|---|
| 10 | |
| 11 | |
| 12 | |
| 13 | |
| 14 | |

**6** 100 102 106 107 110 113 113 115 121 125 126 126 132 132 143 147 147 149

**7** 105 140 103 136 111 142 124 127 128 136 125 110 132 133 103 124 143

**8** 118 121 139 117 145 156 153 126 133 152 152 110 140 134

GRAPHS AND DATA

# Drawing dot plots

Draw a dot plot for each set of scores:

**1** 1 1 3 3 1 2 2 3 4 1 4 4

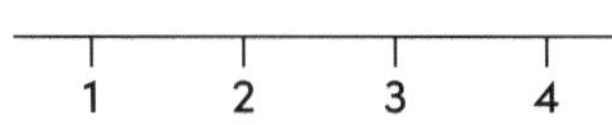


**2** 6 7 6 7 7 8 9 9 7 6

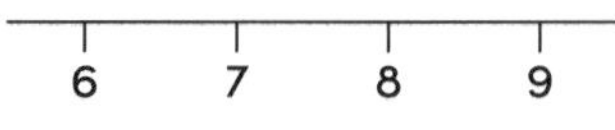


**3** 3 3 4 6 4 4 5 3 4 5 4 6 3

**4** 13 10 11 12 11 13 13 13 12 10

**5** Draw a dot plot for the following frequency table:

| Score ($x$) | Frequency ($f$) |
|---|---|
| 1 | 3 |
| 2 | 1 |
| 3 | 1 |
| 4 | 2 |

**6** Draw a dot plot for the following frequency table:

| Score ($x$) | Frequency ($f$) |
|---|---|
| 10 | 3 |
| 11 | 2 |
| 12 | 1 |
| 13 | 4 |

**7** The table shows the ratings viewers gave for a TV show:

| Score ($x$) | Frequency ($f$) |
|---|---|
| 2.0 | 1 |
| 2.5 | 2 |
| 3.0 | 4 |
| 3.5 | 3 |
| 4.0 | 3 |

Draw a dot plot displaying the table above.

UNIT 6

GRAPHS AND DATA

# Divided bar graphs

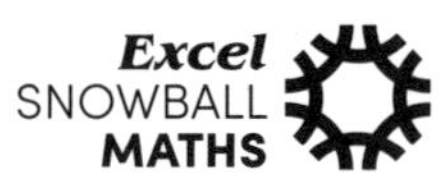

**1** The table shows the number of different-coloured marbles in a bag:

| Blue | Red | Yellow | Green |
|---|---|---|---|
| 4 | 2 | 3 | 1 |

Complete the divided bar graph:

Blue

4 cm 2 cm 3 cm 1 cm

**2** Students were asked what pet they own. The table shows the results:

| Cat | Dog | Bird | Rabbit |
|---|---|---|---|
| 3 | 4 | 2 | 1 |

Complete the divided bar graph:

3 cm 4 cm 2 cm 1 cm

**3** The table shows which electronic device people prefer:

| Tablet | Phone | Laptop | Desktop |
|---|---|---|---|
| 2 | 4 | 3 | 1 |

Complete the divided bar graph:

2 cm

**4** A survey shows what people prefer to watch on the television:

| News | Cartoons | Cooking | Sport |
|---|---|---|---|
| 1 | 5 | 2 | 2 |

Complete the divided bar graph:

1 cm

GRAPHS AND DATA

# Divided bar graphs

**1** A survey shows what drink people prefer:

| Water | Juice | Coffee | Milk |
|---|---|---|---|
| 2 | 3 | 2 | 3 |

Construct a divided bar graph:

**2** The table shows what subject students prefer in school:

| Maths | English | Science | PE |
|---|---|---|---|
| 4 | 2 | 3 | 1 |

Construct a divided bar graph:

**3** Draw a divided bar graph for the results below:

| A | B | C | D |
|---|---|---|---|
| 3.4 | 1.5 | 2.2 | 2.9 |

Complete the divided bar graph:

3.4 cm 1.5 cm 2.2 cm 2.9 cm

**4** Construct a divided bar graph for the results below:

| A | B | C | D |
|---|---|---|---|
| 2.7 | 1.4 | 3.6 | 2.3 |

UNIT 8

GRAPHS AND DATA

# Divided bar graphs

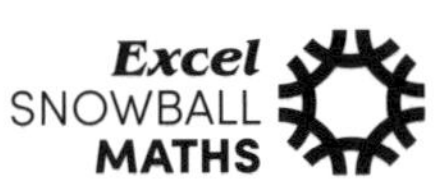


**1** A sample of 50 people were surveyed on their exercise and the results are shown:

| Swimming | Running | Yoga | Cycling |
|---|---|---|---|
| 15 | 17 | 8 | 10 |

Complete:
(correct to one decimal place)

$\frac{15}{\square} \times 10 =$ ______ cm　　$\frac{8}{\square} \times 10 =$ ______ cm

$\frac{17}{\square} \times 10 =$ ______ cm　　$\frac{10}{\square} \times 10 =$ ______ cm

Construct a divided bar graph:

**2** The table shows the languages other than English spoken from a sample of 45 people:

| Korean | Arabic | Vietnamese | Italian |
|---|---|---|---|
| 12 | 16 | 11 | 6 |

Complete:
(correct to one decimal place)

$\frac{\square}{\square} \times 10 =$ ______ cm　　$\frac{\square}{\square} \times 10 =$ ______ cm

$\frac{\square}{\square} \times 10 =$ ______ cm　　$\frac{\square}{\square} \times 10 =$ ______ cm

Construct a divided bar graph:

**3** A sample of 70 people were surveyed on their jobs and the results are shown:

| Teacher | Nurse | Doctor | Police |
|---|---|---|---|
| 19 | 25 | 10 | 16 |

Complete:
(correct to one decimal place)

$\frac{\square}{\square}$ ______ = ______ cm　　$\frac{\square}{\square}$ ______ = ______ cm

$\frac{\square}{\square}$ ______ = ______ cm　　$\frac{\square}{\square}$ ______ = ______ cm

Construct a divided bar graph:

GRAPHS AND DATA

# Divided bar graphs

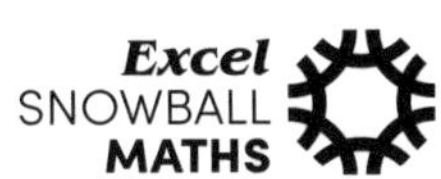


1 A sample of 100 people were surveyed on their age:

| 0–19 | 20–39 | 40–59 | 60–79 |
|---|---|---|---|
| 23 | 34 | 31 | 12 |

Display this as a divided bar graph.

2 A sample of 200 people were surveyed on the pet they have:

| Dogs | Cats | Fish | Birds | Reptiles |
|---|---|---|---|---|
| 88 | 62 | 18 | 20 | 12 |

Display this as a divided bar graph.

3 A sample of 350 children were surveyed on their favourite school subject:

| English | Maths | Science | PE | Art | Other |
|---|---|---|---|---|---|
| 28 | 60 | 40 | 91 | 76 | 55 |

Display this as a divided bar graph.

**GRAPHS AND DATA**

# Drawing sector graphs

**1** Display the following data in a sector graph:

| A | B | C | D | E |
|---|---|---|---|---|
| **30°** | 50° | 60° | 100° | 120° |

Complete:

30 + 50 = ________

30 + 50 + 60 = ________

30 + 50 + 60 + 100 = ________

30 + 50 + 60 + 100 + 120 = ________

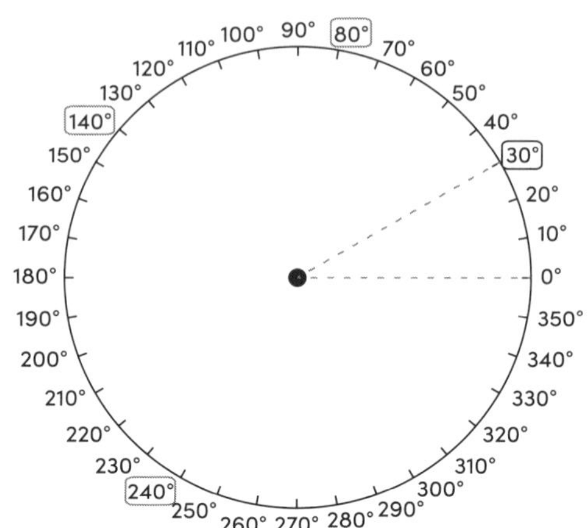


**2** Display the following data in a sector graph:

| A | B | C | D | E |
|---|---|---|---|---|
| **70°** | 120° | 90° | 40° | 40° |

Complete:

70 + 120 = ________

70 + 120 + 90 = ________

70 + 120 + 90 + 40 = ________

70 + 120 + 90 + 40 + 40 = ________

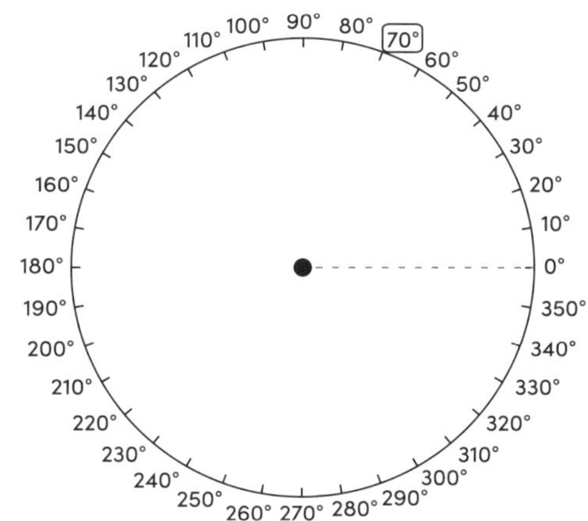


**3** Display the following data in a sector graph:

| A | B | C | D | E |
|---|---|---|---|---|
| **105°** | 40° | 70° | 85° | 60° |

Complete:

105 + 40 = ________

105 + 40 + ________ = ________

105 + 40 + ________ + ________ = ________

105 + 40 + ________ + ________ + ________ = ________

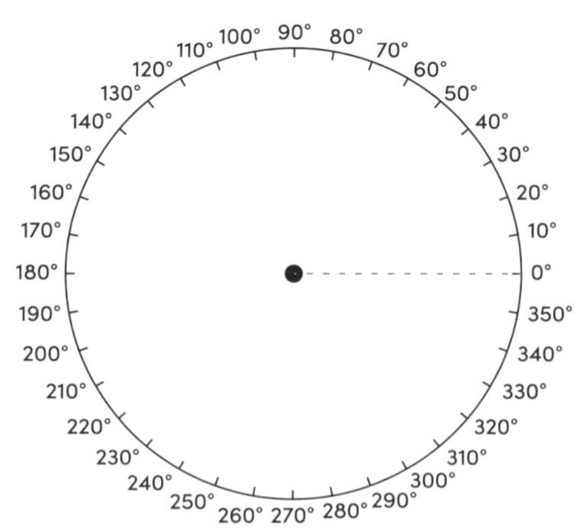

UNIT 11

GRAPHS AND DATA

# Drawing sector graphs

1 Display the following data in a sector graph:

| A | B | C | D | E |
|---|---|---|---|---|
| **125°** | 50° | 75° | 65° | 45° |

Complete:

125 + ________ = ________

125 + ________ + ________ = ________

125 + ________ + ________ + ________ = ________

125 + ________ + ________ + ________ + ________ = ________

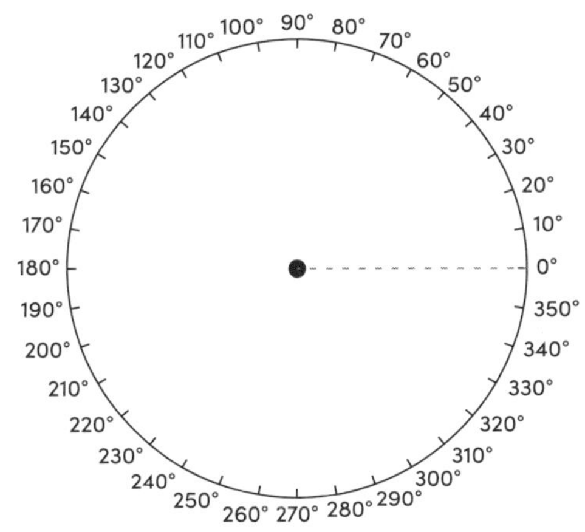

2 Display the following data in a sector graph:

| A | B | C | D | E |
|---|---|---|---|---|
| **72°** | 126° | 94° | 46° | 22° |

Complete:

________ + ________ = ________

________ + ________ + ________ = ________

________ + ________ + ________ + ________ = ________

________ + ________ + ________ + ________ + ________ = ________

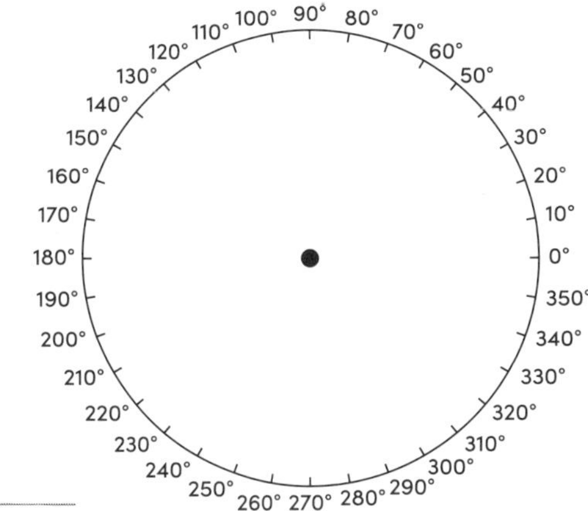

3 Display the following data in a sector graph:

| A | B | C | D | E |
|---|---|---|---|---|
| 55° | 36° | 108° | 87° | 74° |

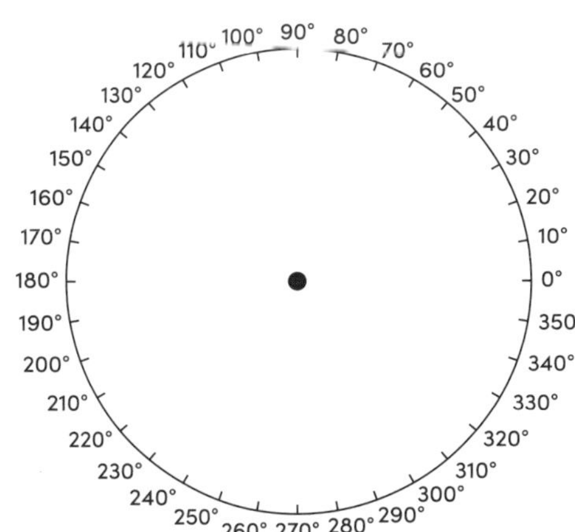

## GRAPHS AND DATA

# Drawing sector graphs

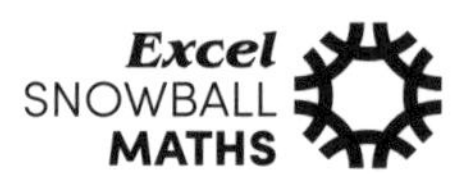

1 The table shows the number of hours studied each day.

Display this in a sector graph.

| Mon. | Tue. | Wed. | Thu. | Fri. |
|---|---|---|---|---|
| 3 | 2 | 1 | 2 | 4 |

Total = ________

$\frac{3}{\square} \times 360 =$ ________ °

+

$\frac{2}{\square} \times 360 =$ ________ ° = 150°

+

$\frac{1}{\square} \times 360 =$ ________ ° = 180°

+

$\frac{2}{\square} \times 360 =$ ________ ° = ________ °

+

$\frac{4}{\square} \times 360 =$ ________ ° = ________ °

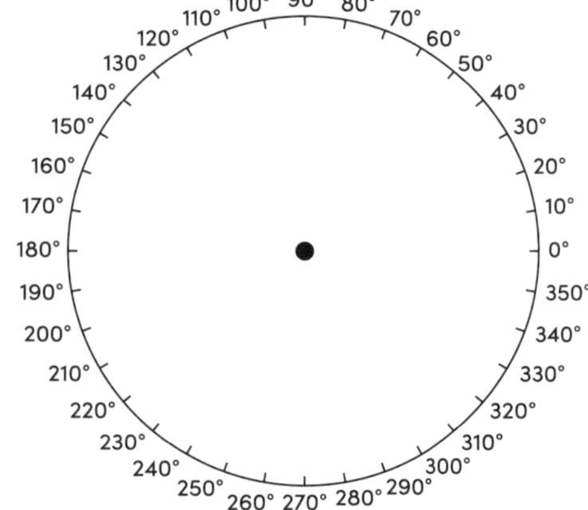

2 The table shows the number of pieces of furniture in a garage sale.

Display this in a sector graph.

| Chairs | Desks | Bookshelves | Beds | Cabinets |
|---|---|---|---|---|
| 8 | 4 | 1 | 2 | 3 |

Total = ________

$\frac{8}{\square} \times 360 =$ ________ °

+

$\frac{4}{\square} \times 360 =$ ________ ° = 240°

+

$\frac{\square}{\square} \times 360 =$ ________ ° = ________ °

+

$\frac{\square}{\square} \times 360 =$ ________ ° = ________ °

+

$\frac{\square}{\square} \times 360 =$ ________ ° = ________ °

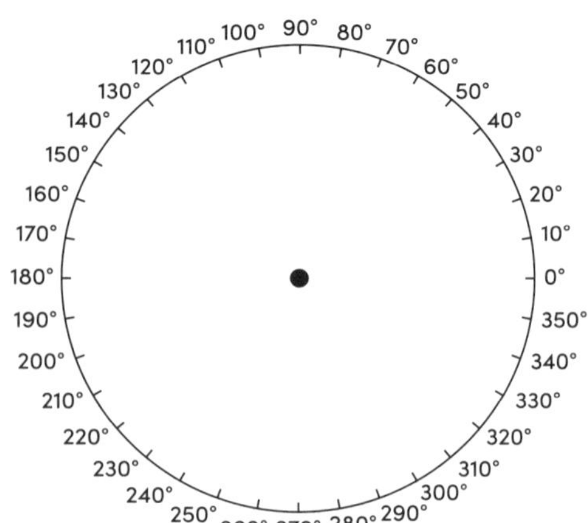

GRAPHS AND DATA

# Drawing sector graphs

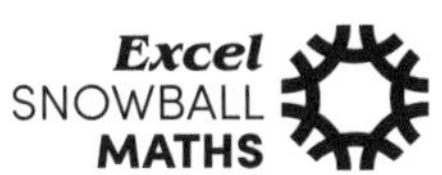


**1** The table shows how people travel to work.

Display this in a sector graph.

| Car | Train | Bus | Walk | Cycle |
|---|---|---|---|---|
| 40 | 15 | 25 | 10 | 10 |

$\frac{\square}{\square}$ × ______ = ______

\+

$\frac{\square}{\square}$ × ______ = ______ = ______

\+

$\frac{\square}{\square}$ × ______ = ______ = ______

\+

$\frac{\square}{\square}$ × ______ = ______ = ______

\+

$\frac{\square}{\square}$ × ______ = ______ = ______

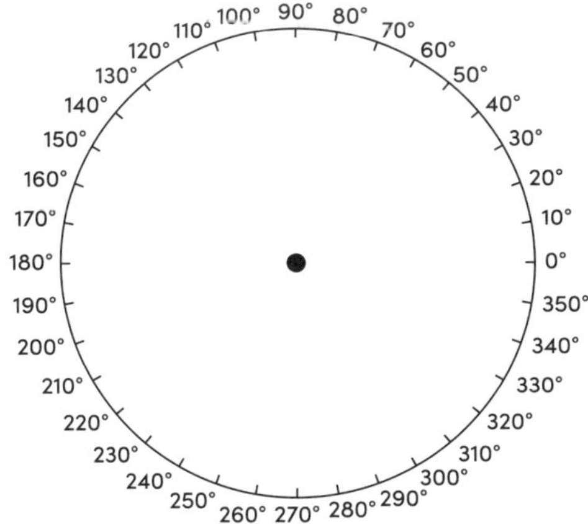

**2** The table shows the favourite sports of students.

Display this in a sector graph.

| Soccer | Basketball | Cricket |
|---|---|---|
| 20 | 35 | 15 |

| Netball | Volleyball |
|---|---|
| 13 | 17 |

Total = ______

$\frac{\square}{\square}$ × ______ = ______

\+

$\frac{\square}{\square}$ × ______ = ______ = ______

\+

$\frac{\square}{\square}$ × ______ = ______ = ______

\+

$\frac{\square}{\square}$ × ______ = ______ = ______

\+

$\frac{\square}{\square}$ × ______ = ______ = ______

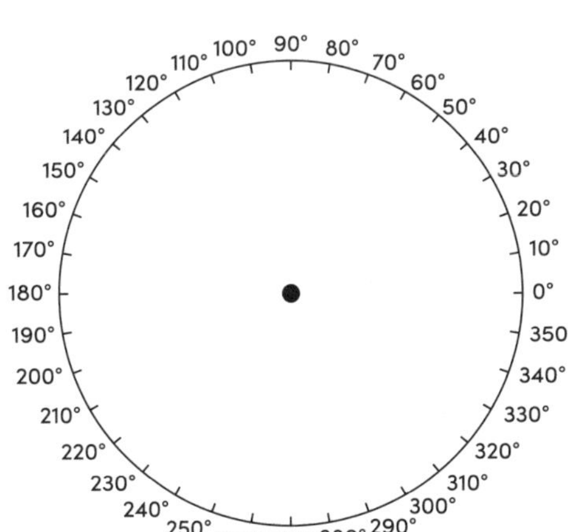

UNIT 14

GRAPHS AND DATA

# Drawing sector graphs

**1** The table shows the number of coloured marbles in a bag.

Display this in a sector graph.

| Red | Blue | Green | White | Black |
|---|---|---|---|---|
| 7 | 6 | 3 | 4 | 4 |

Total = ________

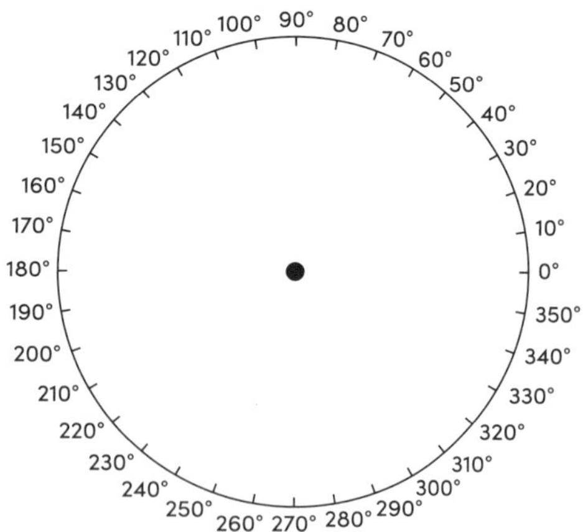


**2** The table shows the student grades in Mathematics.

Display this in a sector graph.

| A | B | C | D | E |
|---|---|---|---|---|
| 3 | 6 | 15 | 5 | 1 |

UNIT 15

# GRAPHS AND DATA
## Analysing graphs

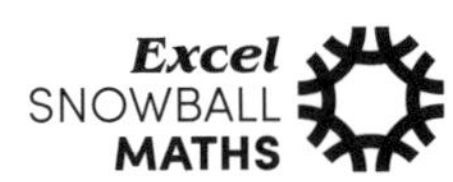


**1** The column graph below shows the percentage of the population in different generations.

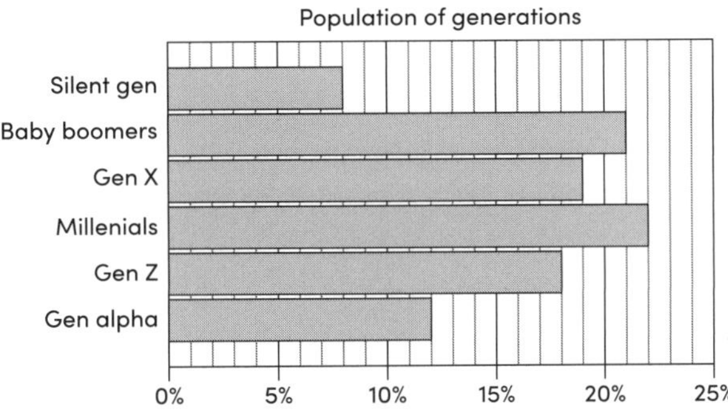


**a** Complete the table below:

| | Gen alpha | Gen Z | Millennials | Gen X | Baby boomers | Silent gen |
|---|---|---|---|---|---|---|
| Percentage % | 12 | 18 | | | | |

**b** Which generation has the greatest population? ______________________

**c** Which generation has the least population? ______________________

---

**2** The column graph shows the population of Australian cities, in millions:

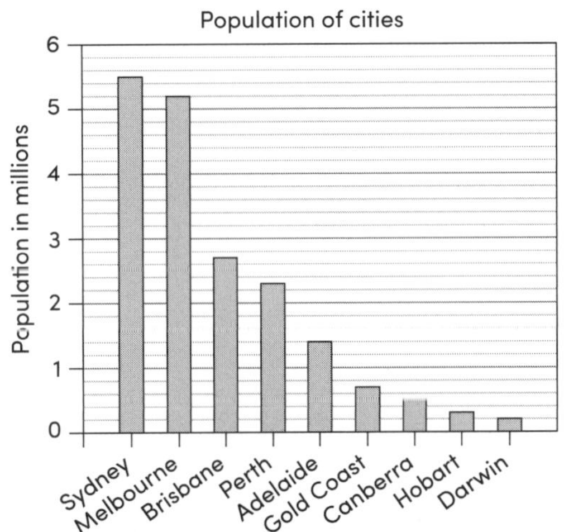


**a** Complete the table below:

| | Sydney | Melbourne | Brisbane | Perth | Adelaide | Gold Coast | Canberra | Hobart | Darwin |
|---|---|---|---|---|---|---|---|---|---|
| Population (millions) | 5.5 | 5.2 | | | | 0.7 | | | |

**b** How many times greater is Sydney's population than Adelaide's population?

______________________

UNIT 16

GRAPHS AND DATA

# Analysing graphs

**1** The clustered column graph below shows how students travel to school.

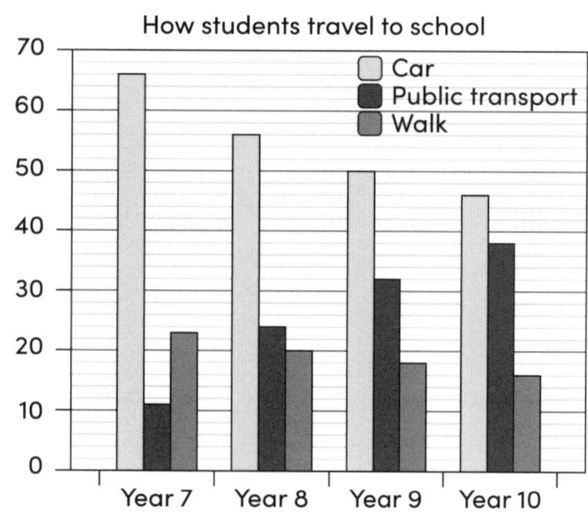


**a** Complete the table below:

| | Year 7 | Year 8 | Year 9 | Year 10 |
|---|---|---|---|---|
| Car | 66% | | | |
| Public transport | 11% | | | |
| Walk | | | | |

**b** Which method of travel is the most popular overall?

______________________

**c** Which method of travel is the least popular overall?

______________________

**d** Give a reason why the number of students travelling by car has decreased each year.

______________________

______________________

**2** The pie chart below shows the percentage of pets in Australian households.

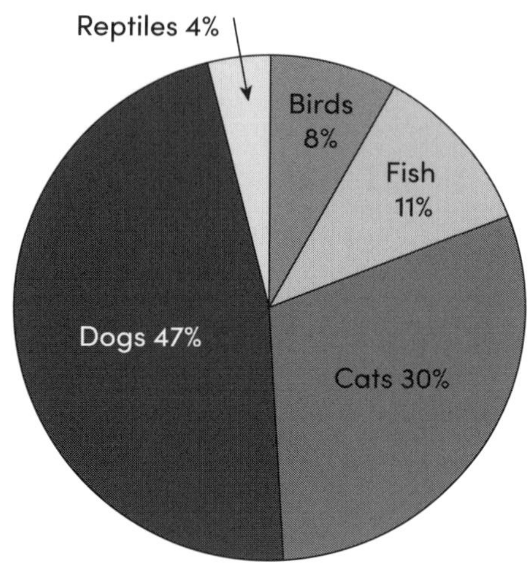


**a** What is the most popular pet?

______________________

______________________

**b** What is the least popular pet?

______________________

______________________

**c** If there were 5000 pets altogether, how many pets would be:

**i** dogs?

$\frac{\square}{100} \times$ 5000

= ________

**ii** cats?

$\frac{\square}{100} \times$ 5000

= ________

**iii** fish?

________

________

**iv** birds?

________

________

**GRAPHS AND DATA**

# Analysing graphs

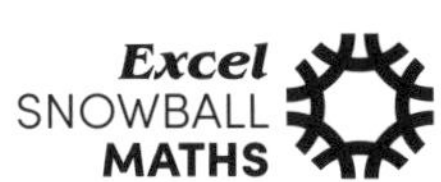


**1** Students were surveyed about their favourite sport and their results were displayed as a divided bar graph.

| Hockey | Soccer | Football | Basketball |
|---|---|---|---|
| 2.8 cm | 4.2 cm | 2.1 cm | 2.9 cm |

12 cm

**a** What was the most popular sport? ______________________

**b** What was the least popular sport? ______________________

**c** What percentage of the students chose hockey? $\frac{\square}{12} \times \underline{100} = \underline{\quad\quad}\%$
(Give correct to one decimal place.)

**d** What percentage of the students chose soccer? $\frac{\square}{12} \times \underline{100} = \underline{\quad\quad}\%$

**e** What percentage of the students chose football? ______________________
(Give correct to one decimal place.)

**f** What percentage of the students chose basketball? ______________________
(Give correct to one decimal place.)

---

**2** The graph shows the students' grade distribution in a maths class.

**a** What was the most common grade? ______________________

**b** What was the least common grade? ______________________

**c** What percentage of the students got a C? $\frac{\square}{\square} \times \underline{\quad\quad} = \underline{\quad\quad}\%$

**d** What percentage of the students got an A? $\frac{\square}{\square} \times \underline{\quad\quad} = \underline{\quad\quad}\%$

**e** What percentage of the students got a B? ______________________

**f** What percentage of the students got an E? ______________________

UNIT 18

# GRAPHS AND DATA
## Analysing graphs

Excel SNOWBALL MATHS

**1** The misleading graph below shows the percentage of people who voted for two different political parties.

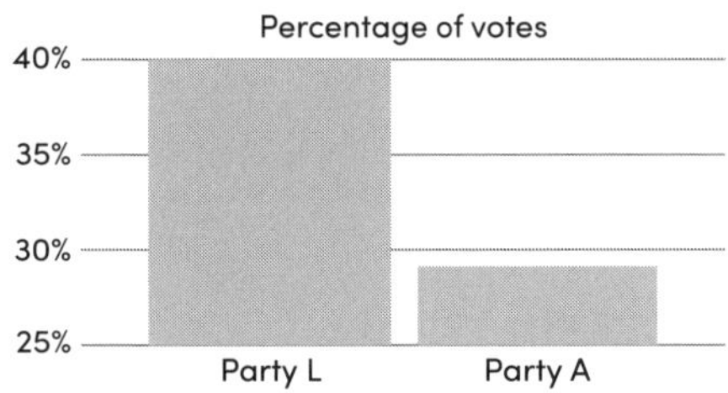


Explain how this graph is misleading.
(Does the graph start at 0?)

**2** The graph compares the median house prices in Melbourne and Sydney. Explain two ways in which this graph is misleading.

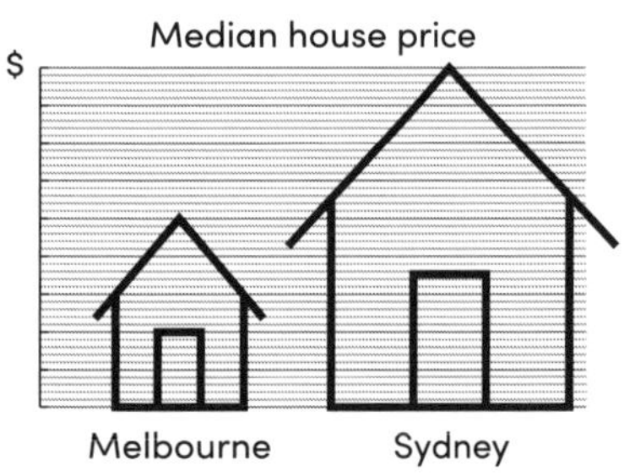


**3** The misleading graph on the right shows the percentage of people who voted for different political parties. Explain two ways in which this graph is misleading.

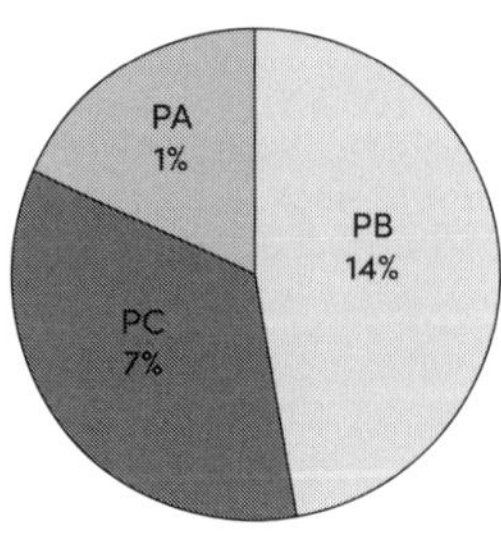


**4** The first graph below has been manipulated to produce the second graph in order to mislead the audience on the number of COVID tests done in different countries.

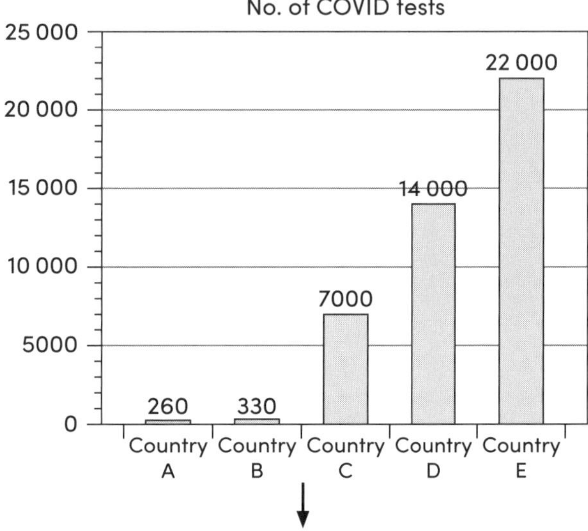


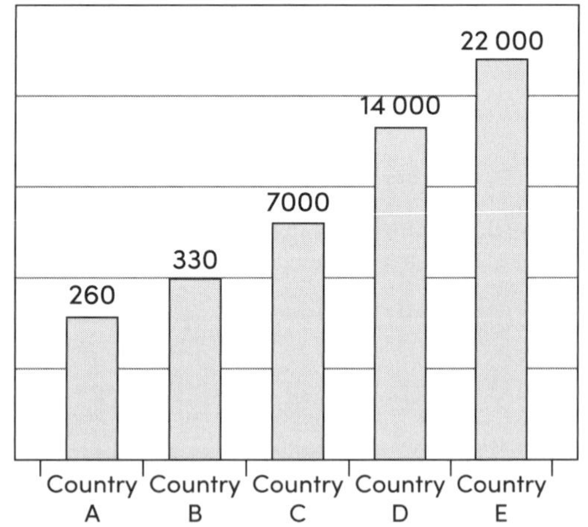


Explain two ways in which it has been manipulated to mislead the audience.

## GRAPHS AND DATA

# Line graphs

**1** The table shows the weight of a baby girl recorded each month from birth. Display this as a line graph:

| Month(s) | 0 | 1 | 2 | 3 | 4 | 5 | 6 | 7 | 8 | 9 | 10 |
|---|---|---|---|---|---|---|---|---|---|---|---|
| Weight (kg) | 3 | 4 | 5 | 6 | 6.5 | 7 | 7.5 | 8 | 8.2 | 8.5 | 9 |

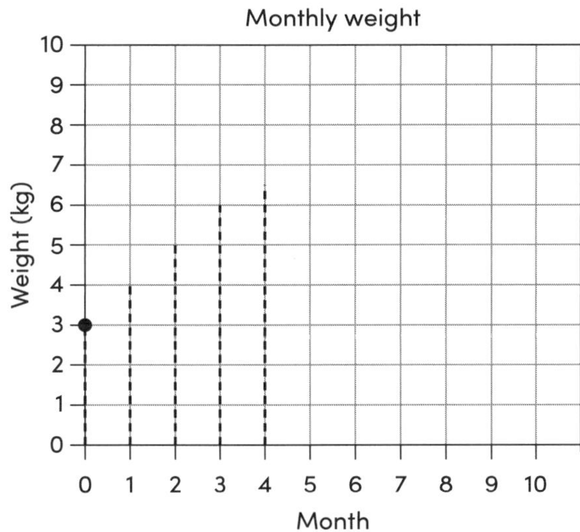


**2** The table shows the weight of a baby boy recorded each month from birth. Display this as a line graph:

| Month(s) | 0 | 1 | 2 | 3 | 4 | 5 | 6 | 7 | 8 | 9 | 10 |
|---|---|---|---|---|---|---|---|---|---|---|---|
| Weight (kg) | 3 | 4.5 | 5.6 | 6.4 | 7 | 7.5 | 8 | 8.3 | 8.6 | 8.9 | 9 |

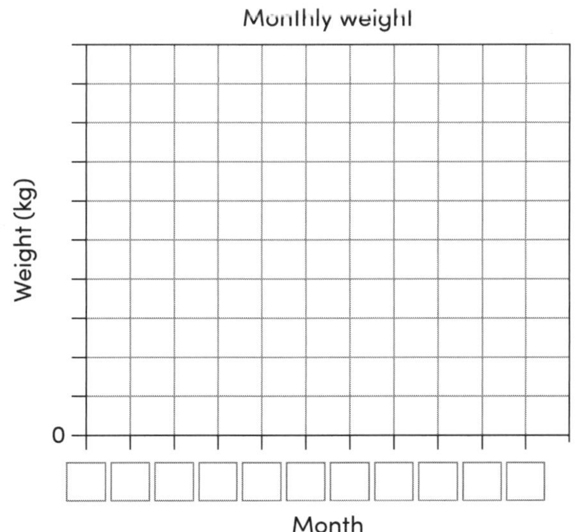

**GRAPHS AND DATA**

# Line graphs

**1** The table shows the height of a baby girl recorded each month from birth. Display this as a line graph:

| Month(s) | 0 | 1 | 2 | 3 | 4 | 5 | 6 | 7 | 8 | 9 | 10 |
|---|---|---|---|---|---|---|---|---|---|---|---|
| Height (cm) | 49 | 54 | 56 | 60 | 62 | 64 | 64 | 67 | 69 | 70 | 72 |

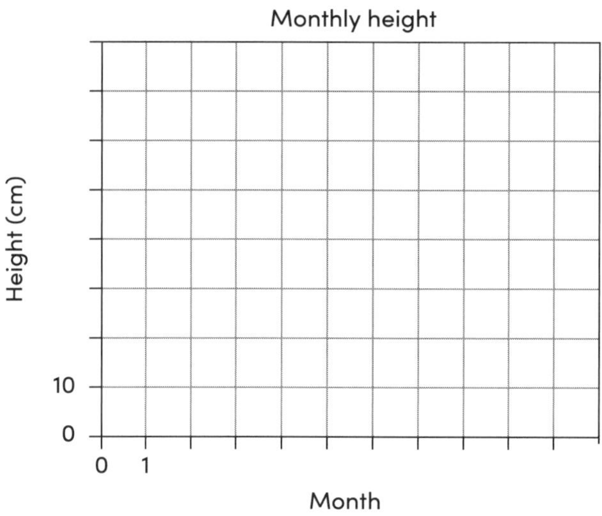


**2** The table shows the height of a baby boy recorded each month from birth. Display this as a line graph:

| Month(s) | 0 | 1 | 2 | 3 | 4 | 5 | 6 | 7 | 8 | 9 | 10 |
|---|---|---|---|---|---|---|---|---|---|---|---|
| Height (cm) | 50 | 55 | 58 | 61 | 64 | 66 | 68 | 69 | 70.5 | 72 | 73 |

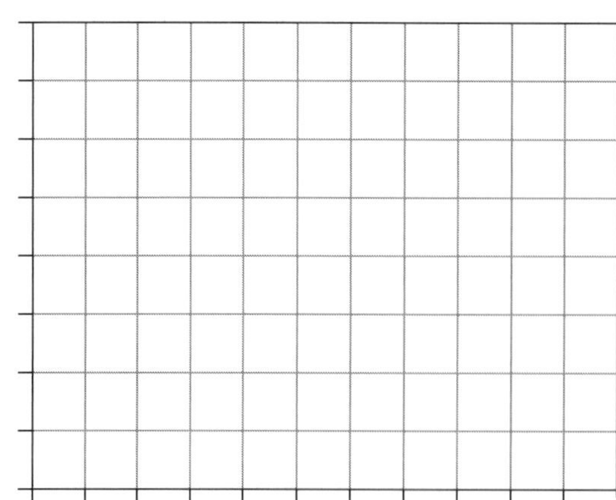

## GRAPHS AND DATA
# Line graphs

1 The table shows the average maximum temperature recorded each month in Sydney. Display this as a line graph:

| Month | Jan. | Feb. | Mar. | Apr. | May | Jun. | Jul. | Aug. | Sep. | Oct. | Nov. | Dec. |
|---|---|---|---|---|---|---|---|---|---|---|---|---|
| Temp. (°C) | 26 | 26 | 25 | 23 | 20 | 17 | 17 | 18 | 21 | 22 | 23 | 25 |

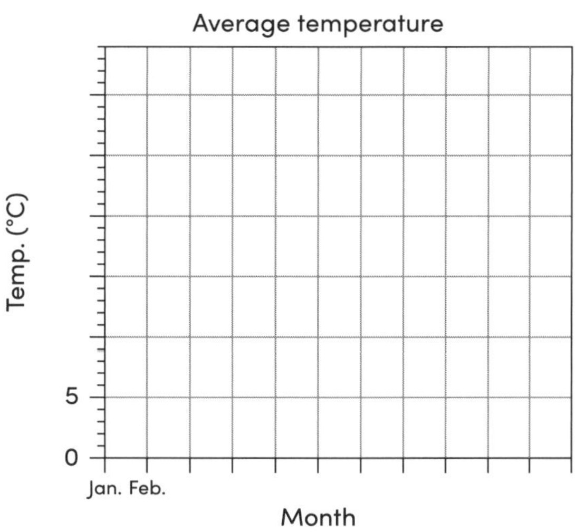


2 The table shows the average minimum temperature recorded each month in Sydney. Display this as a line graph:

| Month | Jan. | Feb. | Mar. | Apr. | May | Jun. | Jul. | Aug. | Sep. | Oct. | Nov. | Dec. |
|---|---|---|---|---|---|---|---|---|---|---|---|---|
| Temp. (°C) | 19 | 19 | 18 | 15 | 12 | 9 | 8 | 9 | 11 | 14 | 16 | 18 |

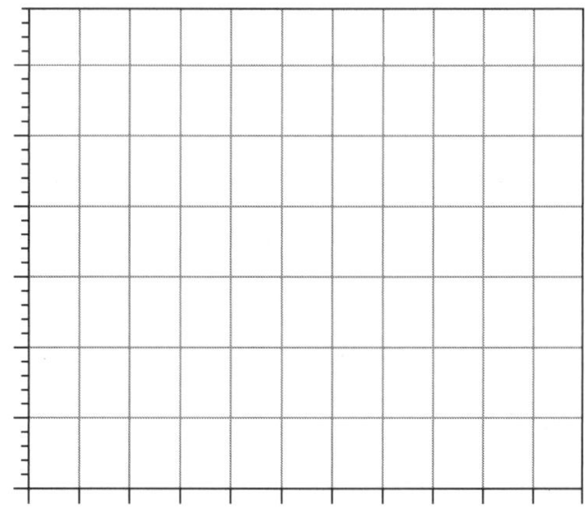

UNIT 22

# GRAPHS AND DATA
## Line graphs

**1** The graph shows Lisa's height recorded each year since she was 2 years old.

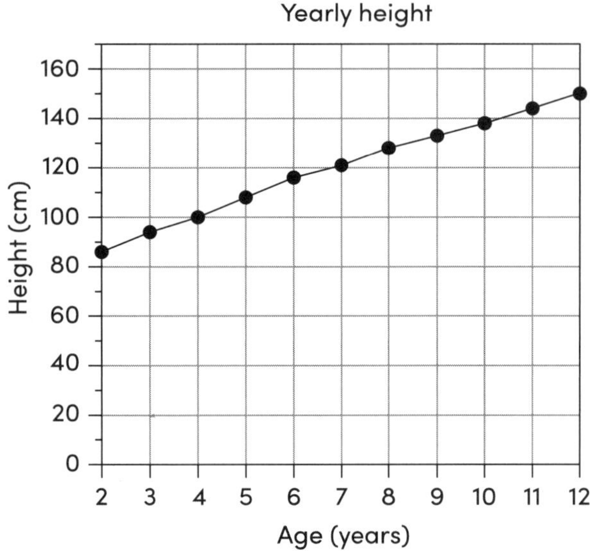


**a** Estimate Lisa's height when she was:

**i** 2 years old? ________ cm

**ii** 7 years old? ________ cm

**iii** 9 years old? ________

**iv** 12 years old? ________

**b** Estimate the year when Lisa's height was:

**i** 100 cm? ________ years

**ii** 121 cm? ________ years

**iii** 138 cm? ________

**iv** 144 cm? ________

**c** How much did Lisa grow between 4 years and 7 years?

________ – ________ = ________
7 years    4 years

**d** How much did Lisa grow between 4 years and 10 years?

________ – ________ = ________
10 years    4 years

**e** How much did Lisa grow between 7 years and 12 years?

**f** How much did Lisa grow from 7 years to 8 years old?

**g** How much did Lisa grow in the year when she was 4 years old?

**h** How much did Lisa grow in the year when she was 11 years old?

**i** Based on this trend, predict how tall Lisa will be when she is 13 years old.

**j** Will the graph continue to increase indefinitely? Explain your answer.

UNIT 23

GRAPHS AND DATA

# Types of data

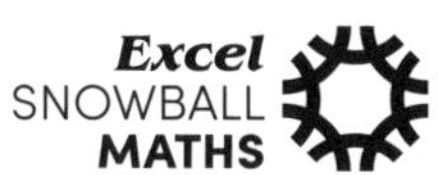

1 Classify each type of data as numerical (N) or categorical (C):

a Heights of Year 9 students ________

b Classroom temperature ________

c Brands of phone ________

d Ice-cream flavour ________

e Language spoken at home ________

f Favourite sport team ________

g Foot length ________

h Arm span ________

i Time to race 100 m ________

j Blood pressure of a patient ________

k Birth date ________

l Year level ________

m Number of students absent at school ________

n Distance travelled by car ________

o Suburb you live in ________

p Postcode ________

2 Classify each numerical data as discrete (D) or continuous (C):

a Number of siblings in a family ________

b Number of birds at a park ________

c Heights of students ________

d Classroom temperature ________

e Number of guests ________

f Number of pets ________

g Time spent doing homework ________

h Blood pressure of a patient ________

i Amount of rainfall ________

j Speed of a car ________

k Number of phones in a household ________

l Population of Sydney ________

3 Classify each categorical data as nominal (N) or ordinal (O):

a High-school year level ________

b School grade ________

c Hair colour ________

d Country ________

e Swimming level ________

f Ratings of a restaurant ________

g Types of pets ________

h Types of coats ________

i Education level ________

j Music genre ________

k Car brands ________

l Customer satisfaction ________

CHAPTER 8 REVIEW

# Graphs and data

Use your own paper for this page.

**1** The frequency table shows the number of pets owned. Represent this data as a frequency histogram and polygon.

| No. of pets | Frequency |
|---|---|
| 0 | 5 |
| 1 | 10 |
| 2 | 4 |
| 3 | 2 |
| 4 | 1 |

**2** Display the scores below as a stem-and-leaf plot.

1 2 2 4 10 13 21 23 27 28 30 34

**3** The heights, in centimetres, of a group of Year 7 students are recorded below.
Draw a stem-and-leaf plot to display this data.

140 151 139 162 147 153 138 159 144
148 150 141 152 156

**4** The table shows the number of students in 20 different classes at a high school. Display this as a dot plot.

| No. of students | Frequency |
|---|---|
| 20 | 2 |
| 21 | 4 |
| 22 | 1 |
| 23 | 6 |
| 24 | 7 |

**5** Display the data below in a 10-cm divided bar graph.

| *W* | *X* | *Y* | *Z* |
|---|---|---|---|
| 4 | 2.6 | 1.4 | 2 |

**6** A sample of 140 people were surveyed on what job industry they were in and the results are shown. Display this in a 10-cm divided bar graph, correct to 1 decimal place for each segment.

| Retail | Healthcare | Hospitality | Construction |
|---|---|---|---|
| 37 | 50 | 28 | 25 |

**7** Display the data below as a sector graph.

| *Q* | *R* | *S* | *T* |
|---|---|---|---|
| 60° | 110° | 75° | 115° |

**8** A sample of 200 people were surveyed on what occupation they held and the results are shown. Display this as a pie chart.

| Managers | Technicians | Sales workers | Machinery operators |
|---|---|---|---|
| 58 | 77 | 26 | 39 |

**9** The table shows daily maximum temperatures throughout the week. Display this as a line graph.

| Day | Mon. | Tue. | Wed. | Thu. | Fri. | Sat. | Sun. |
|---|---|---|---|---|---|---|---|
| (°C) | 17 | 14 | 14 | 16 | 16 | 18 | 19 |

**10** The column graph shows the number of different mammals at a zoo.

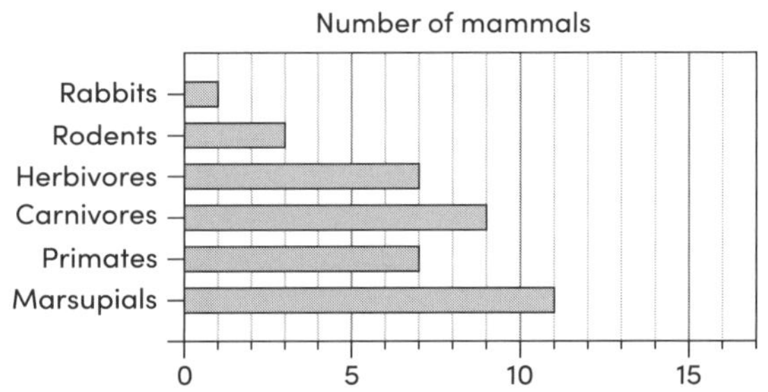


**a** Complete the table below:

| | Marsupials | Primates | Carnivores | Herbivores | Rodents | Rabbits |
|---|---|---|---|---|---|---|
| Number | | | | | | |

**b** Which type of mammal had the greatest number?

**11** The graph shows the net worth of two companies, in millions. Explain why the graph is misleading.

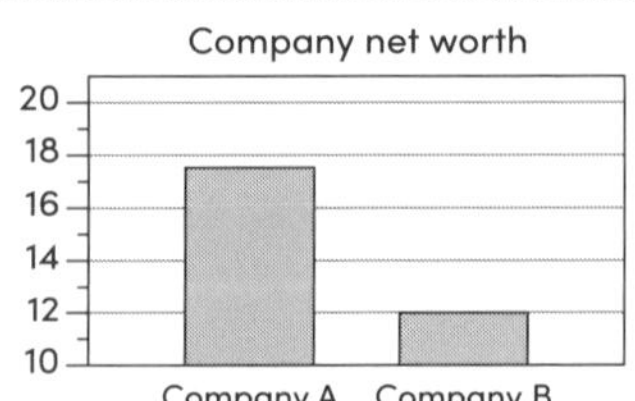


**12** A person weighs 74.6 kg. What type of data is this classified as?

**A** Numerical discrete **B** Numerical continuous

**C** Categorical nominal **D** Categorical ordinal

# CHAPTER 9

UNIT 1

## PROBABILITY
# Introduction to probability

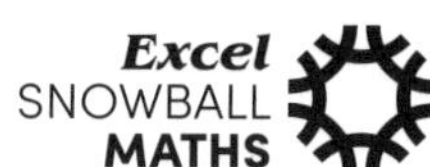


| 0% | 25% | 50% | 75% | 100% |
|---|---|---|---|---|
| Impossible | Unlikely | Even chance | Likely | Certain |

**1** What are the chances of each event occurring?

**a** I will be getting Maths homework. ______

**b** The sun will rise tomorrow. ______

**c** I will be a year older next year. ______

**d** I will be a year younger next year. ______

**e** If I flip a coin, it will land on tails. ______

**f** I will go to school on the weekend. ______

**2** The spinner is spun once. What is the chance that the arrow will land on:

**a** blue? ______

**b** red? ______

**c** yellow? ______

**d** Are red and blue equally likely? ______ (Do they have the same size?)

**3** The spinner is spun once. What is the chance that the arrow will land on:

**a** blue? ______

**b** red? ______

**c** green? ______

**d** Are red and blue equally likely? ______

**4** The spinner is spun once. What is the chance that the arrow will land on:

**a** red? ______

**b** blue? ______

**c** Are red and blue equally likely? ______

**5** The spinner is spun once. What is the chance that the arrow will land on:

**a** blue? ______

**b** red? ______

**c** green? ______

**d** Are red and green equally likely? ______

**6** The spinner is spun once. What is the chance that the arrow will land on:

**a** red? ______

**b** green? ______

**c** yellow? ______

**d** Are green and red equally likely? ______

**7** The spinner is spun once. What is the chance that the arrow will land on:

**a** red? ______

**b** black? ______

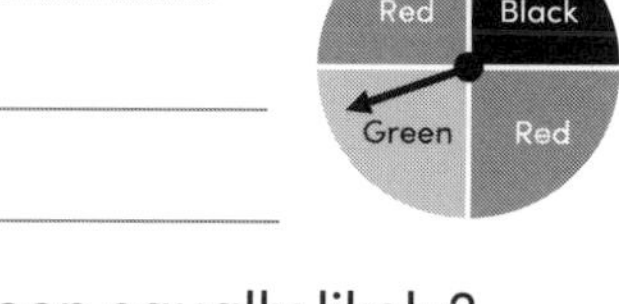

**c** green? ______

**d** Are black and green equally likely? ______

**e** Are red and green equally likely? ______

**PROBABILITY**

# Introduction to probability

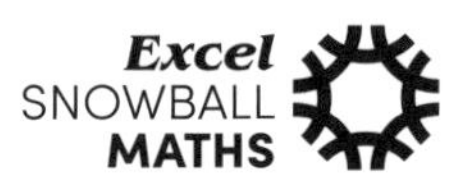


| 0% | 25% | 50% | 75% | 100% |
|---|---|---|---|---|
| Impossible | Unlikely | Even chance | Likely | Certain |

**1** For the spinner on the right:

**a** list the sample space

(What colours are there?)

**b** how many possible outcomes are there?

(How many colours?)

**c** are the outcomes equally likely?

**2** For the spinner on the right:

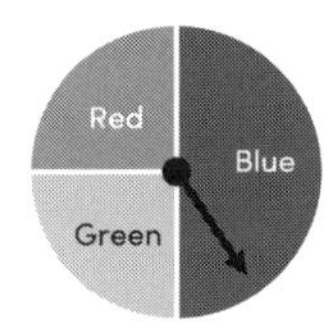


**a** list the sample space

**b** how many possible outcomes are there?

**c** are the outcomes equally likely?

**3** For the spinner on the right:

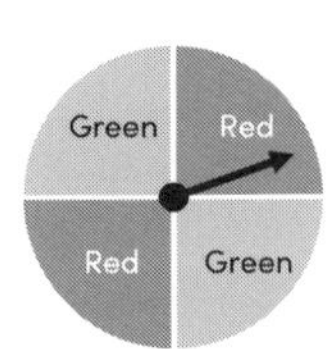


**a** list the sample space

**b** how many possible outcomes are there?

**c** are the outcomes equally likely?

**4** For the spinner on the right:

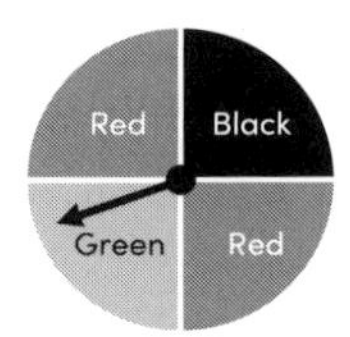


**a** list the sample space

**b** how many possible outcomes are there?

**c** are the outcomes equally likely?

**5** For the spinner on the right:

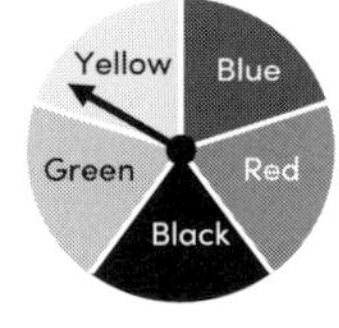


**a** list the sample space

**b** how many possible outcomes are there?

**c** are the outcomes equally likely?

**6** Which outcomes are equally likely in the spinner shown below?

UNIT 3

**PROBABILITY**

# Introduction to probability

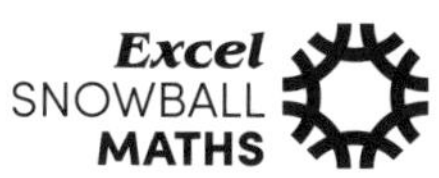

| 0% | 25% | 50% | 75% | 100% |
|---|---|---|---|---|
| Impossible | Unlikely | Even chance | Likely | Certain |

**1** There are 3 white balls and 3 grey balls in a bag.

**a** List the sample space.

(What coloured balls are there?)

**b** How many possible outcomes are there?

(How many colours?)

**c** Are the outcomes equally likely?

**d** Describe the chance of picking a white ball.

**2** There are 2 white balls and 3 black balls in a bag.

**a** List the sample space.

**b** How many possible outcomes are there?

**c** Are the outcomes equally likely?

**3** There are 5 grey balls and 2 black balls in a bag.

**a** List the sample space.

**b** How many possible outcomes are there?

**c** Are the outcomes equally likely?

**d** Describe the chance of picking a grey ball?

**4** A bag contains 2 white balls, 5 black balls and 3 red balls.

**a** What is the sample space?

**b** How many possible outcomes are there?

**c** Are the outcomes equally likely?

**d** Which coloured ball is most likely to be chosen?

**e** Which coloured ball is least likely to be chosen?

**f** How many balls are there in total?

**g** Describe the chance of picking a black ball?

**PROBABILITY**

# Introduction to probability

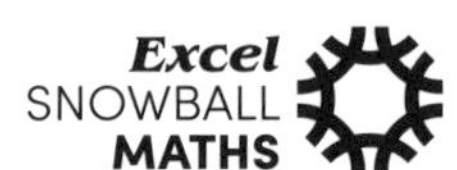


| 0% | 25% | 50% | 75% | 100% |
|---|---|---|---|---|
| Impossible | Unlikely | Even chance | Likely | Certain |

**1** A dice shows the numbers 1, 2, 3, 4, 5, 6.

**a** List the sample space.

(Which numbers are there?)

**b** How many possible outcomes are there?

(How many different numbers?)

**c** Are the outcomes equally likely?

(Are all numbers different?)

**d** Describe the chance of rolling a 2.

**2** A dice shows the numbers 1, 2, 3, 3, 3, 3.

**a** List the sample space.

**b** How many possible outcomes are there?

**c** Are the outcomes equally likely?

**d** Describe the chance of rolling a 3.

**3** A dice has the numbers 1, 2, 3, 4, 4, 4.

**a** List the sample space.

**b** How many possible outcomes are there?

**c** Are the outcomes equally likely?

**d** Describe the chance of rolling a 4.

**e** Describe the chance of rolling a 2.

**f** Describe the chance of rolling an even number?

**4** A bag contains four $1 coins, three 20-cent coins and four 50-cent coins.

**a** Draw a diagram showing the different coins.

**b** What is the sample space?

**c** How many possible outcomes are there?

**d** Are the outcomes equally likely?

**e** Which two coins are equally likely to be chosen?

**f** Which coin is least likely to be chosen?

PROBABILITY

# Theoretical probability

Note: Answers should be correct to one decimal place.

**1** This spinner is spun once. What is the probability that the pointer will land on:

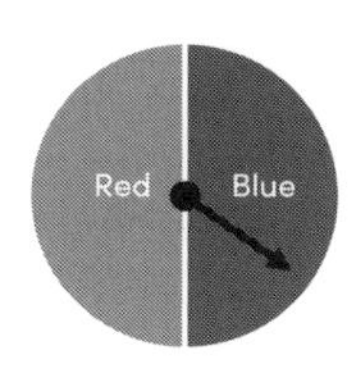


**a** blue? $\frac{\square}{2}$ **b** red? $\frac{\square}{2}$

**c** Are the outcomes equally likely? ______

**d** Describe the chance of it landing on blue. ______

**2** This spinner is spun once. What is the probability that the pointer will land on:

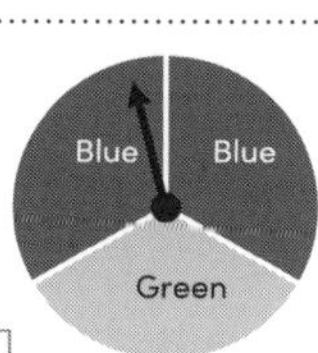


**a** blue? $\frac{\square}{3}$ **b** green? $\frac{\square}{3}$

**c** Are the outcomes equally likely? ______

**d** Describe the chance of it landing on blue. ______

**3** This spinner is spun once. What is the probability that the pointer will land on:

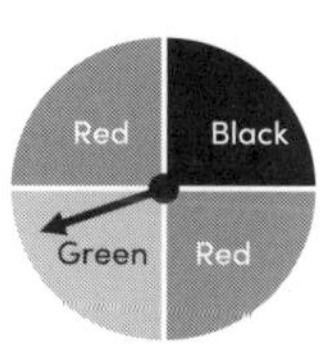


**a** green? ______ **b** black? ______ **c** red? ______

**d** Which two outcomes are equally likely? ______

**e** Describe the chance of it landing on red. ______

**4** This spinner is spun once. What is the probability that the pointer will land on:

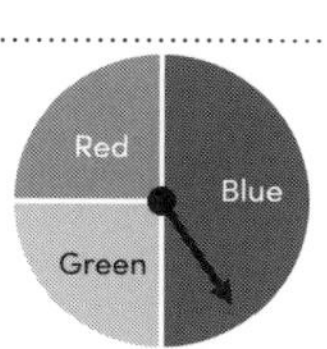


**a** red? ______ **b** green? ______ **c** blue? ______

**d** Which two outcomes are equally likely?

______

**e** Describe the chance of landing on red.

______

**5** This spinner is spun once. What is the probability that the pointer will land on:

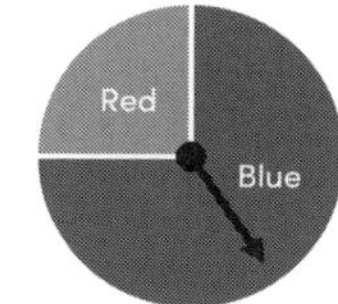


**a** red? ______ **b** blue? ______

**c** Are the outcomes equally likely?

______

**d** Describe the chance of landing on blue.

______

**6** From the letters of the word MATHEMATICS, what is the probability, as a percentage, of picking:

**a** the letter A?

$\frac{\square}{11} \times$ ______ = ______ %

**b** the letter S?

$\frac{\square}{11} \times$ ______ = ______ %

**c** the letter T?

$\frac{\square}{\square} \times$ ______ = ______

**d** the letter I? ______ **e** the letter M? ______

**f** A or E? ______ **g** T or M? ______

UNIT 6

**PROBABILITY**

# Theoretical probability

Note: Answers should be correct to one decimal place.

**1** From the letters of the word ALLITERATION, what is the probability, as a percentage, of picking:

**a** the letter A?

$\frac{\square}{\square}$ × ______ = ______ %

**b** the letter L?

$\frac{\square}{\square}$ × ______ = ______ %

**c** the letter N?

$\frac{\square}{\square}$ × ______ = ______

**d** the letter O? ______________

**e** A or E? ______________

**f** I or L? ______________

**g** A, E, I, O and U? ______________

**2** From the letters of the word EXTRAORDINARY, what is the probability, as a percentage, of picking:

**a** the letter I? ______________

**b** the letter A? ______________

**c** the letter R? ______________

**d** A, R or Y? ______________

**e** X, T or R? ______________

**f** a vowel? ______________

**3** A dice shows the numbers 1, 1, 2, 3, 3, 5. What is the probability, as a fraction and a decimal, of rolling:

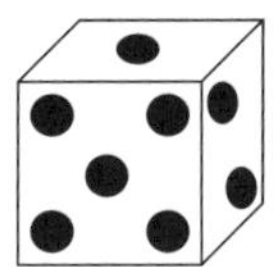

**a** 1? $\frac{\square}{6}$ = ______

**b** 3? $\frac{\square}{6}$ = ______

**c** 5? $\frac{\square}{\square}$ = ______

**d** 1 or 2? $\frac{\square}{\square}$ = ______

**e** 3 or 5? ______

**f** 1, 3 or 5? ______

**4** A fair 6-sided dice shows the numbers 1, 2, 3, 4, 5, 6. What is the probability, as a fraction and a decimal, of rolling:

**a** 2? $\frac{\square}{6}$ = ______

**b** 3? $\frac{\square}{6}$ = ______

**c** 5? ______________

**d** 2, 4 or 6? ______________

**e** 1, 3 or 5? ______________

**f** an even number? ______________

**5** At a service centre, a machine generates a random ticket number from 1 to 9.
What is the probability, as a fraction and a decimal, of getting:

**a** 1? ______________

**b** 4? ______________

**c** a number more than 5? ______________

**d** a number less than 3? ______________

**e** an even number? ______________

**f** an odd number? ______________

**PROBABILITY**

# Theoretical probability

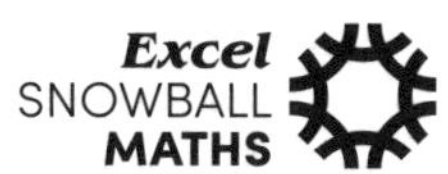


1 A deck of cards contains 52 cards with 4 different suits as shown below:

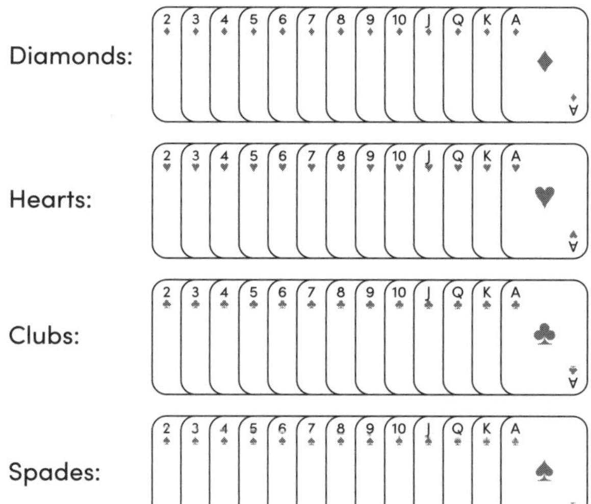


From the deck of cards above if one card is chosen, what is the probability of picking:

a a diamond? $\frac{\square}{52} = \frac{\square}{\square}$

b a club? $\frac{\square}{52} = \frac{\square}{\square}$

c a red card? (diamonds & hearts) $\frac{\square}{\square} = \frac{\square}{\square}$

d a black card? (clubs & spades) $\frac{\square}{\square} = \frac{\square}{\square}$

e a 7? ______

f a 3? ______

g an ace? ______

h a king? ______

i a picture card? (jack, queen, king) ______

j a black 10? ______

k a 5 or 6? ______

l not a picture card? ______

2 A bag contains 3 blue, 2 red and 5 white marbles.

a How many marbles are there in total? ______

b What is the probability of picking:

i a blue marble?

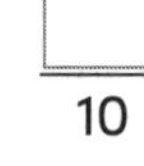

ii a red marble?

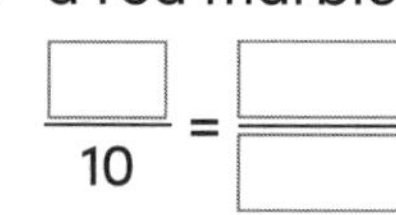

iii a white marble?

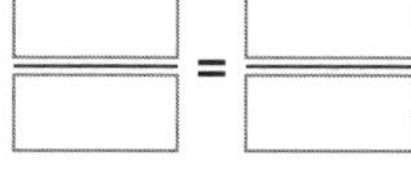

iv not a white marble?

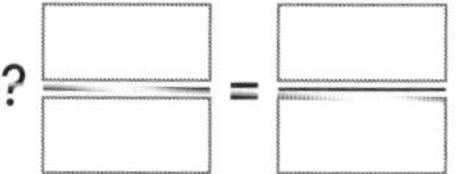

3 In a classroom of 24 students, 10 catch the bus to school, 8 walk and 6 are driven. Find:

a P(bus) ______

b P(driven) ______

c P(walk) ______

d P(not walk) ______

4 In a piggybank, there are five 20-cent coins, seven 10-cent coins and six 50-cent coins.

a How many coins are there in total?

______

b Find:

i P(10-cent) ______

ii P(20-cent) ______

iii P(50-cent) ______

iv P(not 50-cent) ______

v P(not 20-cent) ______

vi P(not 10-cent) ______

## PROBABILITY

# Experimental probability

Note: Answers should be correct to two decimal places.

**1** A spinner was spun 50 times and the results were recorded in the table below.

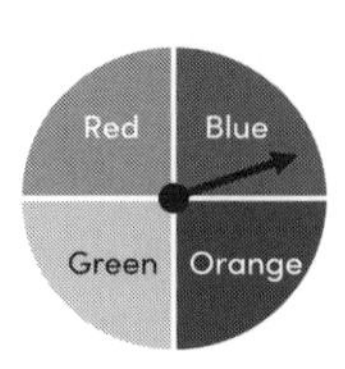


| Colour | Frequency |
|---|---|
| Red | 13 |
| Blue | 11 |
| Green | 9 |
| Orange | 17 |

**a** What is the experimental probability, as a fraction and a decimal, that the spinner lands on:

**i** red?

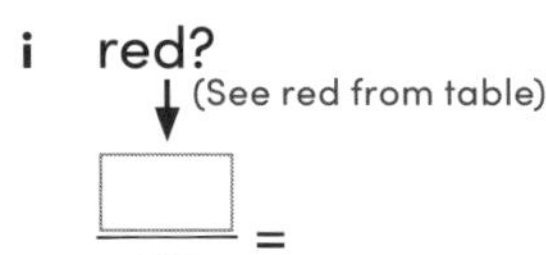


**ii** blue?

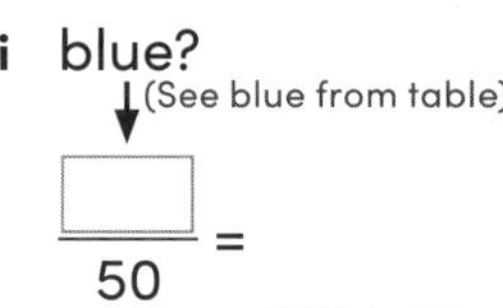


**iii** green?

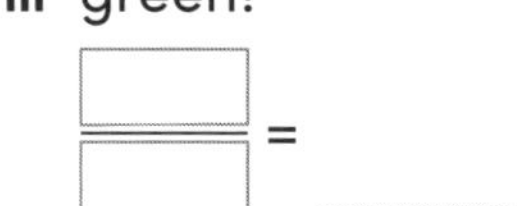

**iv** orange?

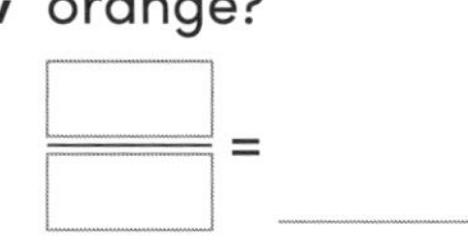

**v** red or blue?

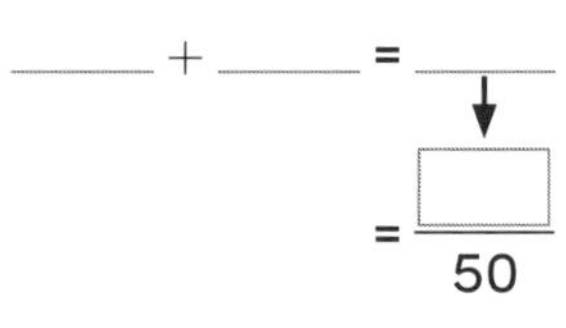


**vi** green or orange?

_____ + _____ = _____

= $\frac{\square}{\square}$

= _____

**b** What is the theoretical probability, as a fraction and a decimal, that the spinner lands on:

**i** red? (See red from spinner)

$\frac{\square}{4}$ = _____

**ii** blue? (See blue from spinner)

$\frac{\square}{4}$ = _____

**iii** green?

**iv** orange?

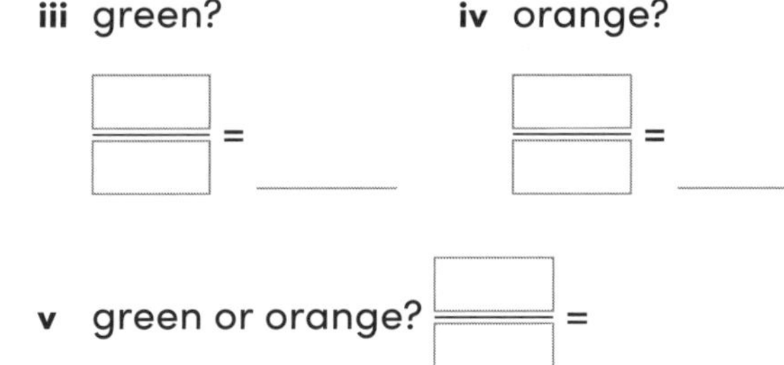

**v** green or orange? $\frac{\square}{\square}$ = _____

**2** A spinner was spun 120 times and the results were recorded in the table below.

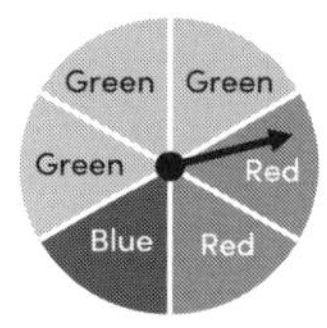


| Colour | Frequency |
|---|---|
| Green | 72 |
| Red | 34 |
| Blue | 14 |

**a** What is the experimental probability, as a fraction and a decimal, that the spinner lands on:

**i** red? _____

**ii** green? _____

**iii** blue? _____

**iv** red or blue? _____

**b** What is the theoretical probability, as a fraction and a decimal, that the spinner lands on:

**i** red? _____

**ii** blue? _____

**iii** green? _____

**iv** yellow? _____

**v** red or green? _____

**vi** red or blue? _____

**PROBABILITY**

# Experimental probability

Note: Answers should be correct to one decimal place.

**1** A dice was rolled and the results were recorded in the table below.

| Number | Frequency |
|---|---|
| 1 | 18 |
| 2 | 12 |
| 3 | 10 |
| 4 | 17 |
| 5 | 13 |
| 6 | 20 |

**a** How many times was it rolled?

18 + 12 + ____ + ____ + ____ + ____

= ____

**b** What is the experimental probability, as a percentage, that the dice rolls:

**i** a 1? $\frac{\square}{90}$ × ____ = ____%

**ii** a 3? $\frac{\square}{90}$ × ____ = ____%

**iii** a 4? $\frac{\square}{\square}$ × ____ = ____

**iv** a 6? $\frac{\square}{\square}$ × ____ = ____

**v** a 5?

______

**vi** a 1 or 4?

______

**vii** a 2 or 6?

______

**viii** an even number?

______

**ix** an odd number?

______

**2** Your turn. Roll a dice 20 times and complete the table below:

| Number | Tally | Frequency |
|---|---|---|
| 1 | | |
| 2 | | |
| 3 | | |
| 4 | | |
| 5 | | |
| 6 | | |

**a** What is the experimental probability, as a percentage, that the dice rolls:

**i** a 1?

______

**ii** a 4?

______

**iii** an even number?

______

**iv** a number less than 4?

______

**b** What is the theoretical probability, as a percentage, of rolling:

**i** a 1?

______

**ii** a 4?

______

**iii** an even number?

______

**iv** a number less than 4?

______

# PROBABILITY
## Complementary events

**1** Write the complementary event for the following:

**a** flipping a head on a coin

______________________

**b** winning a game

______________________

**c** picking a red card from a deck of cards

______________________

**d** rolling a number 5 on a dice

______________________

**e** arriving late to class

______________________

**2** There are 2 white balls and 3 grey balls in a bag. What is the probability of:

**a** picking a white ball? $\frac{\square}{5}$

**b** not picking a white ball? $\frac{\square}{5}$

**c** picking a grey ball? ______

**d** not picking a grey ball? ______

**3** From the picture, what is the probability of:

**a** picking a square? $\frac{\square}{7}$

**b** not picking a square? ______

**c** picking a cross? ______

**d** not picking a cross? ______

**4** The picture shows white and grey tiles. What is the probability of:

**a** picking a grey tile? $\frac{\square}{12}$

**b** not picking a grey tile? ______

**c** picking a white tile? ______

**d** not picking a white tile? ______

**5** From the letters of the word DAFFODILLY, what is the probability of:

**a** picking the letter L? $\frac{\square}{\square} = \frac{\square}{\square}$

**b** not picking the letter L? ______

**c** not picking the letter A? ______

**6** This spinner is spun once. What is the probability that the pointer will land on:

Purple
Purple
Purple
Green
Grey

**a** purple? ______ **b** not purple? ______

**c** green? ______ **d** not green? ______

**e** grey? ______ **f** not grey? ______

**7** This spinner is spun once. What is the probability that the pointer will land on:

Blue
Green
Orange

**a** blue? ______ **b** not blue? ______

**c** green? ______ **d** not green? ______

**e** orange? ______ **f** not orange ______

UNIT 11

**PROBABILITY**

# Complementary events

1 A deck of cards contains 52 cards with 4 different suits as shown below:

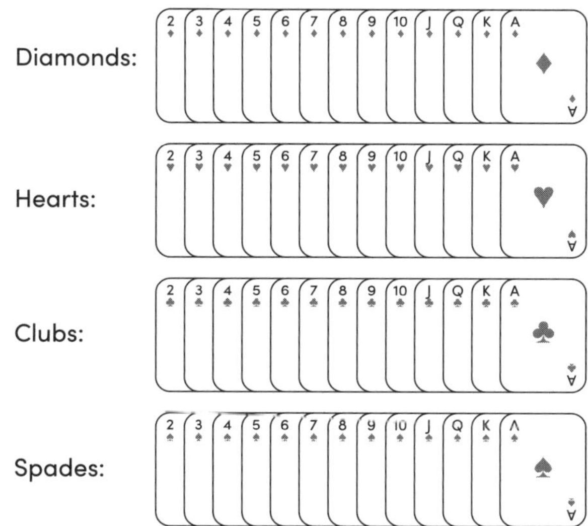

From the deck of cards above, what is the probability of picking:

a a spade?

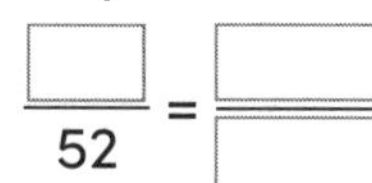

b not a spade?

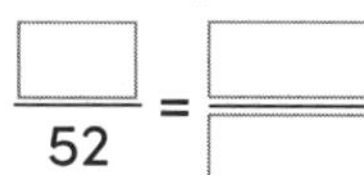

c a queen? ____________

d not a queen? ____________

e a black card? (clubs & spades) ____________

f not a black card? ____________

g a black jack? ____________

h not a black jack? ____________

i a 2 or 3? ____________

j not a 2 or 3? ____________

k not a 7? ____________

l not a 10? ____________

2 A fair 6-sided dice shows the numbers 1, 2, 3, 4, 5, 6. What is the probability of rolling:

a 4? ____________

b not 4? ____________

c an even number? ____________

d not an even number? ____________

3 In a bag of 50 tickets, there are 25 General Admission, 5 VIP, 12 Reserved Seating and 8 Group Package tickets.
Find:

a P(VIP) ________ b P(not VIP) ________

c P(GA) ________ d P(not GA) ________

4 On a bookshelf, there are 12 fiction books, 16 history books and 8 magazines.
Find:

a P(fiction) ________

b P(not fiction) ________

c P(magazine) ________

d P(not magazine) ________

5 If the probability of winning a game is 42%, what is the probability of not winning?

____________________

____________________

6 If the probability of rain tomorrow is 0.7, what is the probability it will not rain?

____________________

____________________

UNIT 12

CHAPTER 9 REVIEW

# Probability

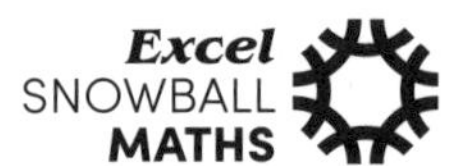


Note: Answers should be correct to one decimal place.

**1** For this spinner:

White, White, Blue, Blue, Blue

**a** list the sample space. ______

**b** how many possible outcomes are there? ______

**c** are the outcomes equally likely? ______

**d** describe the chance of the spinner landing on blue ______

**e** describe the chance of the spinner landing on white ______

**f** what is the probability it will land on blue? ______

**g** what is the probability it will land on white? ______

**2** A fair dice is rolled.

**a** What is the sample space? ______

**b** How many possible outcomes are there? ______

**c** What is the probability of rolling a 2? ______

**d** What is the probability of rolling a 3? ______

**e** Are all the outcomes equally likely? Explain your answer.

______

**f** What is the probability of rolling an even number?

______

**3** From the letters of the word EXCELLENT, what is the probability, as a percentage, of picking the letter:

**a** E? ______ **b** L? ______ **c** E or L? ______

**d** Which letter is the most likely to be chosen?

______

**4** A bag contains 8 blue, 12 black and 4 white buttons. What is the probability of picking a:

**a** blue button? ______ **b** not a blue button? ______

**c** white button? ______ **d** not a white button? ______

**e** Which coloured button is least likely to be chosen? ______

**5** A bag contains 120 tickets. Fifty of these tickets are coloured green and the rest are coloured yellow. What is the probability of selecting a yellow ticket?

______

**6** What is the probability for the following chances:

**a** certain ______ **b** fifty-fifty ______

**c** impossible ______ **d** even chance ______

**7** A spinner was spun, and the results were recorded in the table below.

| Colour | Frequency |
|---|---|
| Green | 29 |
| Red | 15 |
| Orange | 6 |

Green, Green, Green, Red, Red, Orange

**a** What is the experimental probability, as a decimal, of the spinner landing on:

**i** green? ______ **ii** red? ______ **iii** orange? ______

**b** What is the theoretical probability, as a decimal, of the spinner landing on:

**i** green? ______ **ii** red? ______ **iii** orange? ______

**8** A ticket generates a random number from 1 to 30. What is the probability of getting:

**a** 11? ______ **b** 25? ______

**c** a number more than 20? ______

**d** a number less than 20? ______

**e** a number divisible by 5? ______

**f** a two-digit number? ______

**9** Bill bought 5 raffle tickets. If a total of 600 tickets were sold, what is the probability Bill wins first prize?

______

# Answers

## CHAPTER 1

### PAGE 1

**1** a 1 < 4 b 3 > 1 c 4 < 9 d 5 < 6 e 5 > 2 f 3 < 7 g 0 < 5 h 1 > 0 i 4 > 0 j 6 > 0 k 3 > 0 l 0 < 2 m 0 > −5 n −1 < 0 o −4 < 0 p −6 < 0 q −3 < 0 r 0 > −2 s −4 < 3 t 5 > −7 u −6 < 9 v −2 < 4 w 5 > −4 x −8 < 3 **2** a −4 < 4 b −5 < 5 c 1 > −1 d 3 > −3 e −2 < 2 f 8 > −8 g −1 > −7 h −5 > −9 i −6 < −4 j −3 < −2 k −4 > −5 l −7 > −10 m 2 < 5 n 7 > 0 o −4 < 0 p −6 < 5 q −2 < 2 r −9 < −2 s −11 > −12 t 5 > −6 u 0 > −3 **3** a 9, 5, 1 b 5, 3, 0 c 7, 4, 2 d 9, 5, 3 e 6, 4, 0 f 7, 3, 1 g 6, 3, −1 h 5, 1, −2 i 3, 0, −3 j 5, 2, −2 k 4, 0, −3 l 7, 3, −6 m 6, −1, −3 n 4, −2, −5 o 3, −1, −6 p 2, −2, −7 q 1, −1, −7 r 0, −3, −8 s −2, −4, −6 t −1, −3, −5 u −2, −5, −6 v 0, −2, −4 w −2, −5, −7 x −4, −5, −6 y 0, −9, −10 z 8, −6, −7 za −8, −9, −10

### PAGE 2

**1** a 7 b 0 c 6 d 8 e −3 f 3 g 4 h 2 **2** a 1 b −5 c 2 d −1 e −10 f −6 g −5 h −5 **3** a 7 b 1 c 6 d 8 e −1 f −3 g 1 h 2 **4** a 1 °C b −7 °C c 2 °C d −5 °C e 11 °C f −5 °C g 4 °C h 0 °C

### PAGE 3

**1** a 5 b −2 c 3 d −5 e −2 f 4 g −9 h −7 **2** a −$10 b $2 c −$9 d $7 e −$11 f −$8 g −$11 h −$4 **3** a 5 b −1 c 6 d −2 e −1 f 5 g −2 **4** a −1 b 5 c −4 d −7 e −2 f −2 g −5

### PAGE 4

**1** a 1 b 1 c −2 d 2 e −2 f 4 g 2 h −2 i 3 j 0 k −1 **2** a −4 b 1 c 6 d 6 e 4 f −5 g 2 h 2 i −5 j 3 k −4 l 4 m 0 n 0 o 3 p 4 q 7 r 7 s −2 t −4 u 8 v 1

### PAGE 5

**1** a −1 b −1 c 2 d −3 e −4 f 1 g −3 h −2 i 1 j −5 k −5 **2** a −4 b −3 c −4 d 0 e 0 f −6 g −8 h 1 i −6 j −7 k −2 l 1 m 3 n −2 o −5 p −6 q 0 r −1 s 0 t 5 u 5 v −4

### PAGE 6

**1** a −6 b −7 c −7 d −4 e −8 f −9 g −8 h −9 i −8 j −6 k −8 **2** a −6 b −7 c −6 d −6 e −4 f −8 g −8 h −7 i −8 j −5 k −10 l −10 m −9 n −11 o −15 p −11 q −14 r −13 s −13 t −21 u −15 v −10

### PAGE 7

**1** a −5 b −1 c −2 d −2 e −3 f −6 g −4 h −5 i −7 j −5 k −4 **2** a −6 b −2 c −2 d −6 e −3 f −5 g −5 h −3 i −1 j −4 k −7 l −4 m −2 n −3 o −1 p −8 q −1 r −2 s −9 t −3 u −3 v −3

### PAGE 8

**1** a −7 b −6 c −5 d −7 e −7 f −9 g −8 h −7 i −6 j −10 k −6 **2** a −6 b −4 c −4 d −8 e −4 f −8 g −9 h −5 i −5 j −9 k −8 l −5 m −3 n −8 o −2 p −9 q −12 r −10 s −11 t −20 u −16 v −19

### PAGE 9

**1** a 7 b 7 c 6 d 9 e 13 f 5 g 10 h 13 i 16 j 19 **2** a 2 b 3 c 1 d 2 e −2 f −4 g −3 h −3 i 2 j −5

### PAGE 10

**1** a 2 °C b −20 °C c −19 °C d 2 °C e 7 °C f −21 °C **2** a 5th above ground b 4th above ground c 2nd below ground d 6th above ground e 3rd below ground f 5th below ground **3** a 16 h b 6 h c 15 h

### PAGE 11

**1** a 7 h b 14 h c 3 h d 7 h e 4 h **2** a −$50 b $140 c $250 d −$20

### PAGE 12

**1** a −15 b −8 c −36 d −48 e −15 f −14 g −36 h −30 i −30 j −28 k −72 l −40 **2** a 14 b 20 c 40 d 30 e 55 f 60 g 72 h 63 i 24 j 49 k 16 l 25 **3** a 25 b 9 c 36 d 49 e 4 f 64 g 100 h 81 i 16 j 1 k 121 l 144 **4** a −5 b −2 c −9 d −6 e −3 f −7 g −9 h −6 i −3 j −4 k −9 l −5 **5** a 9 b 1 c 7 d 4 e 5 f 5 g 8 h 9 i 8 j 7 k 4 l 5

### PAGE 13

**1** a −3 b −8 c −10 d −4 e −5 f −4 g −3 h −7 i −3 j 4 k 5 l 3 m 6 n 9 o 8 **2** a 3 b 4 c 4 d 8 e 9 f 2 g 9 h 3 i 5 j 4 k −5 l −3 m −6 n −6 o 7 p 9 q −5 r −9 s 5 t −8

### PAGE 14

**1** a −5 b −2 c −2 d −3 e −2 f −4 g −1 h −3 i −4 j −1 k −5 l −1 m −8 n −8 **2** a 5 b 1 c 6 d 3 e 4 f 4 g 2 h 4 i 5 j 4 k 2 l 5 m 4 n 0 **3** a −5 b −2 c −3 d −10 e −7 f −5 g −2 h −3 i −3 j −4 k −3 l −4 m −8 n −6

### PAGE 15

**1** a 7 b 4 c 7 d 11 e 11 f 10 g 11 h 17 i 14 j 20 k 16 l 9 m 15 n 28 **2** a −7 b −14 c −9 d −13 e −12 f −9 g −18 h −8 i −15 j −12 k −14 l −13 m −10 n −26 **3** a −3 b −2 c −3 d −2 e −1 f −3 g −2 h −2 i −4 j −4 k −1 l −1

### PAGE 16

**1** a −15 b −6 c −12 d −12 e −9 f −10 g −7 h −14 i −12 j −20 k −28 l −8 m −25 n −20 **2** a 2 b 8 c 6 d 5 e 8 f 15 g 6 h 6 i 9 j 10 k 12 l 12 m 9 n 30 **3** a 2 b 4 c 2 d 3 e 2 f 5 g 5 h 2 i 7 j 2

### PAGE 17

**1** a 4 b 3 c 5 d 3 e 2 f 2 g 4 h 2 i 3 j 6 k 6 l 7 m 2 n 2 **2** a 8 b 10 c 20 d 15 e 2 f 4 g 4 h 2 i 18 j 25 **3** a −14 b −15 c −21 d −12 e −14 f −11 g −10 h −15 i −12 j −11

### PAGE 18

**1** a −3 b −2 c −3 d −5 e −4 f −2 g −4 h −2 **2** a −1 b −4 c −2 d −1 e 0 f −5 g −7 h −6 **3** a −4 b −2 c −10 d −2 e −8 f −6 g −2 h −4 i −3 j −2 k −5 l −4

### PAGE 19

**1** a −1 < 3 b 0 > −8 c −5 < −2 **2** a 6, 3, −8 b 0, −4, −8 c 4, −5, −9 **3** a −3 m b 5 m c −2 m **4** a −6 b −7 c 10 d 11 e −20 f −21 g 5 h 8 i −5 j −8 **5** 9° C **6** $45 **7** a −18 b 54 c 4 **8** a −5 b 7 c 4 **9** a 3 b −4 **10** a −7 b −4 **11** a 3 b 2 **12** a −10 b −9 **13** a 2 b 3 **14** a 15 b 24 **15** a 3 b 7 **16** a −6 b −4 **17** a −4 b −5 **18** a −1 b −4 **19** a −4 b −2

## CHAPTER 2

### PAGE 20

**1** a $2^4$ b $3^5$ c $5^3$ d $4^4$ e $7^2$ f $8^6$ g $9^5$ h $10^4$ i $6^2$ j $12^6$ **2** a 1 b 4 c 9 d 16 e 25 f 36 g 49 h 64 i 81 j 100 **3** a 2 b 3 c 4 d 5 e 6 f 7 g 8 h 9 i 10 j 11 **4** a 4 b 25 c 16 d 36 e 9 f 49 g 2 h 4 i 5 j 3 k 10 l 6 **5** a 8 b 27 c 64 d 125 e 216 f 343 g 512 h 729 i 1000

### PAGE 21

**1** a 2 b 3 c 4 d 5 e 6 f 7 g 8 h 9 i 10 j 11 **2** a 8 b 125 c 27 d 1000 e 3 f 2 g 4 h 5 i 4 j 25 k 9 l 100 m 3 n 7 o 8 p 10 q 10 r 64 **3** a 4 b 9 c 16 d 25 e 36 f 49 g 64 h −8 i −27

j −64 k −125 l −216 m −343 n −512 ❹ a 5 b 8 c 4 d 7 e 9 f 11 g −2 h −1 i −3 j −5 k −4 l −6

**PAGE 22** ❶ a 8 b 20 c 16 d 12 e 24 f 28 g 2 h 4 i 6 j 3 k 8 l 10 ❷ a 4 b 5 c 7 d 9 e 11 f 13 g 1 h 1 i 3 j 3 k 3 l 1 ❸ a 5 b 7 c 10 d 9 e 11 f 14 g 5 h 4 i 3 j 6 k 2 l 1

**PAGE 23** ❶ a 8 b 4 c 2 d 1 e 15 f 12 g 13 h 16 i 13 j 25 k 41 l 61 m 91 n 84 o 8 p 18 q 4 r 2 ❷ a 1 b 20 c 40 d 50 e 2 f 5 g 72 h 28 ❸ a 9 b 4 c 29 d 27 e 34 f 10 g 75 h 12

**PAGE 24** ❶ a prime b prime c composite d composite e prime f prime g composite h prime i composite j composite k prime l composite ❷ a $2^2 \times 3$ b $2^4$ c $2^2 \times 5$ d $2 \times 3^2$ e $2^2 \times 7$ f $2^3 \times 3$ g $2^5$ h $2^2 \times 3^2$ i $2^3 \times 5$ j $2^4 \times 3$

**PAGE 25** ❶ a $3^3$ b $3^2 \times 5$ c $2 \times 3^3$ d $2^2 \times 3 \times 5$ e $2^3 \times 11$ f $2^2 \times 5^2$ g $2 \times 3 \times 5^2$ h $2 \times 3 \times 7$ i $2^6$ j $2^5 \times 3$ k $2 \times 3^3 \times 5$

**PAGE 26** ❶ a $8^4$ b $2^6$ c $7^4$ d $10^3$ ❷ a 16 b 49 c 64 d 25 e 81 f 100 ❸ a 125 b 27 c 64 d 8 ❹ a 4 b 7 c 2 d 4 e 9 f 36 g 216 h 1000 i 9 j 10 k 3 l 6 m 2 n 5 ❺ a 8 b 10 c 6 d 1 e 20 f 32 g 8 h 6 i 2 j 1 k 25 l 40 m 51 n 70 o 27 p 1 ❻ a prime b composite c composite d prime e prime f composite ❼ a $2^2 \times 11$ b $2^3 \times 7$ c $2^4 \times 5$ d $2 \times 3^3$ e $2^3 \times 17$ f $2^4 \times 3 \times 5$ g $2^2 \times 3^2 \times 11$ h $2^5 \times 13$ i $2 \times 3^2 \times 19$ j $2 \times 3 \times 5 \times 23$

## CHAPTER 3

**PAGE 27** ❶ a $\frac{1}{2}$ b $\frac{2}{4}$ c $\frac{4}{8}$ d $\frac{12}{24}$ e $\frac{25}{50}$ f $\frac{50}{100}$ ❷ a $\frac{1}{3}$ b $\frac{2}{6}$ c $\frac{4}{12}$ d $\frac{8}{24}$ ❸ a $\frac{2}{3}$ b $\frac{8}{12}$ c $\frac{16}{24}$ ❹ a $\frac{1}{4}$ b $\frac{4}{16}$ c $\frac{8}{32}$ d $\frac{25}{100}$ ❺ a $\frac{3}{4}$ b $\frac{12}{16}$ ❻ a $\frac{2}{4}$ b $\frac{4}{6}$ c $\frac{10}{8}$ d $\frac{12}{10}$ e $\frac{4}{12}$ f $\frac{10}{16}$ g $\frac{14}{18}$ h $\frac{6}{20}$ i $\frac{8}{12}$ j $\frac{6}{8}$ k $\frac{16}{10}$ l $\frac{18}{14}$

**PAGE 28** ❶ a $\frac{3}{6}$ b $\frac{6}{9}$ c $\frac{15}{12}$ d $\frac{18}{15}$ e $\frac{9}{18}$ f $\frac{15}{24}$ g $\frac{21}{27}$ h $\frac{9}{30}$ i $\frac{12}{18}$ j $\frac{9}{12}$ k $\frac{24}{15}$ l $\frac{27}{21}$ ❷ a $\frac{4}{8}$ b $\frac{8}{12}$ c $\frac{20}{16}$ d $\frac{24}{20}$ e $\frac{12}{24}$ f $\frac{20}{32}$ g $\frac{28}{36}$ h $\frac{12}{40}$ i $\frac{16}{24}$ j $\frac{12}{16}$ k $\frac{32}{20}$ l $\frac{36}{28}$ ❸ a $\frac{5}{10}$ b $\frac{10}{15}$ c $\frac{25}{20}$ d $\frac{30}{25}$ e $\frac{15}{30}$ f $\frac{25}{40}$ g $\frac{35}{45}$ h $\frac{15}{50}$ i $\frac{20}{30}$ j $\frac{15}{20}$ k $\frac{40}{25}$ l $\frac{45}{35}$ ❹ a $\frac{2}{4}$ b $\frac{2}{3}$ c $\frac{5}{4}$ d $\frac{6}{5}$ e $\frac{2}{6}$ f $\frac{5}{8}$

**PAGE 29** ❶ a $\frac{7}{9}$ b $\frac{3}{10}$ c $\frac{4}{6}$ d $\frac{3}{4}$ e $\frac{8}{5}$ f $\frac{9}{7}$ ❷ a $\frac{2}{3}$ b $\frac{1}{3}$ c $\frac{5}{4}$ d $\frac{7}{5}$ e $\frac{3}{8}$ f $\frac{5}{9}$ g $\frac{7}{10}$ h $\frac{3}{11}$ i $\frac{3}{6}$ j $\frac{5}{4}$ k $\frac{8}{6}$ l $\frac{9}{8}$ ❸ a $\frac{2}{4}$ b $\frac{3}{6}$ c $\frac{6}{8}$ d $\frac{7}{10}$ e $\frac{4}{8}$ f $\frac{6}{5}$ g $\frac{8}{10}$ h $\frac{4}{11}$ i $\frac{3}{7}$ j $\frac{6}{5}$ k $\frac{9}{8}$ l $\frac{7}{9}$ ❹ a $\frac{3}{4}$ b $\frac{5}{3}$ c $\frac{2}{4}$ d $\frac{7}{5}$ e $\frac{2}{6}$ f $\frac{5}{8}$ g $\frac{9}{6}$ h $\frac{12}{4}$ i $\frac{7}{10}$ j $\frac{3}{7}$ k $\frac{8}{1}$ l $\frac{9}{3}$

**PAGE 30** ❶ $0, \frac{1}{2}, 1, 1\frac{1}{2}, 2$; $0, \frac{1}{3}, \frac{2}{3}, 1, 1\frac{1}{3}, 1\frac{2}{3}, 2$ ❷ a < b > c > d < e > f < g < h > ❸ $0, \frac{1}{4}, \frac{2}{4}, \frac{3}{4}, 1, 1\frac{1}{4}, 1\frac{2}{4}, 1\frac{3}{4}, 2$; $0, \frac{1}{5}, \frac{2}{5}, \frac{3}{5}, \frac{4}{5}, 1, 1\frac{1}{5}, 1\frac{2}{5}, 1\frac{3}{5}, 1\frac{4}{5}, 2$ ❹ a > b > c < d < e < f > g > h > i > j > k = l < m < n > o < p =

**PAGE 31** ❶ $0, \frac{1}{6}, \frac{2}{6}, \frac{3}{6}, \frac{4}{6}, \frac{5}{6}, 1, 1\frac{1}{6}, 1\frac{2}{6}, 1\frac{3}{6}, 1\frac{4}{6}, 1\frac{5}{6}, 2$; $0, \frac{1}{7}, \frac{2}{7}, \frac{3}{7}, \frac{4}{7}, \frac{5}{7}, \frac{6}{7}, 1, 1\frac{1}{7}, 1\frac{2}{7}, 1\frac{3}{7}, 1\frac{4}{7}, 1\frac{5}{7}, 1\frac{6}{7}, 2$; $0, \frac{1}{8}, \frac{2}{8}, \frac{3}{8}, \frac{4}{8}, \frac{5}{8}, \frac{6}{8}, \frac{7}{8}, 1, 1\frac{1}{8}, 1\frac{2}{8}, 1\frac{3}{8}, 1\frac{4}{8}, 1\frac{5}{8}, 1\frac{6}{8}, 1\frac{7}{8}, 2$ ❷ a > b < c < d > e = f < g < h > i > j < k < l > ❸ $0, \frac{1}{9}, \frac{2}{9}, \frac{3}{9}, \frac{4}{9}, \frac{5}{9}, \frac{6}{9}, \frac{7}{9}, \frac{8}{9}, 1, 1\frac{1}{9}, 1\frac{2}{9}, 1\frac{3}{9}, 1\frac{4}{9}, 1\frac{5}{9}, 1\frac{6}{9}, 1\frac{7}{9}, 1\frac{8}{9}, 2$ $0, \frac{1}{10}, \frac{2}{10}, \frac{3}{10}, \frac{4}{10}, \frac{5}{10}, \frac{6}{10}, \frac{7}{10}, \frac{8}{10}, \frac{9}{10}, 1, 1\frac{1}{10}, 1\frac{2}{10}, 1\frac{3}{10}, 1\frac{4}{10}, 1\frac{5}{10}, 1\frac{6}{10}, 1\frac{7}{10}, 1\frac{8}{10}, 1\frac{9}{10}, 2$ ❹ a < b > c < d > e < f < g > h > i > j < k < l > m = n > o = p >

**PAGE 32** ❶ a < b > c < d < e < f > g < h < i < j > ❷ a > b < c > d > e < f > g > h < i < j > k > l < ❸ a $\frac{3}{4}$ b $\frac{2}{5}$ c $\frac{5}{6}$ d $\frac{3}{4}$ e $\frac{5}{7}$ f $\frac{11}{12}$ g $\frac{17}{20}$ h $\frac{7}{16}$ i $\frac{3}{4}$ j $\frac{12}{15}$

**PAGE 33** ❶ a $\frac{1}{2}$ b $\frac{3}{4}$ c $\frac{5}{2}$ d $\frac{7}{3}$ e $\frac{9}{10}$ f $\frac{7}{12}$ g 8 h 10 ❷ a $\frac{1}{3}$ b $\frac{4}{5}$ c $\frac{7}{4}$ d $\frac{6}{5}$ e $\frac{8}{9}$ f $\frac{3}{10}$ g 5 h 8 i $\frac{5}{7}$ j $\frac{4}{9}$ ❸ a $\frac{1}{2}$ b $\frac{3}{4}$ c $\frac{6}{5}$ d $\frac{5}{3}$ e $\frac{7}{8}$ f $\frac{5}{8}$ g 4 h 10 ❹ a $\frac{1}{3}$ b $\frac{2}{3}$ c $\frac{5}{4}$ d $\frac{7}{2}$ e $\frac{1}{9}$ f $\frac{8}{11}$ g 2 h 10 i $\frac{9}{10}$ j $\frac{5}{12}$ ❺ a $\frac{1}{2}$ b $\frac{1}{3}$ c $\frac{3}{2}$ d $\frac{6}{5}$ e $\frac{4}{5}$ f $\frac{6}{7}$ g 2 h 6 ❻ a $\frac{1}{3}$ b $\frac{3}{4}$ c $\frac{2}{3}$ d $\frac{4}{5}$ e $\frac{3}{7}$ f $\frac{2}{5}$ g 7 h $\frac{1}{7}$

**PAGE 34** ❶ a $\frac{1}{2}$ b $\frac{3}{4}$ c $\frac{5}{4}$ d $\frac{6}{5}$ e $\frac{7}{9}$ f $\frac{4}{7}$ g 3 h 1 ❷ a $\frac{1}{2}$ b $\frac{3}{5}$ c $\frac{4}{5}$ d $\frac{6}{7}$ e $\frac{8}{9}$ f $\frac{7}{10}$ g 4 h $\frac{2}{3}$ i $\frac{3}{4}$ j $\frac{7}{8}$ k $\frac{10}{7}$ l $\frac{9}{8}$ ❸ a $\frac{1}{3}$ b $\frac{2}{3}$ c $\frac{4}{5}$ d $\frac{7}{10}$ e $\frac{9}{10}$ f $\frac{4}{5}$ g $\frac{5}{6}$ h $\frac{3}{10}$ ❹ a $\frac{1}{2}$ b $\frac{3}{5}$ c $\frac{6}{7}$ d $\frac{4}{5}$ e $\frac{2}{9}$ f $\frac{8}{9}$ g $\frac{1}{10}$ h 11 ❺ a $\frac{2}{5}$ b $\frac{7}{12}$ c $\frac{8}{15}$ ❻ a $\frac{4}{5}$ b $\frac{8}{9}$ c $\frac{3}{10}$ ❼ a $\frac{3}{4}$ b $\frac{5}{6}$ c $\frac{9}{10}$

**PAGE 35** ❶ a $\frac{3}{4}$ b $\frac{2}{3}$ c $\frac{4}{5}$ d $\frac{4}{5}$ e $\frac{5}{6}$ f $\frac{6}{7}$ g $\frac{6}{7}$ h $\frac{5}{8}$ i $\frac{5}{9}$ j $\frac{7}{10}$ ❷ a $\frac{1}{2}$ b $\frac{2}{3}$ c $\frac{1}{4}$ d $\frac{3}{4}$ e $\frac{3}{5}$ f $\frac{4}{5}$ g $\frac{5}{6}$ h $\frac{6}{7}$ ❸ a $\frac{1}{3}$ b $\frac{2}{3}$ c $\frac{2}{5}$ d $\frac{3}{5}$ e $\frac{3}{7}$ f $\frac{5}{9}$ g $\frac{6}{7}$ h $\frac{1}{10}$ ❹ a $\frac{1}{3}$ b $\frac{2}{3}$ c $\frac{3}{5}$ d $\frac{4}{5}$ e $\frac{5}{6}$ f $\frac{7}{8}$ g $\frac{4}{7}$ h $\frac{5}{9}$

### PAGE 36

**1** a $\frac{1}{2}$ b $\frac{2}{3}$ c $\frac{3}{4}$ d $\frac{4}{5}$ e $\frac{2}{5}$ f $\frac{5}{6}$ g $\frac{3}{7}$ h $\frac{2}{9}$ **2** a $\frac{1}{2}$ b $\frac{1}{3}$ c $\frac{3}{4}$ d $\frac{3}{5}$ e $\frac{4}{7}$ f $\frac{6}{7}$ g $\frac{3}{8}$ h $\frac{7}{9}$ **3** a $\frac{1}{2}$ b $\frac{2}{3}$ c $\frac{1}{3}$ d $\frac{3}{5}$ e $\frac{5}{7}$ f $\frac{6}{7}$ **4** a $\frac{1}{2}$ b $\frac{2}{3}$ c $\frac{3}{4}$ d $\frac{4}{5}$ e $\frac{2}{5}$ f $\frac{5}{7}$ **5** a $\frac{1}{2}$ b $\frac{2}{3}$ c $\frac{4}{5}$ d $\frac{5}{6}$

### PAGE 37

**1** a $\frac{3}{4}$ b $\frac{5}{6}$ c $\frac{3}{8}$ d $\frac{7}{8}$ e $\frac{3}{10}$ f $\frac{7}{10}$ g $\frac{9}{10}$ h $\frac{7}{12}$ i $\frac{3}{14}$ j $\frac{5}{14}$ k $\frac{9}{14}$ l $\frac{11}{14}$ m $\frac{11}{14}$ n $\frac{13}{14}$ o $\frac{11}{14}$ p $\frac{13}{16}$ q $\frac{13}{18}$ r $\frac{17}{20}$ s $\frac{19}{24}$ t $\frac{23}{36}$

### PAGE 38

**1** a $\frac{5}{6}$ b $\frac{4}{9}$ c $\frac{7}{9}$ d $\frac{8}{9}$ e $\frac{11}{12}$ f $\frac{4}{15}$ g $\frac{7}{18}$ h $\frac{13}{18}$ i $\frac{17}{18}$ j $\frac{11}{18}$ **2** a $\frac{1}{4}$ b $\frac{1}{6}$ c $\frac{1}{8}$ d $\frac{3}{8}$ e $\frac{3}{10}$ f $\frac{7}{12}$ g $\frac{7}{16}$ h $\frac{5}{16}$ i $\frac{11}{18}$ j $\frac{11}{20}$

### PAGE 39

**1** a $\frac{1}{6}$ b $\frac{2}{9}$ c $\frac{5}{9}$ d $\frac{4}{9}$ e $\frac{7}{12}$ f $\frac{2}{15}$ g $\frac{7}{15}$ h $\frac{8}{15}$ i $\frac{5}{18}$ j $\frac{11}{18}$ k $\frac{8}{21}$ l $\frac{13}{30}$ m $\frac{11}{30}$ n $\frac{5}{24}$ o $\frac{7}{24}$ p $\frac{8}{27}$ q $\frac{14}{27}$ r $\frac{17}{33}$ s $\frac{11}{21}$ t $\frac{13}{30}$ u $\frac{5}{36}$ v $\frac{23}{36}$

### PAGE 40

**1** a $\frac{5}{8}$ b $\frac{7}{8}$ c $\frac{5}{12}$ d $\frac{11}{12}$ e $\frac{5}{16}$ f $\frac{13}{16}$ g $\frac{15}{16}$ h $\frac{9}{20}$ i $\frac{11}{20}$ j $\frac{17}{20}$ k $\frac{19}{20}$ l $\frac{5}{24}$ m $\frac{13}{24}$ n $\frac{23}{24}$ o $\frac{17}{32}$ p $\frac{25}{32}$ q $\frac{29}{40}$ r $\frac{31}{40}$ s $\frac{35}{48}$ t $\frac{41}{44}$

### PAGE 41

**1** a $\frac{7}{10}$ b $\frac{11}{15}$ c $\frac{9}{20}$ d $\frac{17}{20}$ e $\frac{18}{25}$ f $\frac{22}{25}$ g $\frac{11}{30}$ h $\frac{17}{30}$ i $\frac{19}{35}$ j $\frac{22}{35}$ k $\frac{31}{35}$ l $\frac{13}{40}$ m $\frac{27}{40}$ n $\frac{33}{40}$ o $\frac{28}{45}$ p $\frac{37}{45}$ q $\frac{29}{50}$ r $\frac{41}{50}$ s $\frac{51}{55}$ t $\frac{49}{60}$ u $\frac{43}{60}$ v $\frac{51}{55}$

### PAGE 42

**1** a $\frac{13}{15}$ b $\frac{7}{12}$ c $\frac{5}{6}$ d $\frac{7}{6}$ e $\frac{11}{12}$ f $\frac{11}{15}$ g $\frac{15}{28}$ h $\frac{13}{12}$ i $\frac{19}{28}$ j $\frac{14}{15}$ k $\frac{15}{14}$ **2** a $\frac{1}{6}$ b $\frac{1}{15}$ c $\frac{1}{28}$ d $\frac{7}{15}$ **3** a $\frac{3}{20}$ b $\frac{7}{30}$ c $\frac{9}{28}$ d $\frac{7}{20}$ e $\frac{23}{40}$ f $\frac{29}{56}$ g $\frac{19}{30}$ h $\frac{19}{30}$ i $\frac{7}{24}$ j $\frac{31}{28}$ k $\frac{29}{24}$ l $\frac{7}{18}$ m $\frac{24}{35}$ n $\frac{39}{54}$ o $\frac{25}{42}$

### PAGE 43

**1** a $\frac{5}{56}$ b $\frac{16}{45}$ c $\frac{11}{56}$ d $\frac{8}{45}$ e $\frac{4}{45}$ **2** a $\frac{47}{56}$ b $\frac{57}{40}$ c $\frac{71}{63}$ d $\frac{37}{56}$ e $\frac{62}{63}$ **3** a $\frac{73}{88}$ b $\frac{62}{55}$ c $\frac{7}{99}$ d $\frac{19}{70}$ e $\frac{83}{90}$ f $\frac{41}{70}$ g $\frac{13}{70}$ h $\frac{9}{110}$ i $\frac{107}{110}$ j $\frac{70}{99}$ k $\frac{71}{99}$ l $\frac{17}{77}$ **4** $\frac{13}{20}$ **5** $\frac{31}{35}$ **6** $\frac{65}{66}$ **7** $\frac{7}{30}$ **8** $\frac{47}{63}$

### PAGE 44

**1** a mixed numeral b mixed numeral c proper d improper e improper f proper **2** a $\frac{5}{3}$ b $\frac{11}{7}$ c $\frac{8}{3}$ d $\frac{11}{4}$ e $\frac{7}{2}$ f $\frac{23}{7}$ g $\frac{23}{5}$ h $\frac{37}{9}$ i $\frac{27}{5}$ j $\frac{41}{7}$ k $\frac{49}{8}$ l $\frac{53}{7}$ m $\frac{53}{6}$ n $\frac{73}{8}$ **3** a $1\frac{2}{3}$ b $1\frac{1}{2}$ c $1\frac{3}{4}$ d $1\frac{3}{8}$ e $1\frac{3}{7}$ f $1\frac{1}{9}$ g $1\frac{1}{6}$ h $1\frac{2}{9}$ i $1\frac{2}{7}$ j $1\frac{1}{3}$ k $1\frac{1}{4}$ l $1\frac{4}{9}$ **4** a $2\frac{2}{3}$ b $2\frac{1}{2}$ c $2\frac{3}{4}$ d $2\frac{3}{8}$ e $2\frac{1}{6}$ f $2\frac{3}{5}$ g $2\frac{1}{5}$ h $2\frac{2}{9}$ i $2\frac{2}{7}$ j $2\frac{4}{5}$ k $2\frac{1}{7}$ l $2\frac{4}{7}$ m $2\frac{2}{5}$ n $2\frac{1}{8}$ o $2\frac{7}{10}$ p $2\frac{4}{9}$

### PAGE 45

**1** a $3\frac{2}{3}$ b $3\frac{1}{2}$ c $3\frac{3}{4}$ d $3\frac{1}{8}$ e $3\frac{1}{6}$ f $3\frac{2}{5}$ g $3\frac{3}{5}$ h $3\frac{1}{4}$ i $3\frac{2}{7}$ j $3\frac{3}{8}$ k $3\frac{1}{5}$ l $3\frac{4}{5}$ m $3\frac{5}{6}$ n $3\frac{4}{7}$ o $3\frac{2}{9}$ p $3\frac{5}{8}$ **2** a $4\frac{2}{3}$ b $4\frac{1}{2}$ c $4\frac{1}{4}$ d $4\frac{3}{8}$ e $4\frac{5}{6}$ f $4\frac{2}{5}$ g $4\frac{3}{4}$ h $4\frac{2}{9}$ i $4\frac{3}{7}$ j $4\frac{1}{3}$ k $4\frac{4}{5}$ l $4\frac{7}{10}$ **3** a $5\frac{1}{2}$ b $5\frac{2}{3}$ c $5\frac{1}{6}$ d $5\frac{3}{8}$ e $5\frac{2}{5}$ f $5\frac{4}{9}$ g $5\frac{3}{5}$ h $5\frac{6}{7}$ i $5\frac{1}{7}$ j $5\frac{3}{10}$ k $5\frac{2}{9}$ l $5\frac{3}{7}$ m $5\frac{1}{8}$ n $5\frac{1}{10}$ **4** a $6\frac{1}{2}$ b $6\frac{1}{4}$ c $6\frac{3}{4}$ d $6\frac{1}{5}$ e $6\frac{4}{5}$ f $6\frac{2}{3}$ g $6\frac{1}{7}$ h $6\frac{2}{9}$ i $6\frac{5}{6}$ j $6\frac{3}{8}$ k $6\frac{2}{7}$ l $6\frac{3}{7}$

### PAGE 46

**1** a $\frac{4}{5}$ b $\frac{4}{5}$ c $\frac{2}{3}$ d $\frac{3}{4}$ e $\frac{3}{5}$ f $\frac{4}{5}$ g $\frac{3}{5}$ h $\frac{4}{7}$ i $\frac{6}{7}$ j $\frac{6}{7}$ k $\frac{7}{9}$ l $\frac{8}{9}$ **2** a $2\frac{1}{5}$ b $2\frac{1}{4}$ c $2\frac{1}{5}$ d $2\frac{2}{5}$ e $2\frac{1}{4}$ f $2\frac{1}{6}$ g $2\frac{1}{7}$ h $2\frac{2}{7}$

### PAGE 47

**1** a $2\frac{2}{3}$ b $2\frac{3}{5}$ c $2\frac{4}{5}$ d $2\frac{3}{5}$ e $2\frac{3}{4}$ f $2\frac{4}{5}$ g $2\frac{5}{6}$ h $2\frac{4}{7}$ i $2\frac{4}{5}$ j $1\frac{3}{4}$ k $1\frac{2}{3}$ l $1\frac{6}{7}$ m $3\frac{2}{3}$ n $3\frac{3}{4}$ o $3\frac{1}{5}$ p $3\frac{1}{6}$ q $3\frac{1}{4}$ r $3\frac{2}{5}$

### PAGE 48

**1** a $3\frac{1}{5}$ b $3\frac{1}{4}$ c $4\frac{1}{5}$ d $5\frac{1}{4}$ e $5\frac{2}{5}$ f $6\frac{1}{5}$ g $1\frac{2}{3}$ h $1\frac{3}{5}$ i $1\frac{2}{3}$ j $1\frac{1}{7}$ **2** a $2\frac{1}{2}$ b $2\frac{2}{3}$ c $2\frac{1}{3}$ d $2\frac{1}{2}$ e $2\frac{1}{2}$ f $2\frac{1}{3}$

### PAGE 49

**1** a $\frac{5}{6}$ b $\frac{7}{8}$ c $\frac{3}{10}$ d $\frac{7}{10}$ e $\frac{9}{10}$ f $\frac{5}{8}$ g $\frac{3}{8}$ h $\frac{9}{10}$ i $\frac{9}{10}$ j $\frac{7}{8}$ k $\frac{9}{10}$ l $\frac{7}{10}$ m $\frac{3}{4}$ n $\frac{9}{10}$ o $\frac{7}{8}$ p $\frac{11}{12}$ q $\frac{13}{14}$ r $\frac{13}{16}$

### PAGE 50

**1** a $2\frac{1}{4}$ b $2\frac{1}{6}$ c $2\frac{1}{8}$ d $2\frac{1}{8}$ e $2\frac{1}{10}$ f $2\frac{3}{8}$ g $2\frac{5}{8}$ h $2\frac{1}{10}$ i $2\frac{7}{10}$ j $2\frac{5}{6}$ k $2\frac{3}{10}$ l $2\frac{7}{10}$ m $2\frac{1}{12}$ n $2\frac{5}{12}$

### PAGE 51

**1** a $1\frac{1}{2}$ b $2\frac{1}{2}$ c $1\frac{1}{2}$ d $2\frac{1}{2}$ e $2\frac{1}{2}$ f $4\frac{1}{2}$ **2** a $1\frac{1}{2}$ b $1\frac{1}{2}$ c $2\frac{1}{2}$ d $2\frac{1}{2}$ e $3\frac{1}{2}$ f $2\frac{1}{2}$

### PAGE 52

❶ a $\frac{11}{12}$ b $\frac{7}{10}$ c $\frac{7}{12}$ d $\frac{5}{6}$ e $\frac{13}{15}$ f $\frac{19}{20}$ g $\frac{8}{15}$ h $\frac{11}{15}$ i $\frac{14}{15}$ j $\frac{17}{20}$ k $\frac{13}{20}$ l $\frac{7}{10}$ ❷ a $1\frac{1}{10}$ b $1\frac{3}{10}$ c $1\frac{1}{20}$ d $1\frac{3}{20}$

### PAGE 53

❶ a $1\frac{1}{15}$ b $1\frac{5}{12}$ c $1\frac{1}{12}$ d $1\frac{2}{15}$ e $1\frac{3}{35}$ f $1\frac{8}{35}$ ❷ a $\frac{1}{10}$ b $\frac{3}{10}$ c $\frac{1}{12}$ d $\frac{5}{12}$ e $\frac{3}{20}$ f $\frac{7}{20}$ g $\frac{11}{20}$ h $\frac{1}{20}$ i $\frac{3}{10}$ j $\frac{2}{15}$ k $\frac{1}{20}$ l $\frac{7}{15}$

### PAGE 54

❶ a $\frac{2}{15}$ b $\frac{2}{21}$ c $\frac{15}{28}$ d $\frac{12}{35}$ e $\frac{4}{15}$ f $\frac{8}{15}$ g $\frac{4}{35}$ h $\frac{5}{16}$ i $\frac{15}{28}$ j $\frac{8}{15}$ ❷ a $\frac{1}{3}$ b $\frac{2}{7}$ c $\frac{1}{3}$ d $\frac{4}{5}$ e $\frac{5}{8}$ f $\frac{3}{7}$ g $\frac{1}{4}$ h $\frac{1}{3}$ i $\frac{1}{2}$ j $\frac{2}{3}$ ❸ a $1\frac{1}{4}$ b $1\frac{4}{5}$ c $1\frac{1}{7}$ d $1\frac{2}{3}$ e $2\frac{2}{3}$ f $2\frac{2}{5}$ g $2\frac{2}{5}$ h $2\frac{4}{5}$ i $2\frac{2}{7}$ j $2\frac{4}{7}$

### PAGE 55

❶ a $\frac{5}{6}$ b $\frac{14}{15}$ c $\frac{11}{15}$ d $\frac{14}{15}$ e $\frac{7}{15}$ f $\frac{11}{15}$ g $\frac{22}{35}$ h $\frac{7}{12}$ ❷ a $1\frac{1}{15}$ b $1\frac{5}{9}$ c $1\frac{1}{9}$ d $2\frac{1}{4}$ e $2\frac{7}{10}$ f $3\frac{3}{10}$ ❸ a $3\frac{3}{10}$ b $3\frac{1}{8}$ c $5\frac{5}{6}$ d $5\frac{5}{8}$

### PAGE 56

❶ a $2\frac{1}{3}$ b $2\frac{3}{4}$ c $2\frac{2}{5}$ d $2\frac{1}{7}$ e $3\frac{2}{3}$ f $3\frac{2}{5}$ g $4\frac{1}{3}$ h $4\frac{1}{4}$ i 4 j 6 k 5 l 6 m 5 n 7 o 8 p 9 q 11 r 12

### PAGE 57

❶ a $\frac{2}{15}$ b $\frac{15}{28}$ c $\frac{12}{35}$ d $\frac{4}{15}$ e $\frac{3}{40}$ f $\frac{14}{15}$ ❷ a $\frac{1}{3}$ b $\frac{2}{7}$ c $\frac{2}{11}$ d $\frac{4}{5}$ e $\frac{5}{8}$ f $\frac{3}{10}$ ❸ a 6 b 10 c 4 d 10 e 10 f 20 g 6 h 12 i 6 j 12

### PAGE 58

❶ a $1\frac{1}{5}$ b $1\frac{1}{5}$ c $2\frac{2}{5}$ d $2\frac{2}{3}$ e $2\frac{2}{5}$ f $3\frac{3}{4}$ ❷ a $\frac{9}{10}$ b $\frac{9}{14}$ c $\frac{14}{15}$ d $\frac{22}{25}$ e $\frac{21}{22}$ f $\frac{20}{21}$ g $\frac{16}{21}$ h $\frac{22}{27}$ ❸ a $1\frac{5}{9}$ b $1\frac{1}{15}$ c $1\frac{1}{21}$ d $2\frac{2}{9}$

### PAGE 59

❶ a $3\frac{1}{3}$ b $4\frac{2}{3}$ ❷ a $1\frac{1}{8}$ b $1\frac{1}{21}$ c $2\frac{1}{10}$ d $2\frac{7}{10}$ e $1\frac{1}{5}$ f $1\frac{3}{7}$ g $1\frac{2}{7}$ h $1\frac{3}{7}$ i $1\frac{1}{5}$ j $1\frac{1}{2}$

### PAGE 60

❶ a 6 b 2 c 6 d 2 e 5 f 5 g 6 h 6 i 3 j 10 k 9 l 3 ❷ a 8 b 6 c 15 d 6 e 24 f 18 g 12 h 10 i 15 j 15 ❸ 4 ❹ 8 ❺ 4 ❻ 8 ❼ 7

### PAGE 61

❶ 3 ❷ 16 ❸ 27 ❹ \$24 ❺ 24 ❻ \$24 ❼ a 12 b 6 ❽ a 15 b 5 ❾ a 18 b 12 ❿ a 8 b 12 ⓫ a 10 b 20

### PAGE 62

❶ a $\frac{2}{5}$ b 16 c 24 ❷ a $\frac{4}{7}$ b 20 c 15 ❸ a $\frac{3}{5}$ b 27 c 18 ❹ a $\frac{3}{7}$ b 9 c 12 ❺ a $\frac{5}{9}$ b 15 c 12

### PAGE 63

❶ a $\frac{3}{4}$ b $\frac{3}{5}$ c $\frac{5}{12}$ d $\frac{3}{4}$ e $\frac{1}{2}$ f $\frac{3}{8}$ g $\frac{1}{5}$ h $\frac{1}{4}$ i $\frac{6}{19}$ j $\frac{9}{50}$ k $\frac{1}{12}$ l $\frac{3}{4}$ m $\frac{2}{5}$ n $\frac{1}{2}$ o $\frac{1}{10}$ p $\frac{2}{5}$ ❷ a $\frac{1}{2}$ b $\frac{9}{20}$ c $\frac{1}{5}$ d $\frac{13}{20}$ e $\frac{17}{20}$ f $\frac{1}{10}$ g $\frac{2}{5}$ h $\frac{1}{5}$ i $\frac{11}{40}$ j $\frac{3}{20}$ k $\frac{7}{20}$ l $\frac{3}{40}$ m $\frac{3}{10}$ n $\frac{1}{6}$ o $\frac{17}{100}$ p $\frac{13}{200}$ ❸ a $\frac{1}{4}$ b $\frac{3}{4}$ c $\frac{1}{2}$ d $\frac{1}{3}$ e $\frac{2}{3}$ f $\frac{5}{6}$ g $\frac{1}{2}$ h $\frac{1}{3}$

### PAGE 64

❶ a $\frac{3}{20}$ b $\frac{2}{15}$ c $\frac{1}{4}$ d $\frac{3}{35}$ e $\frac{1}{6}$ f $\frac{4}{35}$ g $\frac{1}{6}$ h $\frac{3}{10}$ ❷ a $\frac{3}{100}$ b $\frac{7}{40}$ c $\frac{1}{6}$ d $\frac{18}{125}$ e $\frac{3}{40}$ f $\frac{47}{300}$ g $\frac{13}{180}$ h $\frac{17}{200}$ ❸ a $\frac{1}{30}$ b $\frac{1}{3}$ c $\frac{1}{3}$ d $\frac{1}{20}$ e $\frac{1}{20}$ f $\frac{1}{10}$ g $\frac{1}{12}$ h $\frac{1}{42}$ ❹ a $\frac{1}{2}$ b $\frac{13}{60}$ c $\frac{19}{100}$ d $\frac{3}{25}$ e $\frac{3}{25}$ f $\frac{13}{1000}$ g $\frac{1}{3}$ h $\frac{13}{50}$ ❺ a $\frac{1}{3}$ b $\frac{5}{12}$ c $\frac{1}{8}$ d $\frac{3}{8}$ e $\frac{1}{2}$ f $\frac{5}{6}$

### PAGE 65

❶ a $\frac{4}{10}$ b $\frac{16}{12}$ c $\frac{3}{6}$ d $\frac{5}{7}$ ❷ a $\frac{1}{2}$ b $\frac{2}{3}$ c $\frac{2}{5}$ d $\frac{2}{3}$ ❸ a $\frac{9}{16}$ b $\frac{5}{9}$ ❹ a $\frac{13}{21}$ b $\frac{17}{20}$ c $\frac{9}{35}$ d $\frac{3}{14}$ ❺ a $1\frac{2}{5}$ b $2\frac{3}{4}$ c $3\frac{1}{8}$ d $8\frac{3}{10}$ ❻ a $\frac{7}{5}$ b $\frac{11}{4}$ c $\frac{25}{7}$ d $\frac{29}{6}$ ❼ a 2 b 3 ❽ a $2\frac{1}{6}$ b $3\frac{1}{10}$ ❾ a $\frac{6}{35}$ b $\frac{10}{49}$ c $\frac{8}{35}$ d $1\frac{1}{3}$ e $1\frac{2}{5}$ f $2\frac{1}{3}$ ❿ a 9 b 6 c 20 d 12 ⓫ 9 ⓬ a $\frac{2}{7}$ b 4 c 10 ⓭ a $\frac{7}{20}$ b $\frac{3}{10}$ c $\frac{1}{2}$ d $\frac{3}{4}$ e $\frac{5}{7}$ f $\frac{5}{8}$

## CHAPTER 4

### PAGE 66

❶ a 346.8 b 234.9 c 535.1 d 162.3 e 642.9 f 836.3 g 0.39 h 0.62 i 0.15 j 0.27 k 0.83 l 0.35 m 4 n 2 o 5 p 8 q 6 r 9 ❷ a 156 b 342 c 434 d 987 e 642 f 799 ❸ a 1560 b 3420 c 4340 d 9870 e 6420 f 7990 g 5 h 7 i 3 j 9 k 6.3 l 7.8 m 1.5 n 8.4 ❹ a 34 b 52 c 96 d 14 e 2035 f 6072 g 5135 h 8690 i 20 j 50 k 10 l 60 m 4600 n 5300 o 3900 p 1200

### PAGE 67

❶ a 34.68 b 23.49 c 42.65 d 75.64 e 3.15 f 4.27 g 1.03 h 2.04 i 0.3 j 0.5 k 0.2 l 0.7 m 0.8 n 0.1 o 7.6 p 8.9 q 2.3 r 6.5 ❷ a 1.56 b 2.34 c 4.26 d 7.54 e 31.5 f 52.2 g 10.35 h 20.49 i 0.03 j 0.05 k 0.02 l 0.07 m 0.08 n 0.01 o 0.23 p 0.75 q 0.46 r 0.51 s 0.6 t 0.1 ❸ a 0.0239 b 0.0167 c 0.0465 d 0.0764 e 0.0356 f 0.0207 g 0.2034 h 0.5637 i 0.1928 j 0.7452 k 0.1003 l 0.2034 m 0.0049 n 0.0058 n 0.0023 p 0.0075 q 0.0011 r 0.0066

### PAGE 68

❶ a 20.5 b 12.8 c 14.6 d 16.8 e 10.8 f 18.6 g 14.4 h 27.6 i 10.6 j 24.8 k 14.8 l 12.9 m 0.6 n 0.8 o 0.6 p 0.9 q 3.5 r 4.2 ❷ a 38.4 b 49.2 c 39.6 d 75.9 e 109.2 f 153.3 g 130.2 h 137.6 i 4.8 j 6.9 k 8.4 l 8.2 m 6.8 n 9.9

### PAGE 69

❶ a 11.34 b 19.22 c 15.33 d 13.02 e 2.04 f 2.48 g 18.06 h 32.76 i 38.48 j 49.68 ❷ a 6.24 b 5.55 c 18.2 d 19.24 e 40.28 f 28.35 g 41.76 h 80.84 i 20.25 j 34.81

**PAGE 70** ❶ **a** 1.3 **b** 2.2 **c** 2.1 **d** 3.2 **e** 2.1 **f** 1.1 **g** 2.3 **h** 4.1 **i** 2.3 **j** 3.3 ❷ **a** 0.4 **b** 0.6 **c** 0.8 **d** 0.5 **e** 0.2 **f** 0.5 **g** 0.5 **h** 0.7 **i** 0.2 **j** 0.3 **k** 0.5 **l** 0.2 ❸ **a** 3 **b** 2 **c** 3 **d** 2 **e** 5 **f** 6 **g** 5 **h** 9 **i** 4 **j** 5 **k** 6 **l** 5 **m** 5 **n** 2 **o** 20 **p** 50

**PAGE 71** ❶ **a** 0.2 **b** 0.7 **c** 0.3 **d** 0.5 ❷ **a** 0.4 **b** 0.9 **c** 0.5 **d** 0.7 **e** 0.5 **f** 0.4 **g** 0.2 **h** 0.3 ❸ **a** 1.4 **b** 5.8 **c** 8.8 **d** 3.7 **e** 7.5 **f** 2.4 **g** 6.2 **h** 9.3 **i** 13.6 **j** 19.5 **k** 23.2 **l** 44.9 **m** 52.5 **n** 67.4 **o** 70.2 **p** 83.1 **q** 133.4 **r** 154.7 **s** 245.7 **t** 336.9 **u** 560.2 **v** 749.6 ❹ **a** 0.15 **b** 0.47 **c** 0.95 **d** 0.71 ❺ **a** 0.25 **b** 0.18 **c** 0.44 **d** 0.81 **e** 0.76 **f** 0.46 **g** 0.52 **h** 0.71 **i** 0.93 **j** 0.57 ❻ **a** 6.83 **b** 2.35 **c** 8.54 **d** 5.73 **e** 9.46 **f** 1.36 **g** 6.21 **h** 7.32 **i** 10.58 **j** 34.88 **k** 23.22 **l** 62.53 **m** 73.49 **n** 82.17 **o** 59.65 **p** 94.27 **q** 154.36 **r** 365.12 **s** 745.61 **t** 649.34 **u** 926.84 **v** 852.31

**PAGE 72** ❶ **a** 1.0 **b** 0.9 **c** 2.9 **d** 5.9 ❷ **a** 8.0 **b** 8.9 **c** 3.9 **d** 6.0 **e** 7.0 **f** 3.0 **g** 1.9 **h** 5.9 **i** 5.0 **j** 7.9 ❸ **a** 23.9 **b** 53.0 ❹ **a** 88.0 **b** 68.9 **c** 14.0 **d** 25.9 **e** 77.0 **f** 53.0 **g** 85.9 **h** 62.9 **i** 104.0 **j** 157.0 **k** 310.9 **l** 525.9 ❺ **a** 5 **b** 8 **c** 16 **d** 23 **e** 12 **f** 53 ❻ **a** recurring **b** terminating **c** terminating **d** recurring **e** terminating **f** recurring **g** recurring **h** terminating ❼ **a** $0.\dot{3}$ **b** $0.\dot{1}$ **c** $0.\dot{6}$ **d** $0.\dot{4}$ **e** $0.1\dot{6}$ **f** $0.8\dot{3}$ **g** $0.\dot{1}\dot{8}$ **h** $0.\dot{2}4\dot{3}$ ❽ **a** terminating **b** recurring **c** terminating **d** recurring **e** terminating **f** recurring **g** recurring

**PAGE 73** ❶ **a** $\frac{1}{2}$, 0.5 **b** $\frac{1}{4}$, 0.25 **c** $\frac{7}{10}$, 0.7 **d** $\frac{1}{10}$, 0.1 **e** $\frac{1}{20}$, 0.05 **f** $\frac{2}{25}$, 0.08 **g** $\frac{4}{25}$, 0.16 **h** $\frac{9}{20}$, 0.45 **i** $\frac{6}{5}$, 1.2 **j** 1 ❷ **a** $\frac{1}{8}$, 0.125 **b** $\frac{1}{16}$, 0.0625 **c** $\frac{7}{8}$, 0.875 **d** $\frac{1}{3}$, $0.\dot{3}$ **e** $\frac{11}{40}$, 0.275 **f** $\frac{1}{250}$, 0.004 ❸ **a** 50% **b** 25% **c** 75% **d** 40%

**PAGE 74** ❶ **a** 44.4% **b** 37.5% **c** 250% **d** 11% **e** 12% **f** 225% ❷ **a** 25% **b** 60% **c** 17% **d** 32% **e** 140% **f** 3% **g** 4.05% **h** 365% **i** 500% **j** 100% **k** 208% **l** 703% ❸ **a** $\frac{1}{2}$, 0.5, 50%, $\frac{2}{5}$, 0.4, 40%, $\frac{3}{4}$, 0.75, 75%, $\frac{7}{5}$, 1.4, 140% **b** $\frac{3}{20}$, 0.15, 15%, $\frac{1}{5}$, 0.2, 20%, $\frac{1}{20}$, 0.05, 5%, $\frac{43}{50}$, 0.86, 86%, **c** $\frac{3}{25}$, 0.12, 12%, $\frac{3}{4}$, 0.75, 75%, $\frac{5}{4}$, 1.25, 125%, $\frac{1}{8}$, 0.125, 12.5%, $\frac{49}{100}$, 0.49, 49%, $\frac{5}{9}$, $0.\dot{5}$, 55.6%,

**PAGE 75** ❶ **a** $\frac{14}{25}$ **b** 0.79 **c** 0.72 **d** $\frac{9}{15}$ **e** 0.4 **f** $\frac{9}{50}$ ❷ 90% ❸ $\frac{19}{50}$ ❹ 12% ❺ 79%, $\frac{7}{9}$, 0.76 ❻ $\frac{3}{8}$, 0.23, 20% ❼ 0.405, 42%, $\frac{5}{11}$ ❽ 1.5%, $\frac{2}{100}$, 0.021

**PAGE 76** ❶ **a** 40% **b** 60% ❷ **a** 37.5% **b** 62.5% ❸ **a** 66.7% **b** 33.3% ❹ **a** 76% **b** 82% **c** 15% **d** 34% **e** 40% **f** 60% **g** 66.7% **h** 60% ❺ **a** 50% **b** 35% **c** 40% **d** 60% **e** 45% **f** 2% **g** 10% **h** 5%

**PAGE 77** ❶ **a** 41.7% **b** 16.7% **c** 20% **d** 91.7% **e** 33.3% **f** 12.5% **g** 25% **h** 8.3% ❷ **a** 25% **b** 20% **c** 23.5% **d** 6% **e** 12% **f** 12% **g** 12.5% **h** 2.5% ❸ **a** 25% **b** 50% **c** 62.5% **d** 87.5% **e** 50% **f** 75% **g** 62.5% **h** 87.5%

**PAGE 78** ❶ **a** 30% **b** 50% **c** 20% ❷ **a** 33.3% **b** 50% **c** 16.7% ❸ **a** 16 **b** 25% **c** 12.5% **d** 62.5% ❹ **a** 19 **b** 42.1% **c** 26.3% **d** 57.9% ❺ **a** 36% **b** 24% **c** 76% **d** 76%

**PAGE 79** ❶ **a** 5 **b** 7 **c** \$10 **d** 9 mL **e** 8 g **f** 12 h ❷ **a** 6 **b** 8 **c** \$10 **d** 9 L **e** 5 cm **f** 7 min ❸ **a** 5 **b** 2 **c** 4 kg **d** 12 cm **e** 36° **f** \$75 ❹ **a** 3 **b** 9 **c** 6 min **d** 12 min **e** \$20 **f** 110 mL

**PAGE 80** ❶ **a** \$4.80 **b** \$8.75 **c** \$86.40 **d** \$417.60 **e** 12.4 mL **f** 17.5 mL **g** 171.5 g **h** 174.3 g **i** 87 kg **j** 47.5 kg **k** 24 MB **l** 102 GB ❷ 192 ❸ 472 ❹ 147 ❺ **a** 260 **b** 140 ❻ **a** 15 **b** 285

**PAGE 81** ❶ **a** 246.7 **b** 132.5 **c** 13.5 **d** 49.6 **e** 0.4 **f** 0.7 ❷ **a** 583 **b** 649 **c** 8.5 **d** 9.4 **e** 8390 **f** 7260 ❸ **a** 1.58 **b** 3.24 **c** 0.64 **d** 0.72 **e** 1.3 **f** 2.9 ❹ **a** 0.24 **b** 0.52 **c** 0.203 **d** 0.196 **e** 0.042 **f** 0.089 ❺ **a** 12.6 **b** 16.6 **c** 1.6 **d** 2.1 **e** 38.44 **f** 2.84 ❻ **a** 3.1 **b** 4.3 **c** 0.5 **d** 0.5 **e** 3 **f** 7 ❼ **a** 0.74 **b** 0.52 **c** 7.27 **d** 49.61 **e** 37.23 **f** 8.40 ❽ **a** 5.4 **b** 4.2 **c** 16.8 **d** 62.4 **e** 0.6 **f** 1.0 ❾ **a** terminating **b** recurring **c** recurring ❿ $\frac{7}{8}$, 90%, 0.91

⓫

| Fraction | Decimal | Percentage |
|---|---|---|
| $\frac{2}{5}$ | 0.4 | 40% |
| $1\frac{3}{4}$ | 1.75 | 175% |
| $\frac{3}{20}$ | 0.15 | 15% |
| $\frac{1}{4}$ | 0.25 | 25% |

⓬ **a** 80% **b** 65% **c** 40% **d** 58.3% ⓭ **a** 43.8% **b** 56.3% ⓮ **a** 6 cm **b** \$50 **c** 8 kg **d** \$7.50 ⓯ 28 marks ⓰ **a** \$21 **b** \$119

**CHAPTER 5**

**PAGE 82** ❶ **a** $4x$ **b** $3y$ **c** $6a$ **d** $2m$ **e** $4b$ **f** $4c$ **g** $5x^2$ **h** $3y^2$ **i** $2a^2$ **j** $4h^2$ **k** $4ab$ or $4ba$ **l** $3xy$ or $3yx$ **m** $5pq$ or $5qp$ **n** $4vw$ or $4wv$ ❷ **a** $3x + 2y$ **b** $2a + 3b$ **c** $3p + 2q$ **d** $3s + t$ **e** $3x + 2y$ **f** $4a + 2b$ **g** $3e + 2f$ **h** $3n + 3m$ **i** $3x + 3y$ **j** $4v + 3w$ **k** $3r + 4s$ ❸ **a** $2x^2 + 3y^2$ **b** $2a^2 + 3b^2$ **c** $3p^2 + 2q^2$ **d** $2s^2 + 2t^2$ **e** $3x^2 + 2y^2$ **f** $4a^2 + 2b^2$ **g** $3r^2 + 2x^2$ **h** $3y^2 + 3w$ **i** $3k^2 + 3m$ ❹ **a** $2xy + 2y$ **b** $3wd + 4d$ **c** $2ab + 3a$ **d** $3rm + 2m$ **e** $3wt + 3t$ **f** $4ab + 2b$ **g** $4pq + 2q$ **h** $3xy + 3y$ **i** $2t + 4kt$

**PAGE 83** ❶ **a** $7x$ **b** $10y$ **c** $11a$ **d** $11m$ **e** $9x^2$ **f** $8a^2$ **g** $13e^2$ **h** $13w^2$ **i** $8xy$ **j** $8ab$ **k** $12yd$ **l** $14cf$ **m** $7x + 2y$ **n** $7a + 7b$ **o** $8p + 6r$ **p** $9r + 5t$ **q** $9x + 4t$ **r** $6k + 3g$ **s** $8v + 4d$ **t** $8y + 8p$ ❷ **a** $4x + 3$ **b** $4y + 5$ **c** $5a + 4$ **d** $3m + 2$ **e** $8z^2 + 3$ **f** $12h^2 + 4$ **g** $13e^2 + 8$ **h** $11w^2 + 9$ **i** $6xy + 3$ **j** $3ab + 4$ **k** $5x + 4y$ **l** $5p + 3r$

**m** $r+2t$ **n** $6x+2t$ **o** $4k+2g$ **p** $8h+p$ **q** $11s+7q$ **r** $4y+2x$ **s** $9k+12r$ ❸ **a** $-2x-4y$ **b** $-5p-4r$ **c** $-r-2t$ **d** $-3x-2t$ **e** $-3k-3g$ **f** $-4h-p$ **g** $-2s-3q$ **h** $-4y-4x$ **i** $-7k-2r$ ❹ **a** $a$ **b** 7 ❺ **a** $x$ **b** 12 ❻ **a** $x$ and $y$ **b** 5 ❼ **a** $a$ and $b$ **b** −10

## PAGE 84

❶ **a** $7x+3$ **b** $11x+5$ **c** $8x+4$ **d** $10x+8$ **e** $6x+9$ **f** $9x+3$ ❷ **a** $11m+3n$ **b** $7m+7n$ **c** $18m+6n$ **d** $11m+n$ **e** $5m+6n$ **f** $6m+3n$ ❸ **a** $7xy+8$ **b** $14xy+9$ **c** $11xy+5$ **d** $11xy+3$ **e** $10xy+6$ **f** $9xy+4$ **g** $8xy+6$ **h** $8xy+2$ **i** $2xy+5$ ❹ **a** $5ab+6$ **b** $7ab+9$ **c** $8ab+7$ **d** $8ab+9$ **e** $6ab+8$ **f** $5ab+8$ ❺ **a** $11a+7b$ **b** $15a+9b$ **c** $10a+7b$ **d** $7a+6b$ **e** $8a+5b$ **f** $6a+6b$ ❻ **a** $6w+8vw$ **b** $9w+11vw$ **c** $10w+7vw$ **d** $12w+9vw$ **e** $11w+4vw$ **f** $10w+7vw$ ❼ **a** $7x^2+3x$ **b** $10x^2+3x$ **c** $7x^2+6x$ **d** $4x^2+2x$ **e** $6x^2+3x$ **f** $2x^2+x$ ❽ **a** $8a^2+6a$ **b** $9a^2+10a$ **c** $11a^2+6a$ **d** $6a^2+6a$ **e** $9a^2+4a$ **f** $6a^2+6a$

## PAGE 85

❶ **a** $5x+7y+1$ **b** $6x+4y+2$ **c** $9x+7y+5$ **d** $7x+7y+4$ **e** $8x+5y+7$ **f** $8x+7y+9$ ❷ **a** $2a+2b$ **b** $6a+6b$ **c** $4a+5b$ **d** $8a+2b$ **e** $5a+b$ **f** $a+2b$ ❸ **a** $-2x+2y$ **b** $-3x+5y$ **c** $-4x+7y$ **d** $-5x+5y$ **e** $-x+y$ **f** $-5x+2y$ ❹ **a** $3a-2b$ **b** $2a-2b$ **c** $4a-4b$ **d** $a-7b$ **e** $3a-4b$ **f** $2a-3b$ **g** $6a-3b$ **h** $4a-2b$ **i** $5a-7b$ **j** $3a-b$ ❺ **a** $-6h+3j$ **b** $-9h+3j$ **c** $-5h+3j$ **d** $-10h+6j$ **e** $-4h+4j$ **f** $-9h+2j$

## PAGE 86

❶ **a** $3ab-10a$ **b** $2ab-7a$ **c** $5ab-9a$ **d** $4ab-7a$ **e** $8ab-12a$ **f** $5ab-8a$ **g** $8ab-7a$ ❷ **a** $-5pq+2p$ **b** $-7pq+2p$ **c** $-3pq+5p$ **d** $-4pq+3p$ **e** $-6pq+4p$ **f** $-5pq+2p$ **g** $-3pq+p$ ❸ **a** $2p^2-4pq$ **b** $2p^2-2pq$ **c** $4p^2-3pq$ **d** $5p^2-2pq$ **e** $4p^2-pq$ **f** $5p^2-6pq$ **g** $8p^2-6pq$ ❹ **a** $4a^2+2a$ **b** $5a^2+4a$ **c** $a^2+3a$ **d** $6a^2+2a$ **e** $5a^2+4a$ **f** $8a^2+a$ **g** $6a^2+4a$ ❺ **a** $5x^2-9x+4$ **b** $2x^2-11x+9$ **c** $3x^2-7x+4$ **d** $4x^2-13x+3$ **e** $6x^2-8x+7$ **f** $x^2-5x+4$ **g** $3x^2-10x+8$ **h** $4x^2-13x+9$

## PAGE 87

❶ **a** $6a$ **b** $35y$ **c** $12x$ **d** $35w$ **e** $8e$ **f** $18f$ **g** $-24a$ **h** $-30w$ **i** $-56h$ **j** $-20b$ **k** $12a$ **l** $20b$ ❷ **a** $12ab$ **b** $28yw$ **c** $4xy$ **d** $45wk$ **e** $-6at$ **f** $-40wg$ **g** $-21ht$ **h** $-12tp$ **i** $12ab$ **j** $8xy$ ❸ **a** $10a^2$ **b** $21y^2$ **c** $12x^2$ **d** $7w^2$ **e** $12m^2$ **f** $24h^2$ **g** $-15h^2$ **h** $-18t^2$ **i** $25p^2$ **j** $28q^2$ ❹ **a** $18e^2m$ **b** $6f^2g$ **c** $20s^2t$ **d** $36g^2h$ **e** $28x^2y$ **f** $20w^2t$ **g** $-16a^2t$ **h** $-48w^2v$ **i** $-24p^2h$ **j** $-3a^2d$ **k** $-12wv^2$ **l** $24dh^2$

## PAGE 88

❶ **a** $40a^2b^2$ **b** $48y^2t^2$ **c** $90x^2w^2$ **d** $14x^2t^2$ **e** $36y^2m^2$ **f** $20u^2v^2$ **g** $-90n^2t^2$ **h** $-144w^2v^2$ **i** $-36p^2h^2$ **j** $-90p^2q^2$ **k** $24x^2y^2$ **l** $30a^2b^2$ **m** $-192z^2k^2$ **n** $-90ap^2$ **o** $-240v^2w$ ❷ **a** $8a^3$ **b** $16y^3$ **c** $30x^3$ **d** $7w^3$ **e** $16b^3$ **f** $6e^3m$ **g** $10f^3g$ **h** $14z^3t$ **i** $6g^3h$ **j** $-48a^3t$ **k** $-12w^3v$ **l** $16hp^3$ **m** $24ab^3$ **n** $-24y^3p^2$ **o** $-40k^2d^3$ **p** $-18x^2y^3$ ❸ **a** $16a^2b^2c$ **b** $24x^2y^2z$ **c** $15s^2r^2t$ **d** $14a^2c^2d$ **e** $-30k^2h^2j$ **f** $-36cd^2e^2$ **g** $-18mn^2p^2$ **h** $21dg^2h^2$ **i** $24c^2de^2$ **j** $120k^2gw^2$ **k** $-48wp^2z^2$ **l** $-48rs^2t^2$ **m** $-270t^2m^2v$ **n** $168rp^2t^2$ **o** $-140z^2l^2k$ **p** $400p^2q^2r$

## PAGE 89

❶ **a** $3a$ **b** $5y$ **c** $4x$ **d** $7w$ **e** $5e$ **f** $4z$ **g** $-8a$ **h** $-5w$ **i** $8w$ **j** $12a$ ❷ **a** $5a$ **b** $3j$ **c** $5y$ **d** $4t$ **e** $4m$ **f** $2a$ **g** $-9h$ **h** $-4t$ ❸ **a** $\frac{4}{b}$ **b** $\frac{2}{n}$ **c** $\frac{8}{c}$ **d** $\frac{3}{t}$ **e** $\frac{4}{m}$ **f** $\frac{4}{a}$ **g** $-\frac{2}{a}$ **h** $-\frac{3}{c}$ **i** $-\frac{9}{h}$ **j** $-\frac{5}{b}$ ❹ **a** $\frac{a}{5}$ **b** $\frac{j}{6}$ **c** $\frac{y}{5}$ **d** $\frac{t}{6}$ **e** $\frac{m}{4}$ **f** $\frac{a}{2}$ **g** $\frac{2x}{3}$ **h** $\frac{3y}{8}$ **i** $\frac{7b}{4}$ **j** $\frac{4w}{3}$ ❺ **a** 4 **b** 4 **c** 3 **d** 3 **e** $\frac{1}{7}$ **f** $\frac{1}{8}$ **g** $\frac{1}{4}$ **h** $\frac{1}{5}$ **i** $\frac{2}{3}$ **j** $\frac{4}{5}$ **k** $-\frac{5}{3}$ **l** $-\frac{5}{3}$ ❻ **a** $5a$ **b** $3h$ **c** $5b$ **d** $5y$ **e** $4p$ **f** $5r$ **g** $-9d$ **h** $-8h$

## PAGE 90

❶ **a** $\frac{q}{7}$ **b** $\frac{a}{5}$ **c** $\frac{h}{3}$ **d** $\frac{g}{2}$ **e** $\frac{m}{3}$ **f** $\frac{p}{6}$ **g** $\frac{r}{4}$ **h** $\frac{d}{5}$ **i** $\frac{3w}{4}$ **j** $\frac{7l}{8}$ **k** $-\frac{6s}{5}$ **l** $-\frac{4x}{3}$ ❷ **a** $\frac{1}{4c}$ **b** $\frac{1}{4z}$ **c** $\frac{1}{5h}$ **d** $\frac{1}{4r}$ **e** $\frac{1}{7p}$ **f** $\frac{1}{2v}$ **g** $\frac{1}{10w}$ **h** $\frac{1}{3w}$ **i** $-\frac{1}{7h}$ **j** $-\frac{1}{6h}$ ❸ **a** $\frac{4a}{3}$ **b** $\frac{7b}{8}$ **c** $\frac{5x}{9}$ **d** $\frac{7a}{5}$ **e** $\frac{3e}{2}$ **f** $\frac{4a}{9}$ **g** $\frac{b}{5}$ **h** $\frac{x}{6}$ **i** $\frac{w}{4}$ **j** $\frac{b}{7}$ **k** $-\frac{p}{3}$ **l** $-\frac{k}{5}$ ❹ **a** $\frac{3x}{4y}$ **b** $\frac{2d}{3e}$ **c** $\frac{3v}{4w}$ **d** $\frac{2z}{5y}$ **e** $\frac{5p}{3q}$ **f** $\frac{3l}{2p}$ **g** $\frac{5s}{r}$ **h** $\frac{2g}{h}$ **i** $-\frac{8e}{j}$ **j** $-\frac{8y}{x}$ ❺ **a** $\frac{2h}{5n}$ **b** $\frac{2b}{3a}$ **c** $\frac{5d}{6c}$ **d** $\frac{6y}{7x}$ **e** $\frac{6y}{7z}$ **f** $\frac{2b}{3a}$ **g** $\frac{3d}{c}$ **h** $\frac{12p}{a}$ **i** $-\frac{5b}{s}$ **j** $-\frac{9t}{k}$ ❻ **a** $\frac{2xz}{3y}$ **b** $\frac{3ac}{4b}$ **c** $\frac{2mp}{3n}$ **d** $\frac{3ce}{5d}$ **e** $\frac{bd}{5c}$ **f** $\frac{3qr}{p}$

## PAGE 91

❶ **a** $x+5$ **b** $y+4$ **c** $b+3$ **d** $h+7$ **e** $x+y$ **f** $x+3$ **g** $c+5$ **h** $w+2$ **i** $2m+7$ **j** $3p+q$ **k** $3h+9$ **l** $7m+6$ **m** $4g+3$ **n** $6n+2v$ **o** $2a+5r$ **p** $3+x$ **q** $4y+2$ **r** $7g+6$ **s** $9k+5$ ❷ **a** $x-5$ **b** $y-4$ **c** $b-3$ **d** $h-7$ **e** $x-y$ **f** $3-x$ **g** $5-c$ **h** $12-w$ **i** $7b-m$ **j** $5p-q$ **k** $3h-9$ **l** $4m-7$ **m** $6g-4$ **n** $2n-3m$ **o** $8a-4r$ **p** $x-3$ **q** $y-2$ **r** $2g-6$ **s** $10x-7$ ❸ **a** $9h$ **b** $7m$ **c** $3s$ **d** $4d$ **e** $-6a$ **f** $-2t$

## PAGE 92

❶ **a** $-5r$ **b** $-e$ **c** $xy$ **d** $-mn$ **e** $2x$ **f** $2n$ **g** $2p$ **h** $3w$ **i** $a^2$ **j** $x^2$ **k** $n^2$ **l** $r^2$ **m** $m^2$ **n** $\frac{k}{2}$ **o** $\frac{a}{2}$ **p** $\frac{x}{2}$ **q** $\frac{y}{4}$ **r** $\frac{w}{3}$ ❷ **a** $\frac{h}{3}$ **b** $\frac{a}{5}$ **c** $\frac{w}{4y}$ **d** $\frac{t}{6e}$ **e** $\frac{h}{9}$ **f** $\frac{m}{7}$ **g** $\frac{c}{5d}$ **h** $\frac{n}{2p}$ **i** $-\frac{5}{k}$ **j** $-\frac{p}{q}$ ❸ **a** $2x+3y$ **b** $2m+5n$ **c** $2p+4r$ **d** $3y-6z$ **e** $3c-7e$ **f** $\frac{x}{3}-1$ **g** $\frac{y}{5}-3$ **h** $\frac{a}{2}+3b$ **i** $\frac{t}{2}+6s$ **j** $\frac{v}{3}+9w$ **k** $xy+3$ **l** $ab+4$ **m** $2p-5q$

## PAGE 93

❶ **a** $x+y$ **b** $m+n$ **c** $r+c$ **d** $20-y$ **e** $10-x$ **f** $40-w$ **g** $\frac{50}{w}$ **h** $\frac{25}{c}$ **i** $\frac{30}{b}$ **j** $\$10p$ **k** $\$15x$ **l** $\$12e$ **m** $\$9.50x$ **n** $\$2.99y$ **o** $\$5.40x+\$3.10y$ **p** $\$9.30m+\$6n$ **q** $\$60-5c$ **r** $\$100-4j$

## PAGE 94

❶ **a** 4 **b** 2 **c** 4 **d** 5 **e** 6 **f** 5 **g** 8 **h** 9 **i** 8 **j** 12 **k** −3 **l** −1 **m** −5 **n** −4 **o** −1 **p** −3 **q** −6 **r** −6 **s** −5 **t** −2 ❷ **a** −7 **b** −8 **c** −8 **d** −15 **e** −12 **f** −12 **g** −9 **h** −12 **i** −15 **j** −12 **k** −15 **l** −23 **m** −25 **n** −26

## PAGE 95

❶ **a** 7 **b** 6 **c** 8 **d** 9 **e** 13 **f** 14 **g** 13 **h** 15 **i** 13 **j** 25 **k** 3 **l** 1 **m** 2 **n** 5 **o** 1 **p** 3 ❷ **a** 2 **b** 2 **c** 4 **d** 4 **e** 4 **f** 7 **g** 3 **h** 2 **i** $-\frac{1}{2}$ **j** $-\frac{1}{2}$ **k** $-\frac{1}{3}$ **l** $-\frac{1}{5}$ **m** $-\frac{1}{5}$ **n** $-\frac{1}{9}$

**PAGE 96** **1** a 12 b 10 c 14 d 18 e 20 f 36 g –8 h –15 i –12 j –16 k –24 l –35 m –36 n –48 o –50 p –36 **2** a 3 b 3 c 4 d 6 e 4 f 8 g 1 h 5 i 8 j 8 k –5 l –2 m –4 n –6 o –12 p –13

**PAGE 97** **1** a 4 b 5 c 4 d 6 e 3 f 7 g 8 h 7 i 10 j 9 **2** a 2 b 3 c 4 d 1 e 4 f 5 g 3 h 6 i 2 j 7

**PAGE 98** **1** a –2 b –3 c –1 d –4 e –2 f –5 g –3 h –1 i –3 j –4 **2** a –3 b –2 c –1 d –4 e –2 f –5 g –2 h –3 i –3 j –4

**PAGE 99** **1** a 8 b 6 c 12 d 2 e 6 f 14 g 12 h 8 i 22 j 16 **2** a 15 b 24 c 21 d 27 e 30 f 21 g 33 h 33 i 44 j 40

**PAGE 100** **1** a $\frac{10}{3}$ b $\frac{8}{3}$ c $\frac{28}{3}$ d $\frac{40}{3}$ e 3 f 10 g 8 h 24 i 21 j $\frac{20}{3}$ **2** a $\frac{15}{2}$ b $\frac{35}{2}$ c –6 d –21 e $-\frac{35}{3}$ f –16 g –10 h $-\frac{27}{4}$ i $-\frac{24}{5}$ j –14

**PAGE 101** **1** a 11 b 17 c 18 d 11 e 25 f 19 g 23 h 26 i 20 j 44 k –8 l –22 m –10 n –11 o 41 p 19 q 59 r 24 s –22 t –45

**PAGE 102** **1** a $7a$ b $6t$ c $5x^2$ d $9a^2$ e $8xy$ f $7ab$ **2** a $10x + 3$ b $3x + 5$ c $7a + 3d$ d $8a + 4d$ e $-b + 3y$ f $-6b + 5y$ g $4x - y$ h $5x - 3y$ i $-6ab + 2a$ j $-10ab + 3a$ **3** a $h$ b –8 **4** a $10tr$ b $30wv$ c $-12hx$ d $-10pq$ e $35a^2$ f $-32y^2$ **5** a $5m$ b $-4n$ c $5y$ d $-8d$ e $6x$ f $-5a$ **6** a $2 + x$ b $y - 4$ c $6r$ d $2k$ e $\frac{d}{3}$ f $\frac{m}{2}$ g $5x - 10$ h $7s + 6$ **7** a 5 b 7 c 8 d 5 e 5 f 8 g 28 h –24 i 5 j 6 k 4 l 7 m 3 n 5 o 12 p 6 q 27 r 30 s 4 t –10 u 15 v 26 w –11 x –17

**CHAPTER 6**

**PAGE 103** **1** a 12 cm b 4 cm c 8 cm d 6 cm **2** a 6.5 cm b 2.8 cm c 7.9 cm d 11.2 cm e 9.4 cm f 0.7 cm **3** 1 a 120 mm b 40 mm c 80 mm d 60 mm 2 a 65 mm b 28 mm c 79 mm d 112 mm e 94 mm f 7 mm **4** Parent/teacher to check

**PAGE 104** **1** a 12 cm b 10 cm c 10 cm d 6 cm e 8.4 cm f 8 cm g 7.8 cm h 9 cm **2** a 11 cm b 12 cm c 16 cm **3** a Parent/teacher to check b 10 cm

**PAGE 105** **1** a 2 cm b Parent/teacher to check **2** a 5 cm b Parent/teacher to check **3** a 4 cm b Parent/teacher to check **4** a 2.5 cm b Parent/teacher to check **5** a Parent/teacher to check b 14 cm **6** Answers can vary, e.g a rectangle with length 2 cm, width 3 cm.

**PAGE 106** **1** a 8 cm b 14 cm c 20 cm d 16 cm e 2.4 cm f 35.6 cm g 10 cm h 18 cm i 11 cm j 11.6 cm k 14.2 cm l 12 cm m 9.2 cm n 15.4 cm o 14.8 cm p 4.5 cm

**PAGE 107** **1** a 14 cm b 9.6 cm c 22.4 cm d 101 cm e 204 cm f 29.4 cm g 16 cm h 44 cm **2** a 7 cm b 6 cm c 8 cm d 9 cm e 11 cm f 12 cm

**PAGE 108** **1** a 7 cm b 30 cm **2** a 7 cm b 32 cm **3** a 36 cm b 34 cm c 68 cm d 82 cm e 42 cm f 60 cm

**PAGE 109** **1** a mm b mm c cm d cm e m f m g km h km i mm j mm k cm l m m m n km o cm p m **2** a 200 cm b 450 cm c 10 cm d 27 cm e 826 cm f 701 cm g 12.5 cm h 67.8 cm **3** a 50 mm b 60 mm c 8 mm d 2 mm e 76 mm f 101 mm

**PAGE 110** **1** a 2000 m b 5000 m c 3800 m d 4750 m e 6010 m f 900 m g 840 m h 150 m **2** a 3 cm b 5 cm c 6.5 cm d 9.7 cm e 10.5 cm f 12.5 cm **3** a 4 m b 5.2 m c 0.9 m d 0.85 m e 3.4 km f 2.5 km g 72 cm h 20.1 cm **4** a 500 mm b 400 mm c 350 mm d 2000 mm e 6800 mm

**PAGE 111** **1** a 7 m b 5.2 m c 0.6 m d 0.41 m e 0.505 m **2** a 5.2 m b 4.6 m c 3000 m d 7600 m e 3 m f 4.2 m g 13 m h 25.6 m **3** 0.28 cm **4** 15.8 cm **5** a 914 mm b 45 books **6** a 213 cm b 3 people **7** a 1.8 cm b 180 cm

**PAGE 112** **1** a 9 cm$^2$ b 25 cm$^2$ c 64 mm$^2$ d 100 mm$^2$ e 0.36 m$^2$ f 51.8 m$^2$ **2** a 15 cm$^2$ b 54 cm$^2$ c 30.4 m$^2$ d 7.6 m$^2$ e 40.6 m$^2$ f 98.4 mm$^2$

**PAGE 113** **1** 1.4 m$^2$ **2** 1250 m$^2$ **3** 1.4 m$^2$ **4** 0.8 m$^2$ **5** a 3 cm, 20 cm b 4 cm, 20 cm c 6 cm, 30 cm d 5 cm, 30 cm **6** a 5 m b 26 m

**PAGE 114** **1** a 5 m, 25 m$^2$ b 8 m, 64 m$^2$ c 7 m, 49 m$^2$ d 9 m, 81 m$^2$ e 12 m, 144 m$^2$ f 15 m, 225 m$^2$ **2** a 9 m b 81 m$^2$ **3** a 10 m b 100 m$^2$ **4** a 5 m b 20 m **5** a 7 m b 28 m

**PAGE 115** **1** a 12 cm$^2$ b 22.5 cm$^2$ c 24.5 mm$^2$ d 48 mm$^2$ e 14 m$^2$ f 10.7 m$^2$ g 38.5 m$^2$ h 39 m$^2$ i 31.2 cm$^2$ j 42 cm$^2$ k 44 m$^2$ l 10.7 m$^2$ **2** 75 cm$^2$ **3** 0.2 m$^2$

**PAGE 116** **1** 21 505 m$^2$ **2** 4.5 m$^2$ **3** a 17.6 m$^2$ b \$2147.20 **4** a 3 cm b 6 cm c 6 cm d 10 cm e 4 m f 10 m **5** 6 cm

**PAGE 117** **1** a 45 cm$^2$ b 21 cm$^2$ c 18 mm$^2$ d 91 mm$^2$ e 71.3 m$^2$ f 39.8 m$^2$ g 10.1 m$^2$ h 36 m$^2$ **2** 47.4 m$^2$ **3** a 0.55 m$^2$ b \$79.75 **4** a 131.5 m$^2$ b 11 buckets

**PAGE 118** **1** a 4 m b 5 m c 7 m d 18 m e 15 m f 4 m **2** 4 m **3** 8.9 m **4** 14 m **5** 7.7 cm **6** 30 cm

**PAGE 119** **1** a 52 m$^2$ b 94 m$^2$ c 71 m$^2$ d 108 m$^2$ **2** a 68 m$^2$ b 47 m$^2$ c 93 m$^2$ d 107 m$^2$

**PAGE 120** **1** a 131 m$^2$ b 60 m$^2$ **2** a 26 m$^2$ b 60 m$^2$ c 45 m$^2$ d 80.5 m$^2$ e 75 m$^2$ f 30 m$^2$

**PAGE 121** **1** a 19 m$^2$ b 22 m$^2$ c 32 m$^2$ d 69 m$^2$ e 54 m$^2$ f 61 m$^2$ g 179 m$^2$ h 31 m$^2$

**PAGE 122** **1** a 92 m$^2$ b 65 m$^2$ c 107 m$^2$ d 125 m$^2$ e 235 m$^2$ f 192 m$^2$

**PAGE 123** **1** a 99 m$^2$ b 55 m$^2$ c 122 m$^2$ d 264 m$^2$ e 206 m$^2$ f 160 m$^2$

**PAGE 124** **1** a 46 m$^2$ b 61 m$^2$ c 46 m$^2$ d 19.4 m$^2$ **2** a 25 m$^2$ b 19 m$^2$ c 13 m$^2$

**PAGE 125** **1** a mm$^2$ b mm$^2$ c cm$^2$ d cm$^2$ e m$^2$ f m$^2$ g m$^2$ h ha i km$^2$ j km$^2$ k m$^2$ l cm$^2$ m cm$^2$ n m$^2$

**o** km$^2$ **p** mm$^2$ **q** cm$^2$ **r** cm$^2$ ❷ **a** 5.3 cm$^2$ **b** 9.54 cm$^2$ **c** 3.84 cm$^2$ **d** 5.08 cm$^2$ **e** 16.03 cm$^2$ **f** 0.72 cm$^2$ **g** 620 mm$^2$ **h** 954 mm$^2$ **i** 1603 mm$^2$ **j** 72 mm$^2$ **k** 384 mm$^2$ **l** 2410 mm$^2$ **m** 4.5 cm$^2$ **n** 580 mm$^2$

**PAGE 126** ❶ **a** 4.65 m$^2$ **b** 74.2 m$^2$ **c** 16.03 m$^2$ **d** 9.5 m$^2$ **e** 0.21 m$^2$ **f** 0.65 m$^2$ **g** 74.6 cm$^2$ **h** 550 cm$^2$ **i** 1300 cm$^2$ **j** 2100 cm$^2$ ❷ **a** 3.2 ha **b** 9.2 ha **c** 72.4 ha **d** 0.16 ha **e** 920 m$^2$ **f** 640 m$^2$ **g** 415 m$^2$ **h** 3500 m$^2$ **i** 200 m$^2$ **j** 8.6 ha ❸ **a** 0.946 km$^2$ **b** 3.2 km$^2$ **c** 0.16 km$^2$ **d** 7.4 km$^2$ **e** 450 m$^2$ **f** 3200 m$^2$ **g** 15 000 m$^2$ **h** 0.186 km$^2$ **i** 48 m$^2$

**PAGE 127** ❶ 5600 mm$^2$ ❷ 400 mm$^2$ ❸ 40 000 mm$^2$ ❹ 150 000 cm$^2$ ❺ 40 300 cm$^2$ ❻ **a** 21 960 m$^2$ **b** 2.20 ha ❼ **a** 5341 m$^2$ **b** 0.53 ha ❽ **a** 930 000 m$^2$ **b** 620 m

**PAGE 128** ❶ **a** 28 cm **b** 12 cm **c** 20.8 cm **d** 34 cm **e** 34 cm **f** 24 cm **g** 36 cm **h** 50 cm ❷ **a** 50 mm **b** 126 mm **c** 15 cm **d** 7.5 cm **e** 4500 m **f** 950 m **g** 1.8 m **h** 0.6 m **i** 1.2 km **j** 5 km **k** 40 mm **l** 3.5 m ❸ **a** 49 m$^2$ **b** 36.8 m$^2$ **c** 42 m$^2$ **d** 7 m$^2$ **e** 135 m$^2$ **f** 368 m$^2$ ❹ 3.25 m$^2$ ❺ 1.24 m$^2$ ❻ 15 m ❼ **a** 98 m$^2$ **b** 124 m$^2$ **c** 27 m$^2$ **d** 110 m$^2$ ❽ **a** 2.4 cm$^2$ **b** 850 mm$^2$ **c** 5.6 m$^2$ **d** 17 000 m$^2$

**CHAPTER 7**

**PAGE 129** ❶ **a** *ST* **b** *QR* **c** *AB* **d** *DC* **e** *WX, YZ* **f** *CF, DE* ❷ **a** A B **b** P Q

❸ **a** ∠*BAC* or ∠*CAB* **b** ∠*ECD* or ∠*DCE* **c** ∠*PRQ* or ∠*QRP* **d** ∠*KJL* or ∠*LJK* **e** ∠*STU* or ∠*UTS* **f** ∠*FGH* or ∠*HGF* **g** ∠*QPR* or ∠*RPQ* **h** ∠*STU* or ∠*UTS* **i** ∠*LMN* or ∠*NML* **j** ∠*QOP* or ∠*POQ* **k** ∠*TVX* or ∠*XVT* **l** ∠*CDE* or ∠*EDC* ❹ **a** Parent/teacher to check

**PAGE 130** ❶ **a** ∠*ACB* or ∠*BCA*, ∠*BCD* or ∠*DCB* **b** ∠*RTS* or ∠*STR*, ∠*STU* or ∠*UTS* **c** ∠*QPR* or ∠*RPQ*, ∠*RPS* or ∠*SPR* **d** ∠*ABD* or ∠*DBA*, ∠*DBC* or ∠*CBD* **e** ∠*QSR* or ∠*RSQ*, ∠*RST* or ∠*TSR* **f** ∠*DAB* or ∠*BAD*, ∠*BAC* or ∠*CAB* **g** ∠*MOP* or ∠*POM*, ∠*NOP* or ∠*PON* **h** ∠*UTR* or ∠*RTU*, ∠*STU* or ∠*UTS* **i** ∠*HGJ* or ∠*JGH*, ∠*KGH* or ∠*HGK* **j** ∠*RTS* or ∠*STR*, ∠*STU* or ∠*UTS* **k** ∠*LJM* or ∠*MJL*, ∠*KJL* or ∠*LJK* **l** ∠*ZVX* or ∠*XVZ*, ∠*XVY* or ∠*YVX*

**PAGE 131** ❶ **a** ∠*BAC* or ∠*CAB*, ∠*ABC* or ∠*CBA*, ∠*ACB* or ∠*BCA* **b** ∠*QPR* or ∠*RPQ*, ∠*PRQ* or ∠*QRP*, ∠*PQR* or ∠*RQP* **c** ∠*TSU* or ∠*UST*, ∠*STU* or ∠*UTS*, ∠*SUT* or ∠*TUS* **d** ∠*LMN* or ∠*NML*, ∠*MNL* or ∠*LNM*, ∠*MLN* or ∠*NLM* **e** ∠*FDE* or ∠*EDF*, ∠*DFE* or ∠*EFD*, ∠*DEF* or ∠*FED* **f** ∠*DCE* or ∠*ECD*, ∠*CDE* or ∠*EDC*, ∠*DEC* or ∠*CED* ❷ **a** ∠*AED* or ∠*DEA*, ∠*BEC* or ∠*CEB* **b** ∠*SPQ* or ∠*QPS*, ∠*RPT* or ∠*TPR* **c** ∠*LKM* or ∠*MKL*, ∠*JKN* or ∠*NKJ* **d** ∠*EGD* or ∠*DGE*, ∠*FGH* or ∠*HGF* **e** ∠*EBD* or ∠*DBE*, ∠*ABC* or ∠*CBA* **f** ∠*POQ* or ∠*QOP*, ∠*ROS* or ∠*SOR*

**PAGE 132** ❶ **a** ∠*DEG* or ∠*GED*, ∠*EFH* or ∠*DFH* **b** ∠*ABD* or ∠*DBA*, ∠*BCE* or ∠*ACE* **c** ∠*ADB* or ∠*BDA*, ∠*DEC* or ∠*AEC* **d** ∠*QRS* or ∠*QRT*, ∠*PST* or ∠*TSP* ❷ **a** ∠*FDB* or ∠*BDF*, ∠*CBD* or ∠*DBC* **b** ∠*HGD* or ∠*DGH*, ∠*FDG* or ∠*GDF* **c** ∠*XWZ* or ∠*ZWX*, ∠*UZW* or ∠*WZU* **d** ∠*SPV* or ∠*VPS*, ∠*UVP* or ∠*PVU* ❸ **a** ∠*DAB* or ∠*DAC*, ∠*DBC* or ∠*CBD* **b** ∠*EHG* or ∠*EHF*, ∠*EGF* or ∠*FGE* **c** ∠*PTQ* or ∠*QTP*, ∠*RTS* or ∠*STR* **d** ∠*TSY* or ∠*YST*, ∠*TXZ* or ∠*ZXT* **e** ∠*UTV* or ∠*VTU*, ∠*XVT* or ∠*TVX* **f** ∠*MKN* or ∠*NKM*, ∠*LKN* or ∠*NKL*

**PAGE 133** ❶ **a** Acute **b** Obtuse **c** Right **d** Reflex **e** Straight **f** Revolution **g** Obtuse **h** Acute **i** Acute **j** Reflex ❷ **a** Acute, Acute, Acute **b** Acute, Right, Acute **c** Obtuse, Acute, Acute **d** Acute, Acute, Right **e** Acute, Obtuse, Acute **f** Obtuse, Acute, Obtuse **g** Right, Right, Right

**PAGE 134** ❶ **a** Parent/teacher to check

❷

| Angle | Angle type |
|---|---|
| 45° | Acute |
| 60° | Acute |
| 125° | Obtuse |
| 90° | Right |
| 180° | Straight |
| 165° | Obtuse |
| 179° | Obtuse |
| 270° | Reflex |
| 210° | Reflex |
| 360° | Revolution |
| 79° | Acute |
| 190° | Reflex |

❸ 60° ❹ 45° ❺ 30° ❻ 37° ❼ 60° ❽ 100° ❾ 90° ❿ 90° ⓫ 180°

**PAGE 135** ❶ **a** 50° **b** 70° **c** 35° **d** 65° **e** 50° **f** 80° **g** 45° **h** 90° **i** 120° **j** 135° **k** 146° **l** 110° **m** 155°

**PAGE 136** ❶ **a** 50° **b** 35° **c** 60° **d** 75° **e** 130° **f** 135° **g** 165° ❷ **a** 310° **b** 295° **c** 317° **d** 310° **e** 285°

**PAGE 137** ❶ **a** 50° **b** 65° **c** 35° **d** 120° **e** 40° **f** 54° **g** 35° **h** 120° ❷ **a** 70°, 70° **b** 60°, 60° **c** 70°, 110° **d** 120°, 60°

**PAGE 138** ❶ **a** 50° **b** 38° **c** 30° **d** 45° **e** 80° **f** 60° ❷ **a** 138° **b** 116° **c** 50° **d** 65° **e** 90° **f** 60° ❸ **a** Yes **b** Yes **c** No **d** No **e** Yes **f** No ❹ **a** Yes **b** No **c** Yes **d** Yes **e** No ❺ **a** 29° **b** 20° **c** 70° **d** 32° **e** 35° **f** 45° **g** 24° **h** 21°

**PAGE 139** ❶ **a** 55° **b** 105° **c** 113° **d** 126° **e** 50° **f** 75° **g** 37° **h** 84° **i** 45° **j** 40° **k** 10° **l** 45° ❷ **a** 128°, 52° **b** 131°, 49° **c** 148°, 32° **d** 55, 125° **e** 40°, 140° **f** 130°, 50° **g** 132°, 48° **h** 59°, 121°

**PAGE 140** ❶ **a** 150°, 30° **b** 146°, 34° **c** 29°, 151° **d** 52°, 128° **e** 126°, 54° **f** 82°, 98° **g** 79°, 101° **h** 83°, 97° **i** 49°, 131° ❷ **a** 53°, 127° **b** 48°, 132° **c** 34°, 146° **d** 28°, 152° **e** 147°, 33° **f** 139°, 41° **g** 33°, 147° **h** 29°, 151°

**PAGE 141** ❶ **a** 50° **b** 44° **c** 51° **d** 220° **e** 230° **f** 30° **g** 50° **h** 242° **i** 25° **j** 45° ❷ **a** 89° **b** 109° **c** 151° **d** 125° **e** 87° **f** 105° **g** 40° **h** 94°

**PAGE 142** ❶ **a** 107° **b** 110° **c** 109° **d** 112° **e** 68° **f** 70° **g** 71° **h** 69° **i** 72° **j** 68° **k** 108° **l** 112°

**PAGE 143** ❶ **a** 72° **b** 70° **c** 71° **d** 69° ❷ **a** 107°, 73° **b** 102°, 78° **c** 72°, 108° **d** 105°, 75° ❸ **a** 73°, 107° **b** 78°, 102° **c** 79°, 101° **d** 104°, 76° **e** 107°, 73° **f** 104°, 76° **g** 109°, 71° **h** 70°, 110°

**PAGE 144** ❶ **a** 100°, 71° **b** 98°, 75° **c** 108°, 73° **d** 105°, 70° **e** 90°, 65° ❷ **a** 75°, 44° **b** 58°, 47° **c** 49°, 60° **d** 61°, 80° **e** 63°, 90° **f** 59°, 78°

**PAGE 145** ❶ **a** 75° **b** 68° **c** 65° **d** 72° **e** 102° **f** 105° **g** 114° **h** 101° **i** 70°, 70° **j** 69°, 69° **k** 67°, 67° **l** 73°, 73° **m** 112°, 112° **n** 105°, 105° **o** 110°,110° **p** 106°, 106°

**PAGE 146** ❶ **a** 107°, 73° **b** 112°, 68° **c** 65°, 115° **d** 63°, 117° ❷ **a** 69°, 111° **b** 70°, 110° **c** 115°, 65° **d** 112°, 68° **e** 73°, 107° **f** 72°, 108° **g** 103°, 77° **h** 109°, 71° **i** 70°, 110° **j** 69°, 111° **k** 112°, 68°

**PAGE 147** ❶ **a** 50°, 60° **b** 52°, 61° **c** 47°, 63° **d** 51°, 62°, 67° **e** 50°, 61°, 69° **f** 48°, 58°, 74° **g** 49°, 60°, 71° ❷ **a** 40°, 50° **b** 41°, 52° **c** 39°, 48° **d** 44°, 51° **e** 95° **f** 88° **g** 91° **h** 96°

**PAGE 148** ❶ **a** 113° **b** 75° **c** 78° **d** 109° **e** 71° **f** 107° **g** 68° **h** 111° ❷ **a** 72° **b** 107° **c** 104° **d** 70° ❸ **a** 128°, 52° **b** 124°, 56° **c** 131°, 49° **d** 127°, 53° **e** 125°, 55°

**PAGE 149** ❶ **a** 67°, 113° **b** 72°, 108° **c** 78°, 102° **d** 76°, 104° ❷ **a** 116°, 64° **b** 122°, 58° **c** 60°, 120° **d** 62°, 62° ❸ **a** 77°, 111°, 69° **b** 78°, 107°, 73° **c** 106°, 71°, 109° **d** 75°, 98°, 98° **e** 70°, 79°, 79°

**PAGE 150** ❶ **a** Corresponding **b** Alternate **c** Co-interior **d** Alternate **e** Corresponding **f** Co-interior **g** Corresponding **h** Alternate **i** Co-interior **j** Corresponding ❷ **a** Corresponding **b** Alternate **c** Co-interior **d** Alternate **e** Co-interior **f** Alternate **g** Corresponding **h** Co-interior **i** Corresponding **j** Alternate **k** Corresponding **l** Alternate

**PAGE 151** ❶ **a** $x = 104°$, Corresponding **b** $x = 108°$, Alternate **c** $x = 78°$, Corresponding **d** $x = 108°$, Co-interior **e** $x = 104°$, Co-interior **f** $x = 67°$, Co-interior **g** $x = 55°$, Alternate **h** $x = 132°$, Corresponding ❷ **a** $x = 105°$, Corresponding **b** $y = 105°$, Alternate **c** $x = 105°$, Corresponding; $y = 105°$, Alternate **d** $x = 110°$, Corresponding; $y = 110°$, Alternate **e** $x = 72°$, Corresponding; $y = 72°$, Alternate **f** $x = 68°$, Alternate; $y = 68°$, Corresponding

**PAGE 152** ❶ **a** $x = 75°$, Alternate **b** $y = 75°$, Corresponding **c** $x = 75°$, Alternate; $y = 75°$, Corresponding **d** $x = 69°$, Alternate; $y = 69°$, Corresponding **e** $x = 109°$, Alternate; $y = 109°$, Corresponding **f** $x = 113°$, Alternate; $y = 113°$, Corresponding **g** $x = 110°$, Corresponding; $y = 110°$, Alternate **h** $x = 72°$, Corresponding; $y = 72°$, Alternate ❷ **a** $x = 120°$, Alternate **b** $y = 60°$, Co-interior **c** $x = 120°$, Alternate; $y = 60°$, Co-interior **d** $x = 117°$, Alternate; $y = 63°$, Co-interior **e** $x = 124°$, Alternate; $y = 56°$, Co-interior **f** $x = 53°$, Co-interior; $y = 127°$, Alternate

**PAGE 153** ❶ **a** $x = 67°$, Vertically opposite angles **b** $y = 113°$, Straight angle **c** $x = 67°$, Vertically opposite angles; $y = 113°$, Straight angle **d** $x = 78°$, Vertically opposite angles; $y = 102°$, Straight angle **e** $x = 108°$, Vertically opposite angles; $y = 72°$, Straight angle **f** $x = 72°$, Vertically opposite angles; $y = 108°$, Straight angle ❷ **a** $x = 72°$, Co-interior **b** $y = 108°$, Corresponding **c** $x = 72°$, Co-interior; $y = 108°$, Corresponding **d** $x = 68°$, Co-interior; $y = 112°$, Corresponding **e** $x = 105°$, Co-interior; $y = 75°$, Corresponding **f** $x = 71°$, Co-interior; $y = 109°$, Corresponding

**PAGE 154** ❶ **a** Equilateral **b** Equilateral **c** Equilateral **d** Isosceles **e** Isosceles **f** Isosceles **g** Isosceles **h** Scalene **i** Scalene **j** Scalene **k** Scalene **l** Equilateral **m** Isosceles **n** Scalene **o** Isosceles **p** Isosceles **q** Equilateral **r** Scalene ❷ **a** Right **b** Right **c** Acute **d** Acute **e** Acute **f** Obtuse

**PAGE 155** ❶ **a** Obtuse **b** Right **c** Acute **d** Acute **e** Obtuse **f** Obtuse ❷ **a** Scalene, Right **b** Isosceles, Right **c** Isosceles, Acute **d** Isosceles, Obtuse **e** Scalene, Obtuse **f** Scalene, Acute

**PAGE 156** ❶ **a** Equilateral, Acute-angled **b** Isosceles, Acute-angled **c** Scalene, Obtuse-angled **d** Isosceles, Obtuse-angled ❷ Parent/teacher to check

**PAGE 157** ❶ **a** 120° **b** 23° **c** 34° **d** 21° **e** 30° **f** 70° **g** 45° **h** 40° ❷ **a** 36° **b** 90° **c** 110° **d** 52° **e** 42° **f** 130° **g** 98° **h** 54° **i** 20° **j** 122°

**PAGE 158** ❶ **a** 65° **b** 28° **c** 71° **d** 61° **e** 70° **f** 45° ❷ **a** 133°, 47° **b** 148°, 32° **c** 151°, 29° **d** 139°, 41° **e** 147°, 33° **f** 149°, 31°

**PAGE 159** ❶ **a** equal **b** 90° **c** equal **d** bisect **e** 90° **f** bisect ❷ Rewrite the sentences from Q1a–1f. ❸ **a** equal **b** 90° **c** equal **d** bisect

**PAGE 160** ❶ Rewrite the sentences from Q3a–3d in Unit 31. ❷ **a** equal **b** parallel **c** equal **d** bisect **e** 90° **f** bisect ❸ Rewrite the sentences from Q2a–2f.

**PAGE 161** ❶ **a** equal **b** parallel **c** equal **d** bisect ❷ Rewrite the sentences from Q1a–1d. ❸ **a** equal **b** equal **c** 90° ❹ Rewrite the sentences from Q3a–3c.

**PAGE 162** ❶ parallel ❷ Parent/teacher to check ❸ Parent/teacher to check ❹ Parent/teacher to check ❺ Parent/teacher to check ❻ **a** True **b** False **c** True **d** False **e** True **f** True **g** False

**PAGE 163** ❶ **a** 76° **b** 73° **c** 65° **d** 232° **e** 95° **f** 125° **g** 47° **h** 50° ❷ **a** 65°, 122° **b** 77°, 94° **c** 70°, 69° **d** 86°, 79° **e** 110°, 60° **f** 61°

**PAGE 164** ❶ **a** $\angle POQ$ or $\angle QOP$ **b** $\angle YXZ$ or $\angle ZXY$ **c** $\angle QSR$ or $\angle RSQ$, $\angle STU$ or $\angle UTS$ **d** $\angle YVW$ or $\angle WVY$, $\angle YWX$ or $\angle XWY$ ❷ **a** Acute **b** Right **c** Straight **d** Reflex **e** Obtuse **f** Revolution ❸ **a** 30° **b** 120° ❹ 40° ❺ 70° ❻ **a** 26° **b** 75° **c** $x = 30°$, $y = 150°$ **d** $x = 52°$, $y = 128°$ **e** 40° **f** 120° ❼ **a** 72°, Alternate angles **b** $x = 68°$, Corresponding angles; $y = 112°$, Straight angle **c** 70°, Co-interior angles **d** $x = 109°$, Alternate angles; $y = 109°$, vertically opposite ❽ Isoscles, right-angled triangle ❾ **a** 35° **b** 45° **c** 99° **d** 225° **e** $x = 57°$, $y = 123°$ **f** $x = 106°$, $y = 49°$ ❿ Sample answer: Opposite angles are equal and parallel.

CHAPTER 8

## PAGE 165

1
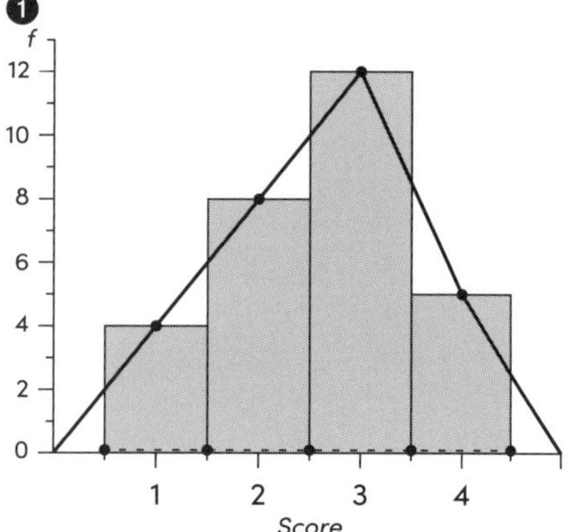


2
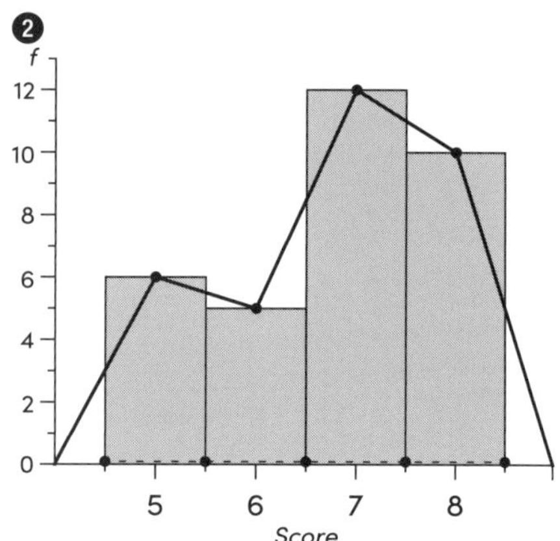


3
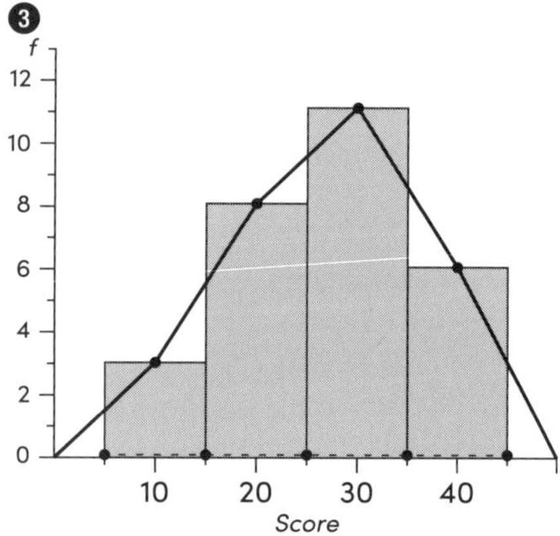


4
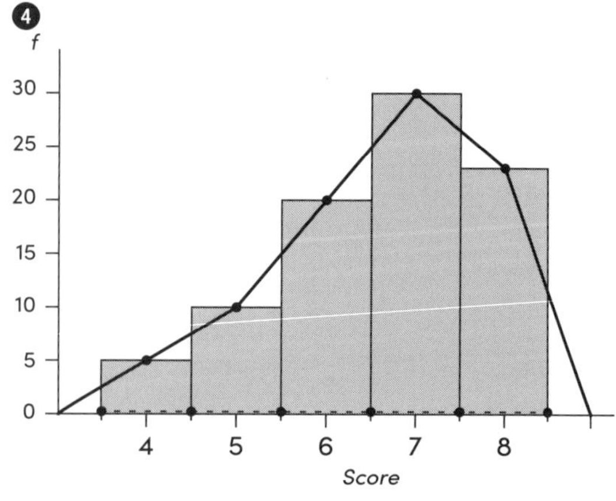


## PAGE 166

1
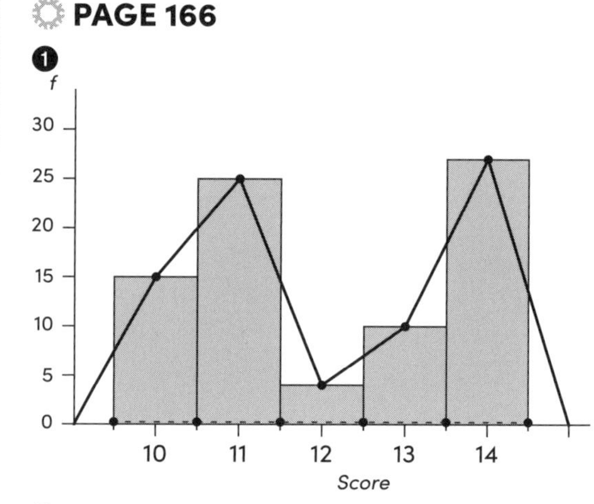


2
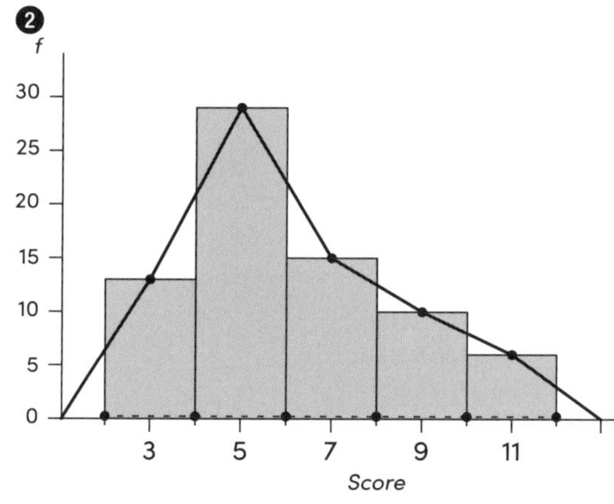


3
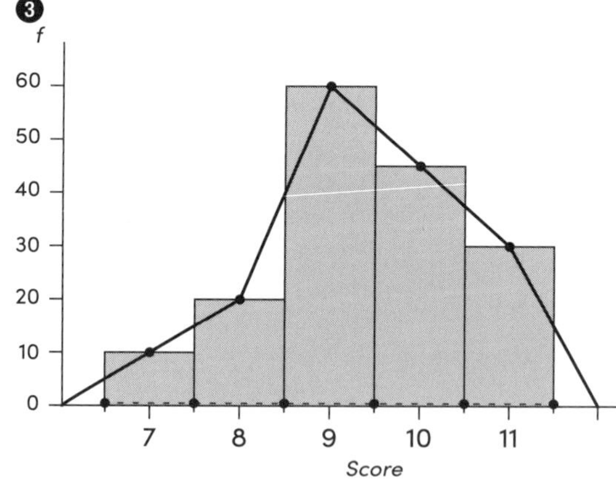


4
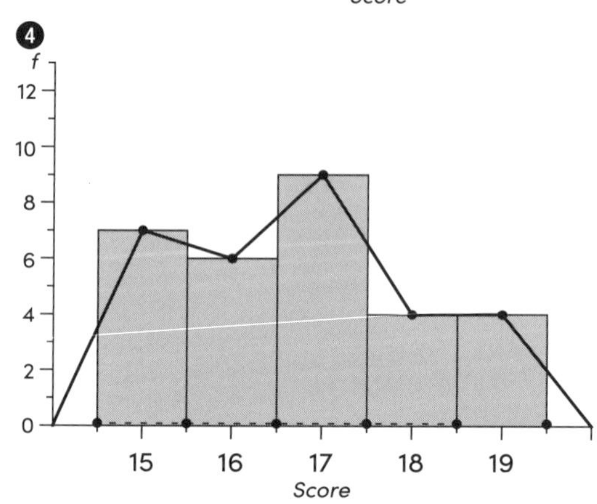

## PAGE 167

**1**

| Stem | Leaf |
|---|---|
| 0 | 1 3 3 3 |
| 1 | 0 2 4 |
| 2 | 1 3 |
| 3 | 2 6 6 9 |

**2**

| Stem | Leaf |
|---|---|
| 0 | 3 4 7 |
| 1 | 2 4 6 6 |
| 2 | 1 3 |
| 3 | 2 5 8 |

**3**

| Stem | Leaf |
|---|---|
| 0 | 5 6 6 7 |
| 1 | 1 3 |
| 2 | 0 4 6 |
| 3 | 4 5 8 9 |

**4**

| Stem | Leaf |
|---|---|
| 0 | 1 5 6 7 |
| 1 | 0 5 8 |
| 2 | 0 4 6 |
| 3 | 1 3 3 |
| 4 | 2 7 |

**5**

| Stem | Leaf |
|---|---|
| 0 | 3 5 5 |
| 1 | 2 3 8 |
| 2 | 6 6 |
| 3 | 0 2 2 5 |
| 4 | 3 4 4 6 |

**6**

| Stem | Leaf |
|---|---|
| 0 | 1 2 2 4 |
| 1 | 2 6 6 |
| 2 | 0 2 4 |
| 3 | 0 3 |
| 4 | 2 2 7 7 |

**7**

| Stem | Leaf |
|---|---|
| 0 | 3 5 |
| 1 | 6 7 7 |
| 2 | 4 5 6 |
| 3 | 2 7 9 |
| 4 | 3 6 7 9 |

**8**

| Stem | Leaf |
|---|---|
| 0 | 2 4 6 9 |
| 1 | 2 2 3 |
| 2 | 4 8 |
| 3 | 2 6 6 |
| 4 | 2 3 7 8 |

## PAGE 168

**1**

| Stem | Leaf |
|---|---|
| 0 | 2 5 6 |
| 1 | 0 1 8 |
| 2 | 0 1 2 4 7 |
| 3 | 2 5 6 |
| 4 | 3 4 4 |
| 5 | 0 1 3 3 |

**2**

| Stem | Leaf |
|---|---|
| 0 | 2 2 2 4 |
| 1 | 3 6 7 |
| 2 | 0 4 6 8 |
| 3 | 2 7 |
| 4 | 1 5 |
| 5 | 0 1 2 6 |

**3**

| Stem | Leaf |
|---|---|
| 6 | 0 2 3 4 |
| 7 | 3 8 |
| 8 | 2 6 |
| 9 | 3 3 4 |
| 10 | 3 5 6 7 |

**4**

| Stem | Leaf |
|---|---|
| 6 | 2 4 |
| 7 | 3 4 5 |
| 8 | 5 6 8 9 9 |
| 9 | 6 6 7 8 |
| 10 | 1 3 3 |

**5**

| Stem | Leaf |
|---|---|
| 10 | 3 5 5 7 |
| 11 | 2 2 5 6 |
| 12 | 4 5 7 |
| 13 | 2 6 9 9 |
| 14 | 1 4 5 |

**6**

| Stem | Leaf |
|---|---|
| 10 | 0 2 6 7 |
| 11 | 0 3 3 5 |
| 12 | 1 5 6 6 |
| 13 | 2 2 |
| 14 | 3 7 7 9 |

**7**

| Stem | Leaf |
|---|---|
| 10 | 3 3 5 |
| 11 | 0 1 |
| 12 | 4 4 5 7 8 |
| 13 | 2 3 6 6 |
| 14 | 0 2 3 |

**8**

| Stem | Leaf |
|---|---|
| 11 | 0 7 8 |
| 12 | 1 6 |
| 13 | 3 4 9 |
| 14 | 0 5 |
| 15 | 2 2 3 6 |

## PAGE 169

**1**

1 2 3 4

**2**

6 7 8 9

**3**

3 4 5 6

**4**

10 11 12 13

**5**

1 2 3 4

**6**

10 11 12 13

**7**

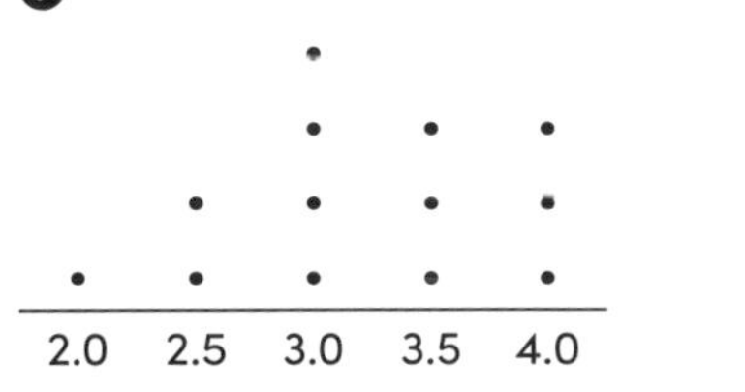


## PAGE 170

**1**

Blue | Red | Yellow | Gr

4 cm | 2 cm | 3 cm | 1 cm

**2**

Cat | Dog | Bird | R

3 cm | 4 cm | 2 cm | 1 cm

**3**

Tablet | Phone | Laptop | D

2 cm | 4 cm | 3 cm | 1 cm

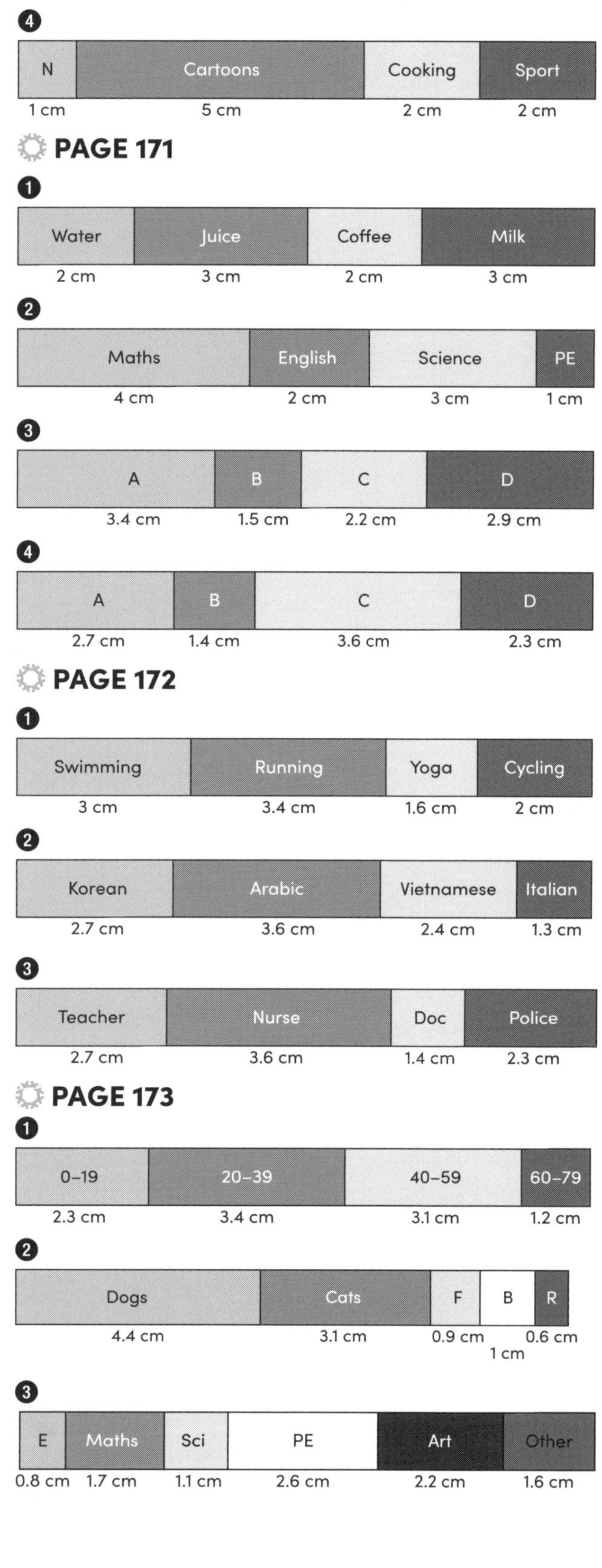


## PAGE 174

**1**

C, B, A, D, E

**2**

B, A, E, C, D

**3**

B, A, C, E, D

## PAGE 175

**1**

A, B, C, E, D

**2**

B, A, E, C, D

**3**

B, C, A, E, D

## PAGE 176

**1**

Tue., Mon., Wed., Thu., Fri.

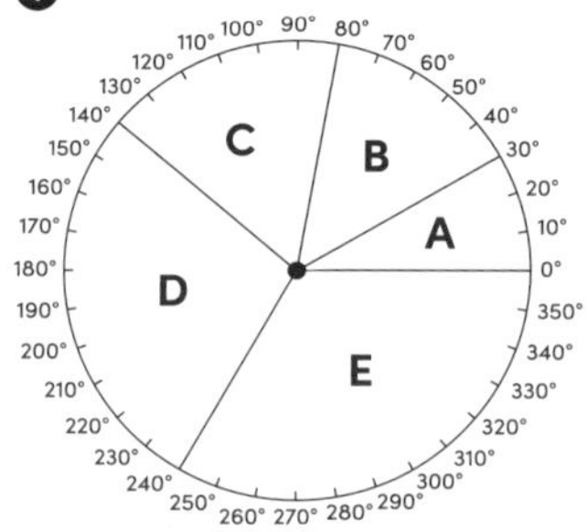

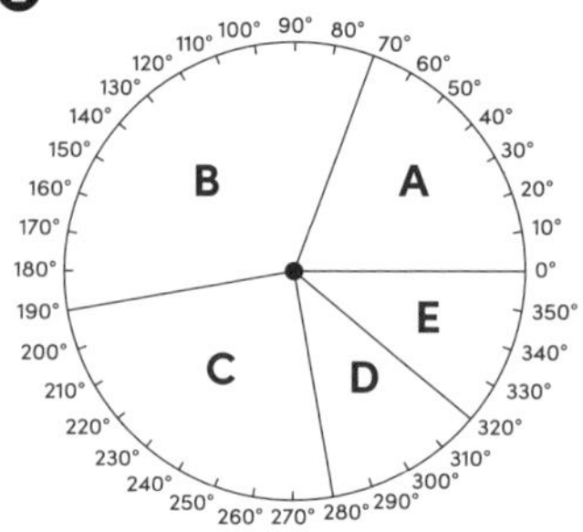

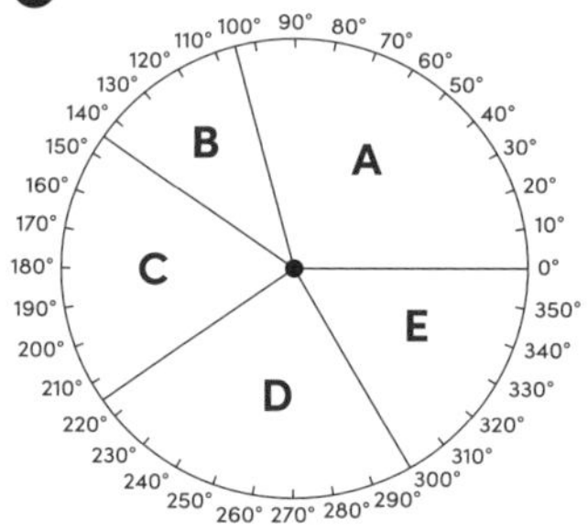

**2**

Chairs, Desks, Cabinets, Beds, Bookshelf

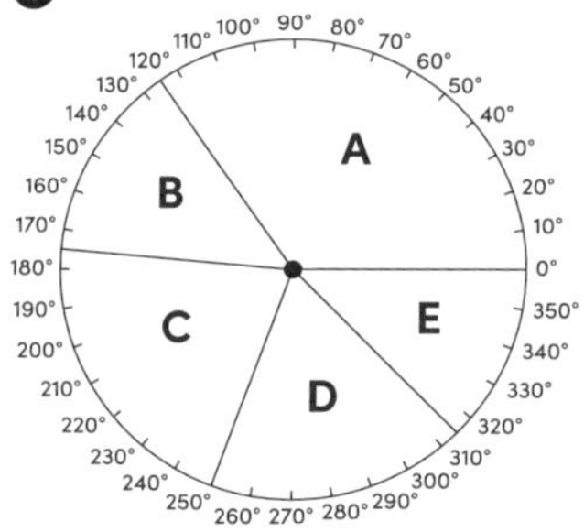

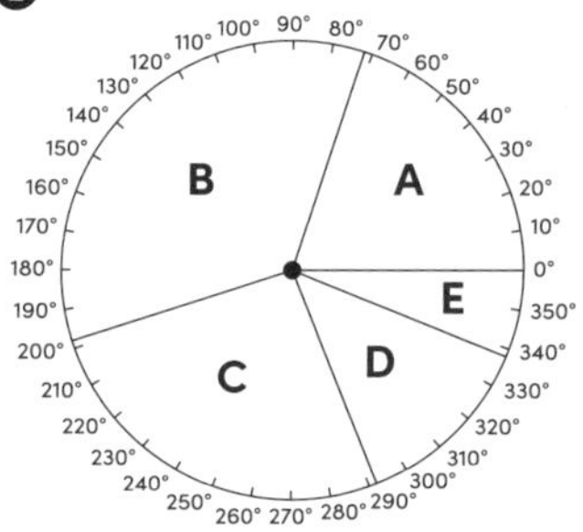

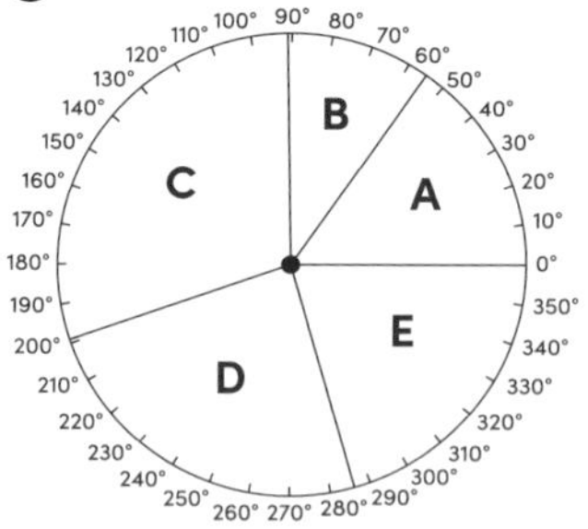

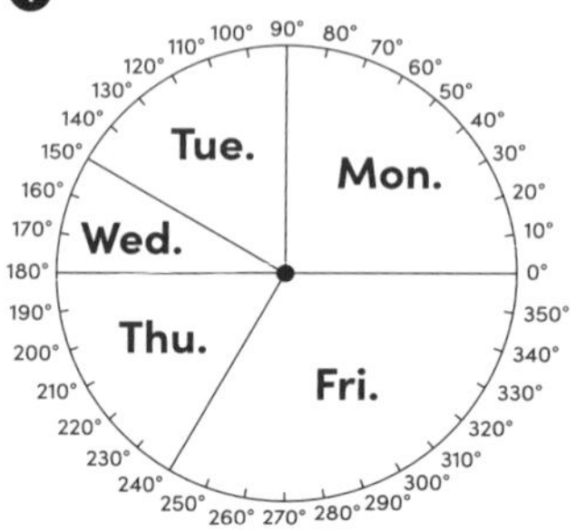

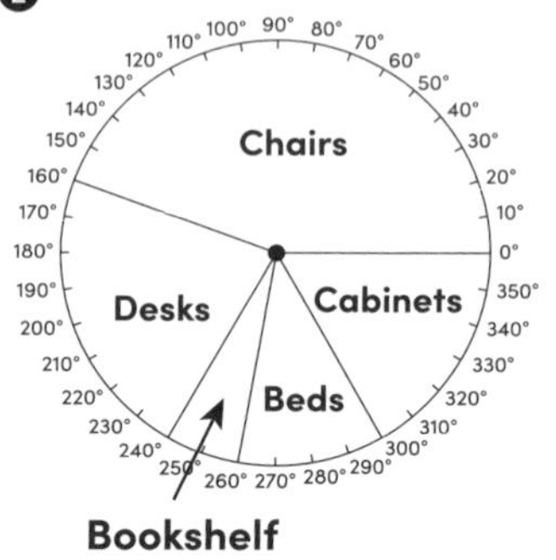

## PAGE 177

❶ ❷

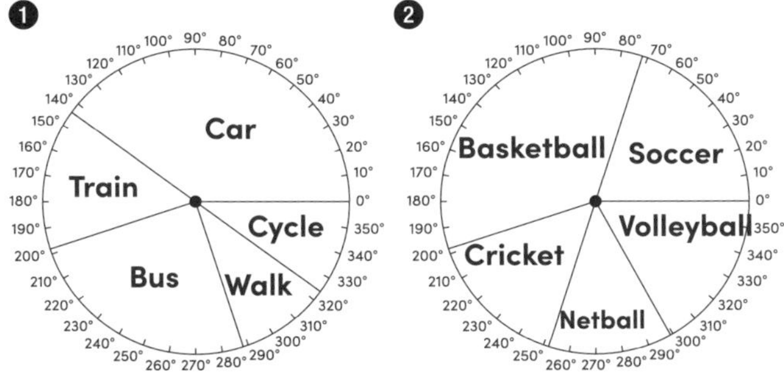


## PAGE 178

❶ ❷

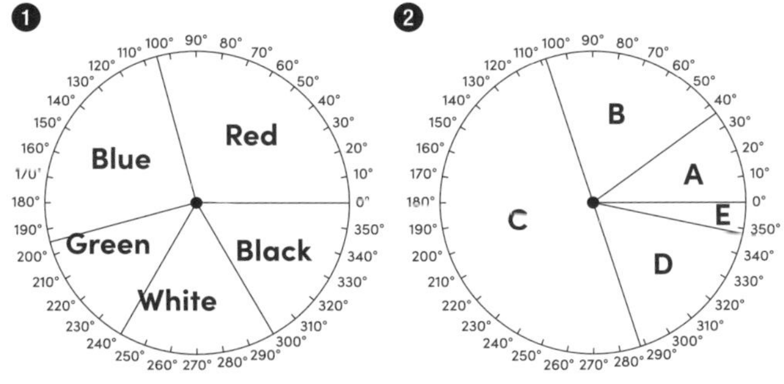


## PAGE 179

❶ a

| | Gen alpha | Gen Z | Millennials | Gen X | Baby boomers | Silent gen |
|---|---|---|---|---|---|---|
| Percentage % | 12 | 18 | 22 | 19 | 21 | 8 |

b Millennials c Silent gen

❷ a

| | Sydney | Melbourne | Brisbane | Perth | Adelaide | Gold Coast | Canberra | Hobart | Darwin |
|---|---|---|---|---|---|---|---|---|---|
| Population (millions) | 5.5 | 5.2 | 2.7 | 2.3 | 1.4 | 0.7 | 0.5 | 0.3 | 0.2 |

b About 4 times

## PAGE 180

❶ a

| | Year 7 | Year 8 | Year 9 | Year 10 |
|---|---|---|---|---|
| Car | 66% | 56 | 50 | 46 |
| Public transport | 11% | 24 | 32 | 38 |
| Walk | 23 | 20 | 18 | 16 |

**b** Car **c** Walk **d** Students can travel to school on their own. ❷ **a** Dogs **b** Reptiles **c i** 2350 **ii** 1500 **iii** 550 **iv** 400

## PAGE 181

❶ **a** Soccer **b** Football **c** 23.3% **d** 35% **e** 17.5% **f** 24.2% ❷ **a** C **b** A **c** 45% **d** 10% **e** 20% **f** 5%

## PAGE 182

❶ **a** The scale does not start at 0% so it makes Party L look a lot higher than Party A. ❷ No scale is given and the pictures make it look like Sydney's house prices are double those of Melbourne's which may not be true. ❸ The percentages do not add up to 100% and the size of each sector is incorrect. ❹ The scale has been removed and the size of the column graphs has been increased.

## PAGE 183

❶

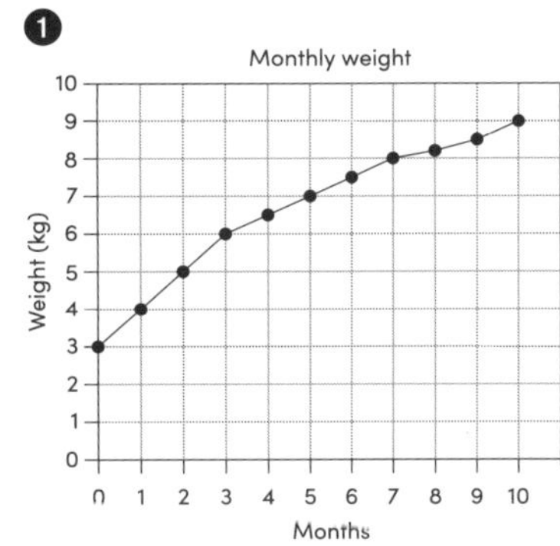


❷

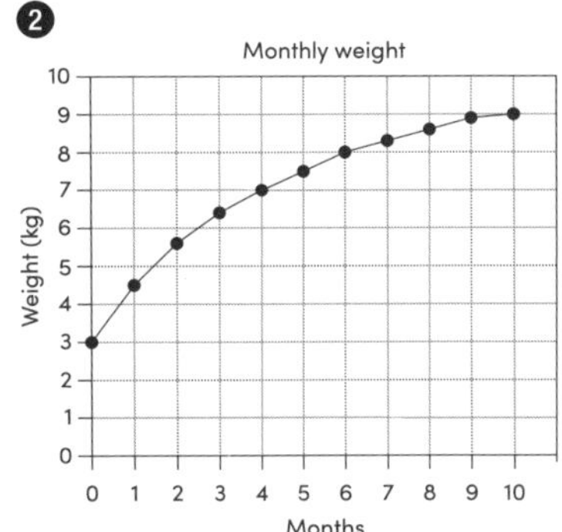


## PAGE 184

❶

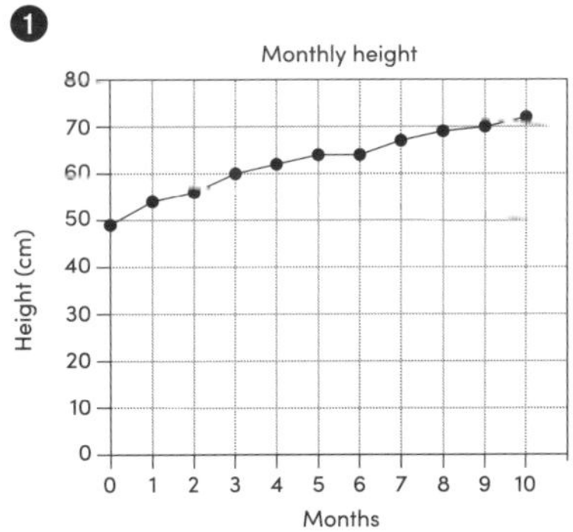


❷

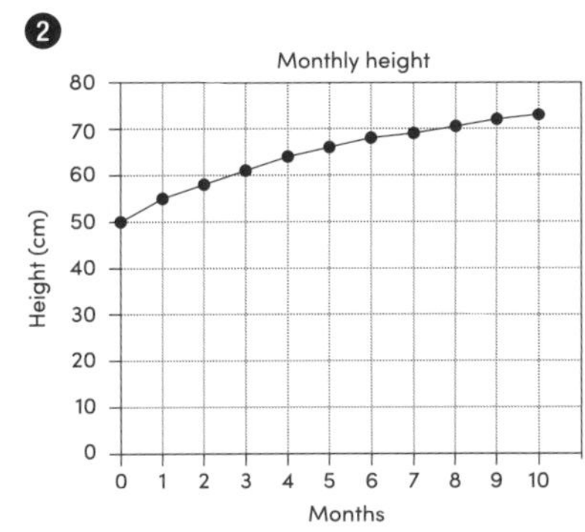

## PAGE 185

**1**

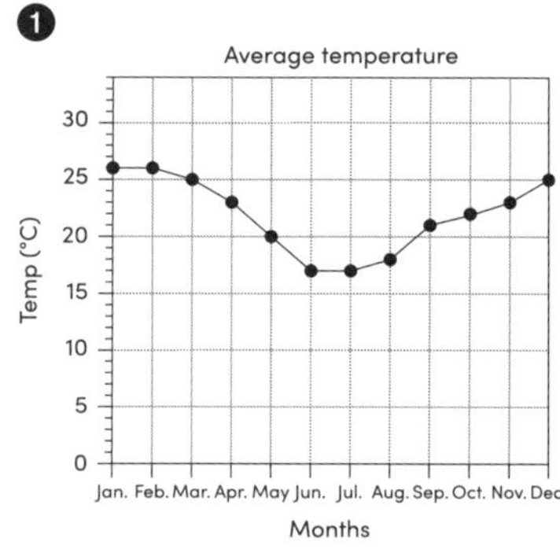


**2**

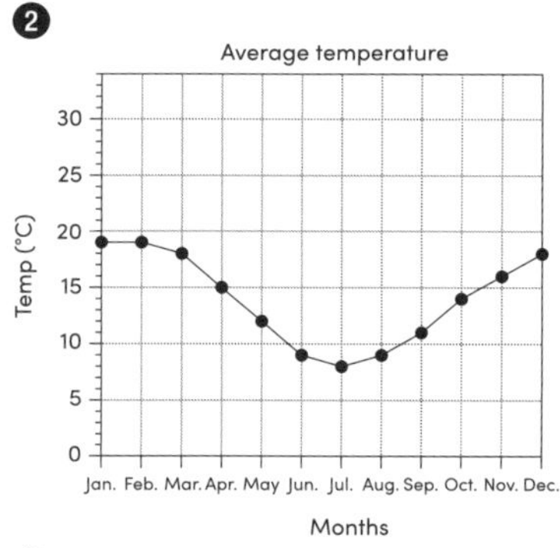


## PAGE 186

**1** Answers may vary. **a** 86 cm, 121 cm, 133 cm, 150 cm **b** 4 yrs, 7 yrs, 10 yrs, 11 yrs **c** 21 cm **d** 38 cm **e** 29 cm **f** 7 cm **g** 8 cm **h** 6 cm **i** 158 cm **j** No, there is a limit to how tall a person can grow.

## PAGE 187

**1** **a** N **b** N **c** C **d** C **e** C **f** C **g** N **h** N **i** N **j** N **k** C **l** C **m** N **n** N **o** C **p** C **2** **a** D **b** D **c** C **d** C **e** D **f** D **g** C **h** C **i** C **j** C **k** D **l** D **3** **a** O **b** O **c** N **d** N **e** O **f** O **g** N **h** N **i** O **j** N **k** N **l** O

## PAGE 188

**1**

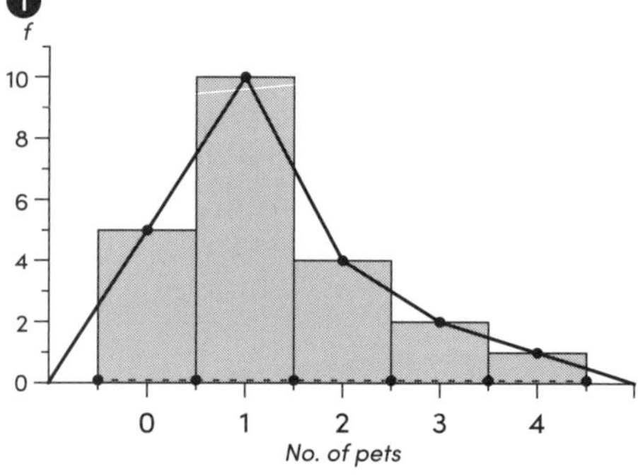


**2**

| Stem | Leaf |
|---|---|
| 0 | 1 2 2 4 |
| 1 | 0 3 |
| 2 | 1 3 7 8 |
| 3 | 0 4 |

**3**

| Stem | Leaf |
|---|---|
| 13 | 8 9 |
| 14 | 0 1 4 7 8 |
| 15 | 0 1 2 3 6 9 |
| 16 | 2 |

**4**

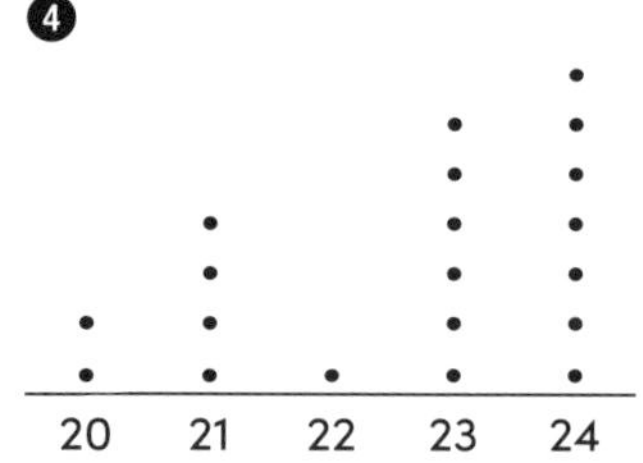


**5**

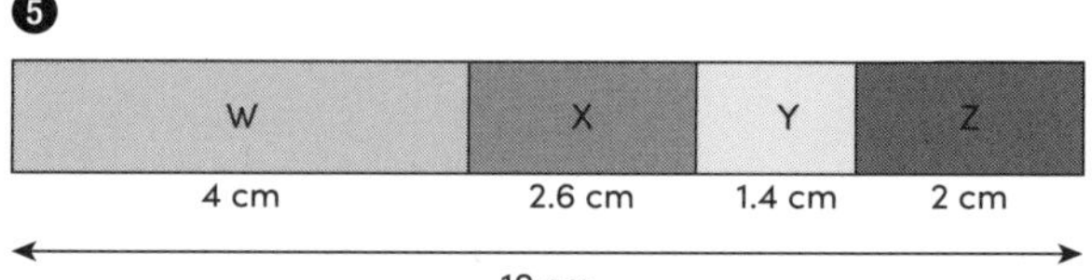


**6**

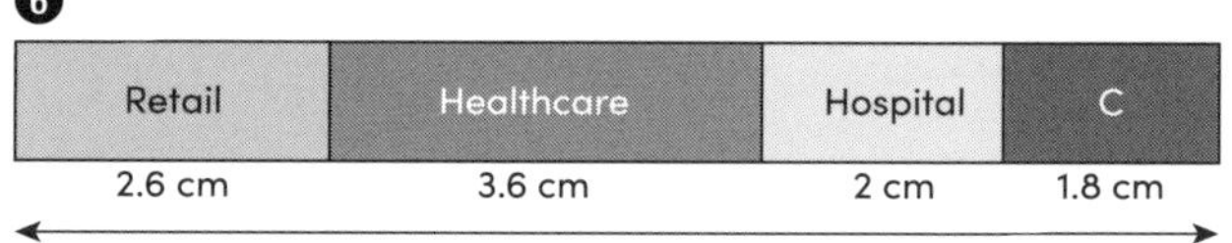


**7** **8**

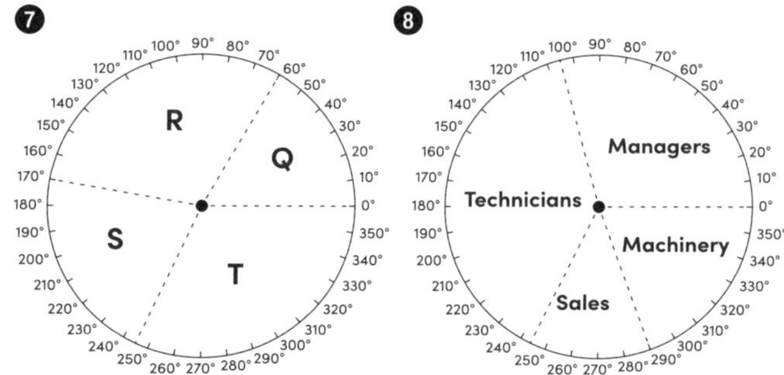


**9**

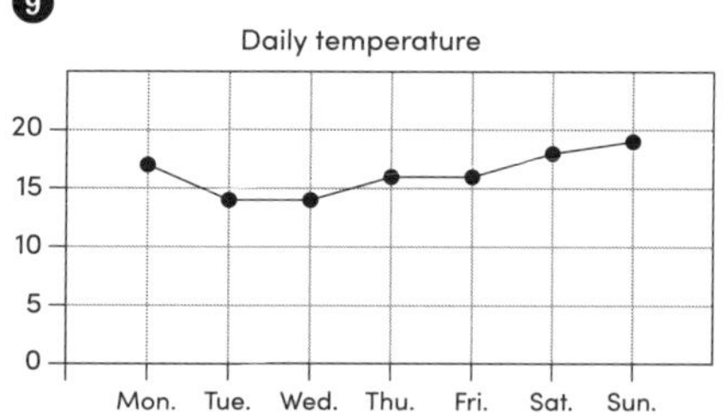


**10** **a**

| | Marsupials | Primates | Carnivores | Herbivores | Rodents | Rabbits |
|---|---|---|---|---|---|---|
| Number | 11 | 7 | 9 | 7 | 3 | 1 |

**b** Marsupials **11** The scale does not start at 0 so it makes Company A's net worth look a lot higher than Company B's. **12** B

### CHAPTER 9

## PAGE 189

**1** **a** Likely **b** Certain **c** Certain **d** Impossible **e** Even chance **f** Unlikely **2** **a** Even chance **b** Even chance **c** Impossible **d** Yes **3** **a** Likely **b** Unlikely **c** Impossible **d** No **4** **a** Less than even chance **b** Less than even chance **c** Yes **5** **a** Even chance **b** Unlikely **c** Unlikely **d** Yes **6** **a** Even chance

**b** Even chance **c** Impossible **d** Yes **7** **a** Even chance **b** Unlikely **c** Unlikely **d** Yes **e** No

**PAGE 190** **1** **a** Blue, red **b** 2 **c** Yes **2** **a** Blue, red, green **b** 3 **c** No **3** **a** Green, red **b** 2 **c** Yes **4** **a** Red, black, green **b** 3 **c** No **5** **a** Blue, red, yellow, green, black **b** 5 **c** Yes **6** 1 and 2

**PAGE 191** **1** **a** white, grey balls **b** 2 **c** Yes **d** Even chance **2** **a** white, black balls **b** 2 **c** No **3** **a** grey, black balls **b** 2 **c** No **d** Likely **4** **a** white, black, red balls **b** 3 **c** No **d** Black **e** White **f** 10 **g** Even chance

**PAGE 192** **1** **a** 1, 2, 3, 4, 5, 6 **b** 6 **c** Yes **d** Unlikely **2** **a** 1, 2, 3 **b** 3 **c** No **d** Likely **3** **a** 1, 2, 3, 4 **b** 4 **c** No **d** Even chance **e** Unlikely **f** Likely **4** **a** Parent/teacher to check **b** $1, 20-cent, 50-cent **c** 3 **d** No **e** $1 and 50-cent **f** 20-cent

**PAGE 193** **1** **a** $\frac{1}{2}$ **b** $\frac{1}{2}$ **c** Yes **d** Even chance **2** **a** $\frac{2}{3}$ **b** $\frac{1}{3}$ **c** No **d** Less than even chance **3** **a** $\frac{1}{4}$ **b** $\frac{1}{4}$ **c** $\frac{1}{2}$ **d** Green and black **e** Even chance **4** **a** $\frac{1}{4}$ **b** $\frac{1}{4}$ **c** $\frac{1}{2}$ **d** Red and Green **e** Unlikely **5** **a** $\frac{1}{4}$ **b** $\frac{3}{4}$ **c** No **d** Likely **6** **a** 18.2% **b** 9.1% **c** 18.2% **d** 9.1% **e** 18.2% **f** 27.3% **g** 36.4%

**PAGE 194** **1** **a** 16.7% **b** 16.7% **c** 8.3% **d** 8.3% **e** 25% **f** 33.3% **g** 50% **2** **a** 7.7% **b** 15.4% **c** 23.1% **d** 46.2% **e** 38.5% **f** 38.5% **3** **a** $\frac{2}{6}$, 0.3 **b** $\frac{2}{6}$, 0.3 **c** $\frac{1}{6}$, 0.2 **d** $\frac{3}{6}$, 0.5 **e** $\frac{3}{6}$, 0.5 **f** $\frac{5}{6}$, 0.8 **4** **a** $\frac{1}{6}$, 0.2 **b** $\frac{1}{6}$, 0.2 **c** $\frac{1}{6}$, 0.2 **d** $\frac{3}{6}$, 0.5 **e** $\frac{3}{6}$, 0.5 **f** $\frac{3}{6}$, 0.5 **5** **a** $\frac{1}{9}$, 0.1 **b** $\frac{1}{9}$, 0.1 **c** $\frac{4}{9}$, 0.4 **d** $\frac{2}{9}$, 0.2 **e** $\frac{4}{9}$, 0.4 **f** $\frac{5}{9}$, 0.6

**PAGE 195** **1** **a** $\frac{1}{4}$ **b** $\frac{1}{4}$ **c** $\frac{1}{2}$ **d** $\frac{1}{2}$ **e** $\frac{1}{13}$ **f** $\frac{1}{13}$ **g** $\frac{1}{13}$ **h** $\frac{1}{13}$ **i** $\frac{3}{13}$ **j** $\frac{1}{26}$ **k** $\frac{2}{13}$ **l** $\frac{10}{13}$ **2** **a** 10 **b** **i** $\frac{3}{10}$ **ii** $\frac{1}{5}$ **iii** $\frac{1}{2}$ **iv** $\frac{1}{2}$ **3** **a** $\frac{5}{12}$ **b** $\frac{1}{4}$ **c** $\frac{1}{3}$ **d** $\frac{2}{3}$ **4** **a** 18 **b** **i** $\frac{7}{18}$ **ii** $\frac{5}{18}$ **iii** $\frac{1}{3}$ **iv** $\frac{2}{3}$ **v** $\frac{13}{18}$ **vi** $\frac{11}{18}$

**PAGE 196** **1** **a** **i** $\frac{13}{50}$, 0.26 **ii** $\frac{11}{50}$, 0.22 **iii** $\frac{9}{50}$, 0.18 **iv** $\frac{17}{50}$, 0.34 **v** $\frac{24}{50}$, 0.48 **vi** $\frac{26}{50}$, 0.52 **b** **i** $\frac{1}{4}$, 25% **ii** $\frac{1}{4}$, 25% **iii** $\frac{1}{4}$, 25% **iv** $\frac{1}{4}$, 25% **v** $\frac{1}{2}$, 50% **2** **a** **i** $\frac{34}{120}$, 0.28 **ii** $\frac{72}{120}$, 0.6 **iii** $\frac{14}{120}$, 0.12 **iv** $\frac{48}{120}$, 0.4 **b** **i** $\frac{1}{3}$, 0.33 **ii** $\frac{1}{6}$, 0.17 **iii** $\frac{1}{2}$, 0.5 **iv** 0 **v** $\frac{5}{6}$, 0.83 **vi** $\frac{1}{2}$, 0.5

**PAGE 197** **1** **a** 90 **b** **i** 20% **ii** 11.1% **iii** 18.9% **iv** 22.2% **v** 14.4% **vi** 38.9% **vii** 35.6% **viii** 54.4% **ix** 45.6% **2** **a** Parent/teacher to check **b** **i** 16.7% **ii** 16.7% **iii** 50% **iv** 50%

**PAGE 198** **1** **a** Flipping a tail **b** Losing or drawing **c** Picking a black card **d** Rolling a 1, 2, 3, 4 or 6 **e** Early arrival or on time to class **2** **a** $\frac{2}{5}$ **b** $\frac{3}{5}$ **c** $\frac{3}{5}$ **d** $\frac{2}{5}$ **3** **a** $\frac{2}{7}$ **b** $\frac{5}{7}$ **c** $\frac{5}{7}$ **d** $\frac{2}{7}$ **4** **a** $\frac{5}{12}$ **b** $\frac{7}{12}$ **c** $\frac{7}{12}$ **d** $\frac{5}{12}$ **5** **a** $\frac{1}{5}$ **b** $\frac{4}{5}$ **c** $\frac{9}{10}$ **6** **a** $\frac{3}{5}$ **b** $\frac{2}{5}$ **c** $\frac{1}{5}$ **d** $\frac{4}{5}$ **e** $\frac{1}{5}$ **f** $\frac{4}{5}$ **7** **a** $\frac{1}{4}$ **b** $\frac{3}{4}$ **c** $\frac{1}{2}$ **d** $\frac{1}{2}$ **e** $\frac{1}{4}$ **f** $\frac{3}{4}$

**PAGE 199** **1** **a** $\frac{1}{4}$ **b** $\frac{3}{4}$ **c** $\frac{1}{13}$ **d** $\frac{12}{13}$ **e** $\frac{1}{2}$ **f** $\frac{1}{2}$ **g** $\frac{1}{26}$ **h** $\frac{25}{26}$ **i** $\frac{2}{13}$ **j** $\frac{11}{13}$ **k** $\frac{12}{13}$ **l** $\frac{12}{13}$ **2** **a** $\frac{1}{6}$ **b** $\frac{5}{6}$ **c** $\frac{1}{2}$ **d** $\frac{1}{2}$ **3** **a** $\frac{1}{10}$ **b** $\frac{9}{10}$ **c** $\frac{1}{2}$ **d** $\frac{1}{2}$ **4** **a** $\frac{1}{3}$ **b** $\frac{2}{3}$ **c** $\frac{2}{9}$ **d** $\frac{7}{9}$ **5** 58% **6** 0.3

**PAGE 200** **1** **a** White, Blue **b** 2 **c** No **d** Likely **e** Unlikely **f** $\frac{3}{5}$ **g** $\frac{2}{5}$ **2** **a** 1, 2, 3, 4, 5, 6 **b** 6 **c** $\frac{1}{6}$ **d** $\frac{1}{6}$ **e** Yes, they all have probability of $\frac{1}{6}$ **f** $\frac{1}{2}$ **3** **a** 33.3% **b** 22.2% **c** 55.6% **d** E **4** **a** $\frac{1}{3}$ **b** $\frac{2}{3}$ **c** $\frac{1}{6}$ **d** $\frac{5}{6}$ **e** White button **5** $\frac{7}{12}$ **6** **a** 1 **b** $\frac{1}{2}$ **c** 0 **d** $\frac{1}{2}$ **7** **a** **i** 0.6 **ii** 0.3 **iii** 0.1 **b** **i** 0.5 **ii** 0.3 **iii** 0.2 **8** **a** $\frac{1}{30}$ **b** $\frac{1}{30}$ **c** $\frac{1}{3}$ **d** $\frac{19}{30}$ **e** $\frac{1}{5}$ **f** $\frac{21}{30}$ **9** $\frac{1}{120}$

ISBN 978 1 74125 737 3

Pascal Press
PO Box 250
Glebe NSW 2037
www.pascalpress.com.au

Publisher: Vivienne Joannou
Edited by Rosemary Peers
Answers checked by Melinda Amaral
Cover and page design by Sonia Woo
Typeset by lj Design (Julianne Billington)
Printed by Vivar Printing/Green Giant Press